Machine Learning in 2D Materials Science

Data science and machine learning (ML) methods are increasingly being used to transform the way research is being conducted in materials science to enable new discoveries and design new materials. For any materials science researcher or student, it may be daunting to figure out if ML techniques are useful for them or, if so, which ones are applicable in their individual contexts, and how to study the effectiveness of these methods systematically.

KEY FEATURES

- Provides broad coverage of data science and ML fundamentals to materials science researchers so that they can confidently leverage these techniques in their research projects.
- Offers introductory material in topics such as ML, data integration, and 2D materials.
- Provides in-depth coverage of current ML methods for validating 2D materials using both experimental and simulation data, researching and discovering new 2D materials, and enhancing ML methods with physical properties of materials.
- Discusses customized ML methods for 2D materials data and applications and high-throughput data acquisition.
- Describes several case studies illustrating how ML approaches are currently leading innovations in the discovery, development, manufacturing, and deployment of 2D materials needed for strengthening industrial products.
- Gives future trends in ML for 2D materials, explainable AI, and dealing with extremely large and small, diverse datasets.

Aimed at materials science researchers, this book allows readers to quickly, yet thoroughly, learn the ML and AI concepts needed to ascertain the applicability of ML methods in their research.

Parvathi Chundi, PhD is Professor of Computer Science, University of Nebraska-Omaha. Prior to Omaha, Dr. Chundi was with Agilent Technologies and HP Labs, both in Palo Alto, CA.

Venkataramana Gadhamshetty, PhD, PE is Professor of Environmental Engineering in Department of Civil and Environmental Engineering, South Dakota School of Mines and Technology. He is a cofounder of 2D materials for Biofilm Science Engineering and Technology (2DBEST) center and 2D materials laboratory (2DML) at SDSM&T.

Bharat K. Jasthi, PhD is Associate Professor, Department of Materials and Metallurgical Engineering, South Dakota School of Mines and Technology. Dr. Jasthi has research expertise in the areas of microstructural modification, structure-property correlation, new alloy development, powder metallurgy, additive manufacturing, and development of engineered surface thin films and coatings for a wide range of applications.

Carol Lushbough, MA is an Emeritus Professor of Computer Science, University of South Dakota.

Machine Learning in 2D Materials Science

Edited by
Parvathi Chundi,
Venkataramana Gadhamshetty,
Bharat K. Jasthi, and Carol Lushbough

CRC Press is an imprint of the
Taylor & Francis Group, an **informa** business

Designed cover image: Sourav Verma

First edition published 2024
by CRC Press
2385 NW Executive Center Drive, Suite 320, Boca Raton FL 33431

and by CRC Press
4 Park Square, Milton Park, Abingdon, Oxon, OX14 4RN

CRC Press is an imprint of Taylor & Francis Group, LLC

ISBN: 978-0-367-67820-3 (hbk)
ISBN: 978-0-367-67821-0 (pbk)
ISBN: 978-1-003-13298-1 (ebk)

DOI: 10.1201/9781003132981

Typeset in Times
by codeMantra

Contents

1 Introduction to Machine Learning for Analyzing Material–Microbe Interactions

Venkataramana Gadhamshetty, Parvathi Chundi, and Bharat K. Jasthi

1.1 INTRODUCTION

Analogous to the silicon revolution in the 1970s where silicon-enabled miniaturized computing has transformed the field of information technology, especially by replacing the obsolete centralized mainframes, advanced materials are also expected to revolutionize emerging industries. Such materials are expected to address health, economic, and environmental challenges facing modern society. They are also expected to transform the performance of construction, defense, energy, environment, mining, healthcare, and manufacturing domains [1]. For instance, advanced materials that feature sustainability benefits can potentially mitigate the negative environmental impacts of current methods of material production. For instance, greenhouse gas emissions from material production using current technologies have reached 11 billion tons of CO_2 equivalent, which represents a 120% increase compared with that in 1995 [2]. Given the predictions of an increase in material consumption (62 Gt/year, current) to 100 Gt by 2030 [3], the financial and environmental burden can also be expected to increase accordingly. Advanced materials are also expected to alleviate recurring issues related to abiotic corrosion and microbiologically influenced corrosion (MIC) of materials. Assuming that advanced materials can significantly improve health, environmental, economic, and performance improvement benefits, there is a compelling need for embracing artificial intelligence (AI) and machine learning (ML) methods for accelerating the discovery and implementation of advanced material systems.

Unusual nanoscale phenomena and associated properties of two-dimensional (2D) materials render them promising advanced materials in many technologically relevant applications. Some of the desirable properties observed so far include low-profile thickness, low permeability, mechanical flexibility, and higher values of carrier mobilities, superconductivity, and optical absorption compared with their bulk counterparts [4–6]. These properties render them candidates for serving high-performance barrier coatings [7–11], energy devices [12–16], catalysts [17–22], biosensors [23–26],

DOI: 10.1201/9781003132981-1

spintronics [16,27–29], and supercapacitors [30–33]. With the successful exfoliation of graphene in 2010 and its implementation in diverse applications, there has been a significant interest in discovering other promising 2D materials including hexagonal boron nitride (hBN), metallic carbides, nitride, and carbonitrides (MXenes) [34–36]. However, such a discovery and implementation process requires significant infrastructure (human and hardware) and extensive efforts for extended periods of time (~10–20 years) [20], all with many odds of failure. ML methods, some of which are discussed in this book, can accelerate the discovery of new 2D materials and our ability to predict their properties without any apriori experiments.

Every material, equipment, and piece of infrastructure exposed to the natural and built environments are subject to degradation by physical, chemical, and biological processes, or their combinations. The prospects for degradation or corrosion are typically amplified in the presence of living microorganisms. Biofilms represent a robust, self-excreted extracellular polymeric substance that is known to encapsulate living microbial cells. Owing to the complex three-dimensional (3D) architecture, multicellular community, and surface colonization lifestyle, biofilms are known to confer protection to the encapsulated cells against any known environmental stressor (e.g., antibiotics, disinfectants, turbulence, nutrient limitations, and extreme physical and geochemical conditions). Thus, biofilms are omnipresent, and they critically influence the performance of materials used in any agricultural, industrial, and human systems. The design and development of advanced materials, especially protective coatings for controlling biofilms and their MIC effects, should consider biological mechanisms along with typical cost and performance metrics. Such noninvasive coatings based on 2D materials can then be used to influence genotypical and subsequently phenotypical responses (e.g., adherence state, quorum sensing, and MIC) in each biological environment. Recent works by the authors and their coworkers have demonstrated the use of graphene and 2D hBN materials for enabling beneficial biofilms in environmental biotechnology applications (e.g., microbial fuel cells [37–41] and bioenergy production [42]) and for controlling the detrimental effects of harmful biofilms (e.g., MIC prevention [43–49]). Further efforts can enable the use of these 2D coatings in a range of practical applications including corrosion prevention in water pipes, oil wells, air conditioners, cooling towers, and other engineering applications.

Considering the rush to accelerate the discovery and deployment of 2D materials, one can expect the use of computer science aspects (e.g., big data and ML) for enabling 2D material innovation. Nearly, 1000 different promising 2D materials have been reported. Considering the interest in functionalizing these materials, for example, by doping with one of the 84 stable elements, one can expect 84 variations of each 2D material when doped with a single element. The variety of these materials will increase to 3486 with two dopants, 95,284 combinations with two dopants, and 2 million combinations with four dopants. One can thus expect 2 billion combinations for 1000 different 2D materials. If one were to explore their performance in biological environments, especially individually with thousands of technologically relevant bacterial species, the estimated efforts and the amount of big data can be overwhelming. Typically, the "big data" at the microbe–material interfaces are characterized in terms of three Vs (greater variety, increasing volumes, and higher velocity). To help readers visualize the gigantic amount of data, ~2314 exabytes of big data have been reported to be generated by health care alone in 2020. Both environmental biotechnology and biomedical fields envision omics to be a cornerstone for big data, which include different modalities at a level of individual gene, protein, and metabolite. It is clear that the growth of biological data can be expected to grow at an unprecedented rate.

Computer science tools will become necessary for handling the big data, when one tries to fuse the materials and biofilm phenotype data when performance of advanced materials is assessed in microbial environments. Many of the state-of-the-art computational tools that predict biofilm phenotypes using gene and genome sequencing data do not necessarily take material properties into account.

To address the above issues, the core contributors to this book, primarily editors and authors, have formed an interjurisdictional consortium that was funded by the National Science Foundation (OIA # 1920954 [50] and 1849206 [51]). This consortium used convergence research with a focus on exploring the use of 2D materials for addressing vexing research problems facing biological applications. In particular, we focused on exploring the use of 2D materials for addressing the growth of biofilms responsible for MIC. This research required a deep integration of knowledge, theories, methods, and data from diverse disciplines (bioscience, computer science, and material science). A goal for this project was Biofilms Data and Information Discovery System (Biofilm-DIDs), which integrates metadata from accessible materials and biofilm data sources. Natural language processing (NLP) queries will allow users to predict biofilm phenotype on a material. Other parallel goals were to develop automated approaches to analyzing the properties of materials, as well as the properties of biofilms grown on these materials. Ultimately, the material–microbe fusion analysis framework developed in this project is expected to assist in accelerating the development of 2D protective coatings for bioengineering applications. This project used copper (Cu) as a model for technologically relevant metals exposed to biological environments. Thin films of graphene and hBN were used to obtain 2D protective coatings on these metal surfaces. Cu was selected for its catalytic effect in synthesizing graphene and hBN using chemical vapor deposition (CVD) methods. *Oleidesulfovibrio alaskensis* G20 (previously known as *Desulfovibrio Alaskensis* G20) was used as a model organism.

Overview of this book: Chapter 2 provides readers with a comprehensive overview of 2D materials, including the classification of 2D materials and their synthesis methods, principles of 2D material design, and examples of applications that leverage the merits of these materials. Chapter 3 provides an overview of different ML approaches (e.g., supervised, unsupervised, semi-supervised, self-supervised, and reinforcement learning) that can be used in both 2D materials and biofilm research. Chapter 4 introduces the ML approaches that can be used to accelerate the discovery process of 2D materials. This chapter discusses supervised learning, unsupervised learning, and reinforcement learning methods that can effectively enable 2D material discovery. Chapter 5 transits into biology domains, where a hybrid U-Net based on convolutional neural networks (CNN)–vision-transformer-based contraction layers was used to analyze scanning electron microscopy (SEM) images of Oleidesulfovibrio alaskensis (OA)- G20 biofilms. Chapter 6 uses deep

CNNs (DCNNs) for enabling the automatic classification and segmentation of biofilm entities, along with the corrosion products from the SEM images of metals exposed to biofilms. Chapter 7 gets into the depth of using ML methods for analyzing underlying biological mechanisms in these cells (e.g., protein–protein interactions). Chapter 8 provides an overview of the Biofilm-DIDs, with the goal of highlighting their features for collecting and combining a large materials and biological data sets and leveraging AI methods for analyzing and predicting gene responses and biofilm characteristics influenced by material surface properties.

Chapter 9 discusses the use of ML approaches for addressing issues with the characterization of 2D materials (e.g., defects) relevant to many biological applications. Traditional methods of 2D material detection can involve hundreds of hours of manual labor. Despite this assiduous investigation, the structure–property relationships of 2D materials are perplexing and inconclusive. We discuss ML methods to analyze the image and spectrum data sets as input features and streamline them to predict the fingerprint features of 2D materials within seconds. Chapter 10 introduces atomistic-level simulation techniques for analyzing microbe–2D material interactions, and the use of bioinformatics and ML tools for this analysis. Chapter 11 discusses futuristic technologies (e.g., alloy development, drug delivery, and quantum materials) that can leverage ML approaches. Chapter 12 discusses needed Research and Development (R&D) efforts to further enable the development of ML-driven frameworks for 2D material discovery.

REFERENCES

1. National Science and Technology Council. Materials Genome Initiative for Global Competitiveness. 2011. Executive Office of the President, National Science and Technology Council, Washington, DC.
2. Hertwich, E.G., Increased carbon footprint of materials production driven by rise in investments. *Nature Geoscience*, 2021. **14**(3): pp. 151–155.
3. (OECD), O.E.C.D. Green Growth and Sustainable Development. April 5, 2023; Available from: https://www.oecd.org/greengrowth/.
4. Berry, V., Impermeability of graphene and its applications. *Carbon*, 2013. **62**: pp. 1–10.
5. Glavin, N.R., et al., Emerging applications of elemental 2D materials. *Advanced Materials*, 2020. **32**(7): p. 1904302.
6. Liu, X. and M.C. Hersam, 2D materials for quantum information science. *Nature Reviews Materials*, 2019. **4**(10): pp. 669–684.
7. Othman, N.H., et al., Graphene-based polymer nanocomposites as barrier coatings for corrosion protection. *Progress in Organic Coatings*, 2019. **135**: pp. 82–99.
8. Bunch, J.S., et al., Impermeable atomic membranes from graphene sheets. *Nano Letters*, 2008. **8**(8): pp. 2458–2462.
9. Chilkoor, G., et al., Hexagonal boron nitride: the thinnest insulating barrier to microbial corrosion. *ACS Nano*, 2018. **12**(3): pp. 2242–2252.
10. Chilkoor, G., et al., Hexagonal boron nitride for sulfur corrosion inhibition. *ACS Nano*, 2020. **14**(11): pp. 14809–14819.
11. Joseph, A., et al., 2D MoS2-hBN hybrid coatings for enhanced corrosion resistance of solid lubricant coatings. *Surface and Coatings Technology*, 2022. **443**: pp. 128612.
12. Zhai, S., et al., 2D materials for 1D electrochemical energy storage devices. *Energy Storage Materials*, 2019. **19**: pp. 102–123.

13. Xue, Y., et al., Opening two-dimensional materials for energy conversion and storage: a concept. *Advanced Energy Materials*, 2017. **7**(19): pp. 1602684.
14. Sahoo, R., A. Pal, and T. Pal, 2D materials for renewable energy storage devices: outlook and challenges. *Chemical Communications*, 2016. **52**(93): pp. 13528–13542.
15. Zhang, X., et al., 2D materials beyond graphene for high-performance energy storage applications. *Advanced Energy Materials*, 2016. **6**(23): p. 1600671.
16. Ahn, E.C., 2D materials for spintronic devices. *Npj 2D Materials and Applications*, 2020. **4**(1): p. 17.
17. Liu, J., et al., Nanostructured 2D materials: prospective catalysts for electrochemical CO2 reduction. *Small Methods*, 2017. **1**(1–2): p. 1600006.
18. Siahrostami, S., et al., Two-dimensional materials as catalysts for energy conversion. *Catalysis Letters*, 2016. **146**: pp. 1917–1921.
19. Zhu, J., et al., Recent advance in MXenes: a promising 2D material for catalysis, sensor and chemical adsorption. *Coordination Chemistry Reviews*, 2017. **352**: pp. 306–327.
20. Hasani, A., et al., Two-dimensional materials as catalysts for solar fuels: hydrogen evolution reaction and CO2 reduction. *Journal of Materials Chemistry A*, 2019. **7**(2): pp. 430–454.
21. Tang, L., et al., Confinement catalysis with 2D materials for energy conversion. *Advanced Materials*, 2019. **31**(50): p. 1901996.
22. Deng, D., et al., Catalysis with two-dimensional materials and their heterostructures. *Nature Nanotechnology*, 2016. **11**(3): pp. 218–230.
23. Li, S., et al., New opportunities for emerging 2D materials in bioelectronics and biosensors. *Current Opinion in Biomedical Engineering*, 2020. **13**: pp. 32–41.
24. Chen, F., et al., Structures, properties, and challenges of emerging 2D materials in bioelectronics and biosensors. *InfoMat*, 2022. **4**(5): p. e12299.
25. Lei, Z.L. and B. Guo, 2D material-based optical biosensor: status and prospect. *Advanced Science*, 2022. **9**(4): p. 2102924.
26. Shavanova, K., et al., Application of 2D non-graphene materials and 2D oxide nanostructures for biosensing technology. *Sensors*, 2016. **16**(2): p. 223.
27. Feng, Y.P., et al., Prospects of spintronics based on 2D materials. *Wiley Interdisciplinary Reviews: Computational Molecular Science*, 2017. **7**(5): p. e1313.
28. Choudhuri, I., P. Bhauriyal, and B. Pathak, Recent advances in graphene-like 2D materials for spintronics applications. *Chemistry of Materials*, 2019. **31**(20): pp. 8260–8285.
29. Han, W., Perspectives for spintronics in 2D materials. *APL Materials*, 2016. **4**(3): p. 032401.
30. Palaniselvam, T. and J.-B. Baek, Graphene based 2D-materials for supercapacitors. *2D Materials*, 2015. **2**(3): p. 032002.
31. Han, Y., et al., Recent progress in 2D materials for flexible supercapacitors. *Journal of Energy Chemistry*, 2018. **27**(1): pp. 57–72.
32. Murali, G., et al., A review on MXenes: new-generation 2D materials for supercapacitors. *Sustainable Energy & Fuels*, 2021. **5**(22): pp. 5672–5693.
33. Forouzandeh, P. and S.C. Pillai, Two-dimensional (2D) electrode materials for supercapacitors. *Materials Today: Proceedings*, 2021. **41**: pp. 498–505.
34. Naguib, M., et al., Two-dimensional nanocrystals produced by exfoliation of Ti3AlC2. *Advanced Materials*, 2011. **23**(37): pp. 4248–4253.
35. Novoselov, K.S., et al., Electric field effect in atomically thin carbon films. *Science*, 2004. **306**(5696): pp. 666–669.
36. Jena, D., et al., The new nitrides: layered, ferroelectric, magnetic, metallic and superconducting nitrides to boost the GaN photonics and electronics eco-system. *Japanese Journal of Applied Physics*, 2019. **58**(SC): p. SC0801.

37. Islam, J., et al., Vitamin-C-enabled reduced graphene oxide chemistry for tuning biofilm phenotypes of methylotrophs on nickel electrodes in microbial fuel cells. *Bioresource Technology*, 2020. **300**: pp. 122642.
38. Islam, J., et al., Graphene as thinnest coating on copper electrodes in microbial methanol fuel cells. *ACS Nano*, 2022. **17**(1): pp. 137–145.
39. Jawaharraj, K., et al., Electricity from methane by Methylococcus capsulatus (Bath) and Methylosinus trichosporium OB3b. *Bioresource Technology*, 2021. **321**: p. 124398.
40. Islam, J., et al., Surface modification approaches for methane oxidation in bioelectrochemical systems. In: Kumar, P., Kuppam, C. (eds) *Bioelectrochemical Systems: Vol. 2 Current and Emerging Applications*, 2020: pp. 343–374. Springer. https://citations.springernature.com/item?doi=10.1007/978-981-15-6868-8_16
41. Jawaharraj, K., et al., Photosynthetic microbial fuel cells for methanol treatment using graphene electrodes. *Environmental Research*, 2022. **215**: p. 114045.
42. Vemuri, B., et al., Enhanced biohydrogen production with low graphene oxide content using thermophilic bioreactors. *Bioresource Technology*, 2022. **346**: p. 126574.
43. Krishnamurthy, A., et al., Passivation of microbial corrosion using a graphene coating. *Carbon*, 2013. **56**: pp. 45–49.
44. Chilkoor, G., et al., Hexagonal boron nitride for sulfur corrosion inhibition. *ACS Nano*, 2020. **14**(11): pp. 14809–14819.
45. Chilkoor, G., et al., Maleic anhydride-functionalized graphene nanofillers render epoxy coatings highly resistant to corrosion and microbial attack. *Carbon*, 2020. **159**: pp. 586–597.
46. Chilkoor, G., et al., Hexagonal boron nitride: the thinnest insulating barrier to microbial corrosion. *ACS Nano*, 2018. **12**(3): pp. 2242–2252.
47. Krishnamurthy, A., et al., Superiority of graphene over polymer coatings for prevention of microbially induced corrosion. *Scientific Reports*, 2015. **5**(1): p. 13858.
48. Chilkoor, G., et al., Atomic layers of graphene for microbial corrosion prevention. *ACS Nano*, 2021. **15**(1): pp. 447–454.
49. Chilkoor, G., et al., Graphene coatings for microbial corrosion applications. *Encyclopedia of Water: Science Technology and Society*. pp. 1–25.
50. (NSF), N.S.F. RII Track-1: Building on the 2020 Vision: Expanding Research, Education and Innovation in South Dakota. October 1, 2019; Available from: https://www.nsf.gov/awardsearch/showAward?AWD_ID=1849206&HistoricalAwards=false.
51. (NSF), N.S.F. RII Track-2 FEC: Data Driven Material Discovery Center for Bioengineering Innovation. August 1, 2019; Available from: https://www.nsf.gov/awardsearch/showAward?AWD_ID=1920954&HistoricalAwards=false.

2 Introduction to 2D Materials

Roberta Amendola and Amit Acharjee

Since the discovery of graphene and its unique properties, interest in materials that are a few atomic layers thick has been quickly increasing. These "sheet-like" materials are currently known as two-dimensional (2D) materials and consist of more than 150 categories beyond graphene.

This chapter provides a comprehensive overview of 2D material classification, synthesis, functionality, and applications, which equips readers with the basic information needed to understand the principles of 2D material design, while laying a foundation for the topics, which are presented in the following chapters of this book.

2.1 CLASSIFICATION OF 2D MATERIALS

The size and dimensionality are fundamental parameters defining a material's properties.

Nanomaterials are currently classified based on the number of dimensions that are outside the nanoscale range, defined as lower than 100 nm. Based on this definition, materials are identified as zero-dimensional (0D) when no dimension is larger than 100 nm such as quantum dots and nanoparticles; one-dimensional (1D) if one dimension is outside the nanoscale range, which includes nanotubes, nanorods, nanowires, and nanoribbons; 2D characterized by sheet-like configurations where two dimensions are outside the nanoscale and one dimension is a single or few atomic layers thick material like graphene; and three-dimensional (3D) when all dimensions are over the nanoscale range (Gupta et al., 2015). Examples of this last group include nanolayered structures and bulk powders. Example configurations of 0D to 2D materials are illustrated in Figure 2.1.

The sheet-like configuration of 2D materials originates from the fact that the in-plane interatomic interactions are stronger than the ones existing among the stacked planes, which are typically found in the bulk material. Novoselov et al. (2004) demonstrated that it is possible to prepare a few atomic layers thick graphitic sheets, including single-layer graphene by mechanical exfoliation (repeated peeling) of highly oriented pyrolytic graphite. Graphene refers to a single layer of carbon atoms densely packed into a benzene ring structure and is widely used to describe the properties of many carbon-based materials, including graphite, fullerenes, and nanotubes (e.g., carbon nanotubes are thought of as graphene sheets rolled up into cylinders) (Dresselhaus & Dresselhaus 2002; Peres et al., 2006; Shenderova et al., 2002; Walker, 1981) as illustrated in Figure 2.2.

DOI: 10.1201/9781003132981-2

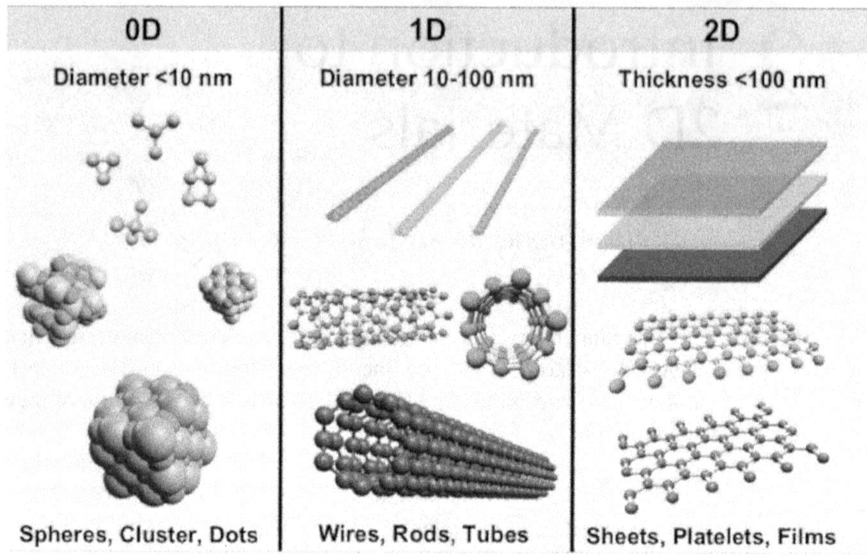

FIGURE 2.1 Common structures for 0D to 2D classification. (Adapted from Shaw, Z.L. et al., Nat. Commun., 12, 1, 2021.)

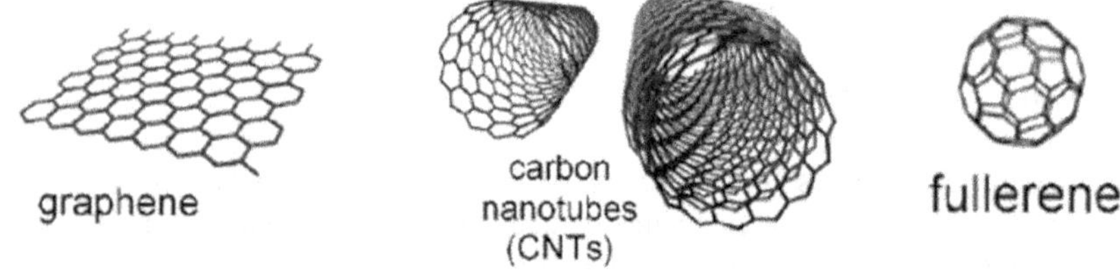

FIGURE 2.2 Schematic diagram of carbon nanostructures. (Adapted from Adorinni, S. et al., Appl. Sci., 11, 2490, 2021.)

The discovery of graphene and its exceptional properties such as high specific surface area, Young modulus, and carrier mobility (Giesbers et al., 2008; Lee et al., 2008; Morozov et al., 2008) revolutionized the last decade leading to significant progress in other 2D materials. In these materials, carbon atoms are bonded by primary intralayer covalent bonds and by weak interlayer secondary van der Waals (VdW) interactions. 2D materials are currently categorized based on their structures (Novoselov et al., 2016) as follows and are also illustrated in Figure 2.3. Such materials can then be combined to form heterostructures known as VdW solids as shown in Figure 2.4.

- Graphene, graphene oxide, and reduced graphene oxide (GO/rGO) (Dreyer et al., 2009; Yan et al., 2010; Zhu et al., 2010)
- Hexagonal boron nitride (*h*-BN) structured like graphene but having boron and nitrogen atoms in place of carbon (Cartamil-Bueno et al., 2017; Gorbachev et al., 2011; Liu et al., 2003)

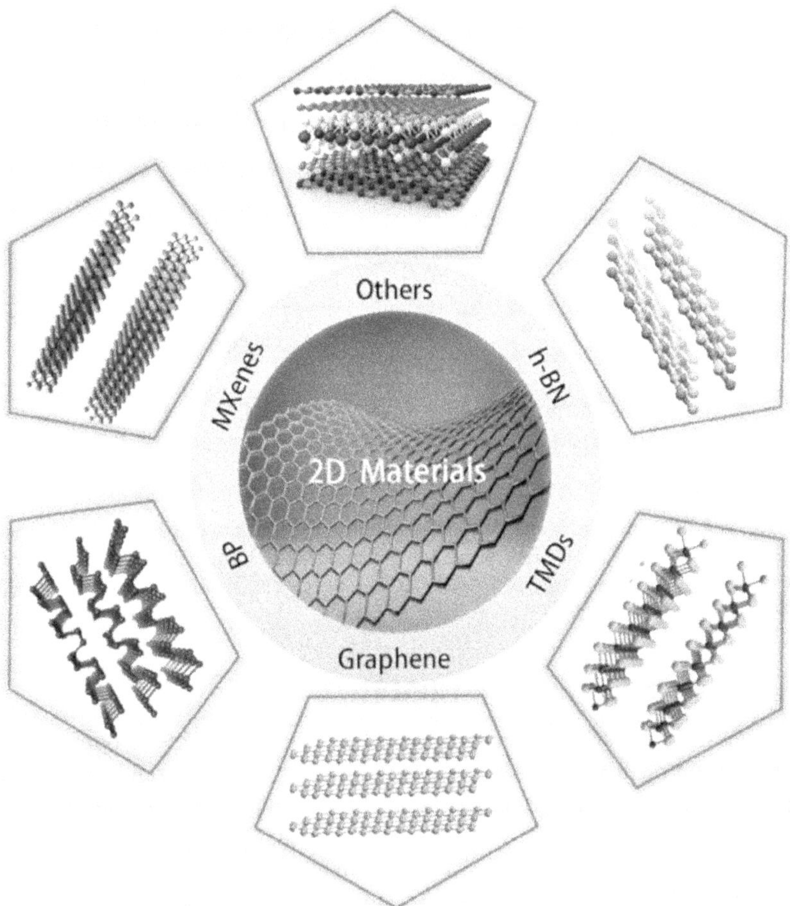

FIGURE 2.3 A schematic diagram of different 2D materials and their structures. (From Dong, Z., Xu, H., Liang, F., Luo, C., Wang, C., Cao, Z. Y., Chen, X. J., Zhang, J., & Wu, X., Molecules, 24(1), 2019, https://doi.org/10.3390/MOLECULES24010088. This figure is reproduced under the terms and conditions of the Creative Commons Attribution (CC BY) license (https://creativecommons.org/licenses/by/4.0/).)

- Transition metal dichalcogenides (TMDCs) with the formula MX_2 (M is a transition metal and X is a chalcogen, typically sulfur, selenium, or tellurium) and hexagonal structure with a tri-layer covalent bonding in the form of X-M-X (Chhowalla et al., 2015; Xiao Li & Zhu, 2015) and metal oxides (e.g., $Bi_2Sr_2CaCu_2O_x$) (Novoselov et al., 2005)
- Black phosphorus (BP) or phosphorene (Jiang & Park, 2014; Tao et al., 2015; Wei & Peng, 2014)
- Metal carbides and nitrides (MXenes) (Anasori et al., 2017)
- 2D metal–organic frameworks (2D MOFs) (Choi et al., 2009)
- 2D covalent–organic frameworks (2D COFs) (Kang et al., 2016)
- 2D perovskite (X. Cai et al., 2018; Lee et al., 2018; Tan et al., 2017; Xu et al., 2013)

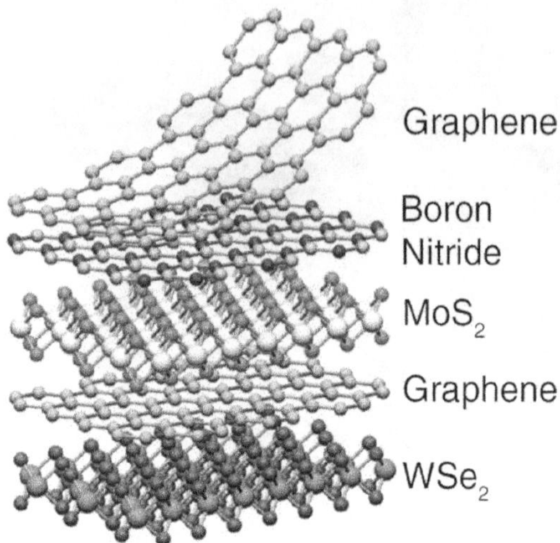

FIGURE 2.4 Example of van der Waals heterostructure that can be made by stacking multiple 2D materials in different ordering. (Adapted from Ramanathan, A. and Aqra, M., *CSAC2021: 1st International Electronic Conference on Chemical Sensors and Analytical Chemistry*, MDPI, Basel, 2021.)

2.2 SYNTHESIS OF 2D MATERIALS

The synthesis methods of 2D materials can be classified into two groups: top-down and bottom-up methods. In the top-down approach, a 2D material is fabricated by exfoliating layers of larger or bulk solid material, while in the bottom-up approach, the 2D material is produced from atomic or molecular precursors, which react and then grow to create a 2D material. The top-down methods include mechanical, liquid-phase, ultrasonic, electrochemical, ion-change, and lithium-intercalated exfoliations, whereas the bottom-up method involves epitaxial growth, chemical vapor deposition (CVD), physical vapor deposition (PVD), wet chemical methods, microwave-assisted method, or topochemical transformation. Among these methods, mechanical exfoliation, liquid-phase exfoliation, CVD, and PVD are commonly used methods for the development of 2D materials (Bian et al., 2022; Shanmugam et al., 2022). Bottom-up and top-down 2D material fabrication methods are schematically summarized in Figure 2.5.

Once the 2D material has been produced, it must be transferred from its growth substrate onto a target substrate, which defines its ultimate application. The ability to transfer large-area 2D materials while avoiding damage is therefore fundamental for preserving their quality. 2D material fabrication and transferring methods are discussed in the following sections.

2.2.1 Top-Down Methods

To obtain monolayer 2D materials, VdW forces must be overcome in a process called exfoliation. Single crystals, grown by chemical vapor transport (CVT) or flux methods, are widely used as bulk material because they provide high-quality mono- or

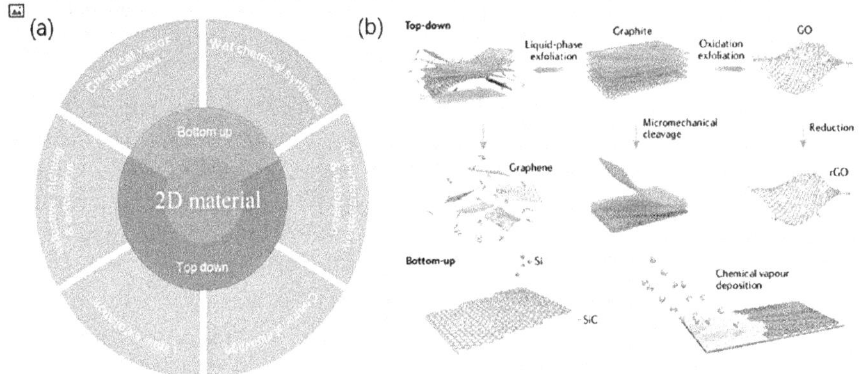

FIGURE 2.5 Summary of top-down and bottom-up 2D material fabrication methods (a) and schematic diagram of the methods using graphene as an example (b) (From Yang, H., Wang, Y., Tiu, Z. C., Tan, S. J., Yuan, L., & Zhang, H., *Micromachines*, *13*(1), 2022, https://doi org/10.3390/MI13010092.)

few-layered 2D materials. Mechanical exfoliation, achieved using adhesive tape, was first introduced to produce graphene and is still used to manufacture high-quality 2D crystals up to hundreds of microns in size (Novoselov et al., 2004). This method is mostly used in the laboratory setting as it has relatively low efficiency and yield. For higher yield, various technologies to assist mechanical exfoliation have been investigated in recent years. Shear force-assisted exfoliation, sonication-assisted exfoliation, and ball-milling exfoliation were all successfully used to produce graphene, *h*-BN, and TMDCs (Bonaccorso et al., 2016; Lei et al., 2015; Nicolosi et al., 2013; Niu et al., 2016; Yi Zhang et al., 2013). Synergistic exfoliation was recently used to exfoliate *h*-BN powders to produce nanosheets. This method, which can also be used with other 2D materials, coupled traditional ball milling with a supplemental vertical load from a weight block and ultrasonication. The vertical load alters the milling ball pattern of motion, which in turn increases the average tangential force and a number of contacts, resulting in a higher exfoliation yield (Wu et al., 2019).

High-quality few-layered graphene flakes, several hundred microns in size, were produced using oxygen plasma cleaning to facilitate the exfoliation process. Compared with traditional mechanical exfoliation, the yield and the area of the transferred flakes were increased 50-fold (Huang et al., 2015). This method is promising and can also be applied to other 2D materials; however, the process efficiency may be radically reduced if the interaction between the selected material (such as TMDCs) and SiO_2 substrate is limited (Bian et al., 2022).

A more efficient top-down method is liquid-phase exfoliation. This process yields a large number of mono- or few-layered 2D flakes from bulk crystals dispersed in a specific solvent. Depending on the nature of the force facilitating the exfoliation process, two main approaches can be identified: direct exfoliation and intercalation-assisted exfoliation.

In direct exfoliation, 2D materials are expanded between their bulk layers when dispersed in the liquid phase with consequent reduction in VdW forces. An ultrasonic wave is then used to disperse the layers (ultrasonic stripping). This methodology is relatively simple and can be used for the preparation of larger quantities of 2D materials. However, the thickness of the resulting flakes is often not even, and impurities are difficult to remove (Hao et al., 2020).

In the intercalation-assisted exfoliation process, small molecules, non-covalently bonded molecules, or polymers are inserted into the bulk material causing expansion of the interlayer spacing, reduction in the VdW forces, and ultimately exfoliation of the 2D layer. Compared to direct exfoliation, larger 2D flakes can be produced. Common intercalation agents are alkali metal atoms or ions in liquid ammonia or naphthalide, or *n*-butyllithium in hexane (Eda et al., 2011; Yin et al., 2016; Zheng et al., 2014). The expansion during the exfoliation processes is achieved through the accumulation of bubbles generated through the hydration of the agents in liquid. Chemical weathering is based on a similar principle and was used for the efficient exfoliation of TDMC nanosheets from bulk material in an alkaline solution. Because of the high chemical potentials of Na^+ and OH^-, interlayer infiltration leads to the accumulation of sodium hydroxide (NaOH) in the bulk material. When its concentration exceeds the critical value of ~7.6% (one NaOH pair per four MX_2 units), exfoliation into ultrathin 2D flakes occurs (Zhao et al., 2015). Due to the chemical processing, both liquid-phase methodologies result in lower quality (i.e., high density of structural defects and lacking in the regulation of sheet size and thickness) 2D materials. Also, the disposal of the products used during processing might be a risk to the environment.

To improve the efficiency of liquid-phase exfoliation, electrochemical intercalation was proposed. In this process, the electric current acts as an attractive driving force to bring foreign molecules or ions into the bulk material, which then causes the exfoliation of mono- or few-layered flakes. This technique has great potential for producing large-area, high-quality atomic thick flakes (He et al., 2019; Howard, 2021; Lin et al., 2018; Wang et al., 2021; Yang et al., 2020; Yu et al., 2020).

2.2.2 Bottom-Up Methods

CVD methods are being widely investigated as efficient bottom-up procedures for producing high-quality large-area 2D materials. In 2009, uniform large-area graphene film was produced for the first time on copper foils by the CVD method (Xuesong Li, Cai, et al., 2009). After that, the process has been successfully applied to produce TMDCs (Gao et al., 2015; Ji et al., 2013; Najmaei et al., 2013; Shi et al., 2015; Zhang et al., 2013, 2019), *h*-BN (Chen et al., 2019; Sun et al., 2018; Sutter et al., 2011), and BP (Smith et al., 2016). In general, the CVD process is a high-temperature chemical synthesis of the 2D material by deposition on a high-purity catalytic substrate such as copper, nickel, or sapphire, where the 2D material is formed. The procedure involved in a CVD process consists of three stages, namely transportation, nucleation, and growth. During transportation, a solid precursor is sublimated at high temperature and delivered to the substrate by an inert carrier gas. Subsequently, the precursor decomposes or diffuses at the hot surface of the substrate (catalytically or non-catalytically), forming the nucleus (nucleation) necessary for the growth of the 2D material. The precursor then continues to react and accumulate in the vicinity of the nucleus growing the 2D material. The produced layer must then be separated from the substrate to obtain the freestanding 2D material.

Graphene is grown on a copper foil substrate using a CH_4:H_2:Ar mixture as the precursor gas. At the hot surface of the substrate up to 1,000°C, catalytic cracking decomposes the mixture. Carbon atoms have low solubility in copper; therefore, atoms assemble in graphene crystals as diffusion does not occur resulting in high-quality graphene crystals and polycrystalline arrangement (Xuesong Li, Cai, et al., 2009).

h-BN was produced by CVD using a thermally active BN precursor (e.g., ammonia borane) over copper, nickel, ruthenium, or rhodium substrate (Sun et al., 2018). The generation of h-BN layers is reported to be obtained at atmospheric pressure CVD (APCVD) and low-pressure CVD (LPCVD) conditions. It was found that, while LPCVD is governed by mass transport control, APCVD relies on surface reaction control (Sun et al., 2018). The pressure, the temperature, and the selected substrate affected the growth of monolayer *h*-BN when borazine is used as a precursor. When the process parameters are set at 10^{-8}Torr of gas pressure, 780°C, and ruthenium Ru(0001) single crystal is used as a substrate, the nucleation process leads to sparse h-BN domains, which then grow to form a closed monolayer film (Sutter et al., 2011). When the parameter is 10^{-7}Torr of gas pressure, 796.8°C, and rhodium Rh(111) single crystals are used as a substrate, the resulting *h*-BN has a highly regular mesh structure and is thermally stable, which makes it a good template to organize molecules (Corso et al., 2004).

As for fabricating 2D TMDCs with CVD methods, various approaches have been implemented such as direct metal sulfurization (Zhan et al., 2012), thermolysis of thiosalts (Sang et al., 2019), and sulfurization of metal oxide (Huang et al., 2014; Shi et al., 2015). Molybdenum disulfide (MoS_2) was produced by direct metal sulfurization using pre-deposited molybdenum on silicon oxide (SiO_2 or silica) as the substrate while sulfur acted as the S source. It was found that the size and thickness of the MoS_2 layer were dependent on the size of the SiO_2 substrate and the thickness of the pre-deposited molybdenum. For thermolysis-based CVD, Ammonium tetrathiomolybdate ($(NH_4)_2MoS_4$) was used as the precursor and thermally decomposed at temperatures in the range of 300°C–900°C. A direct proportionality between the grain size, the number of deposited layers, and the temperature was identified. The sulfurization of metal oxide-based CVD is the most challenging as often thermal decomposition of the precursor is not sufficient because some metal oxides have very high sublimation temperature (i.e., tungsten trioxide, WO_3). For this reason, salt-assisted CVD growth was developed. In this process, alkali metal halides are added to precursors for growing tungsten disulfide (WS_2) or tungsten diselenide (WSe_2) monolayers at moderate temperatures (700°C–850°C) and atmospheric pressure (Li et al., 2015). Until now, a total of 47 compounds and heterostructures were prepared by halide salt-assisted CVD processes (Zhou et al., 2018). Compared to other 2D materials prepared using CVD, TDMCs are more sensitive to the effect of the process variables as the source-to-substrate distance, temperature, gas carrier, precursors, and substrate can all influence the nucleation density and grain growth and ultimately affect the structure of the final material.

In addition to CVD, PVD has also been used for producing graphene, *h*-BN, and TMDCs (Ionescu et al., 2015; Sutter et al., 2011; Wu et al., 2013). This process requires an ultrahigh vacuum and a heated high-purity atomic source to deposit related 2D crystals on a substrate. Molecular beam epitaxy (MBE) and pulsed laser deposition (PLD) are the two widely used PVD approaches. PLD provides fast layer deposition but with uncontrollable thickness, while MBE offers the advantage of precise thickness control and stoichiometric growth (Liu & Hersam 2018).

CVD is regarded as the most utilized method for large-scale and highly efficient production of 2D materials. However, uniform thickness, the ability to reach a wafer-scale product, and the relatively high energy cost due to the elevated temperature process requirement are the major challenges that should be addressed in the near future.

2.2.3 Layer Transfer Methods

The atomic thickness of the 2D layers causes the material to be sensitive to mechanical damage and crumble during transfer, which inevitably compromises their superior properties. Polymer-assisted and polymer-free methods are the main two approaches used for transferring 2D layers.

In polymer-assisted transfer methods, the 2D layer growth on a substrate is coated with a polymer film, usually polymethyl methacrylate (PMMA) due to its flexibility and mechanical strength (Xuesong Li, Zhu, et al., 2009; Ma et al., 2019; Reina et al., 2009). The PMMA/2D layer stack is then delaminated from the substrate using chemical etching, capillary forces, or bubble formation and retrieved using the final application substrate. The PMMA is later removed with a solvent. A schematic diagram of the PMMA-assisted transfer methods of 2D materials is shown in Figure 2.6.

The copper foil used for the growth of graphene is removed using hydrochloric acid (HCl), nitric acid (HNO_3), iron nitrate ($Fe(NO_3)_3$), or copper chloride ($CuCl_2$) (Lee et al., 2017). Strong bases such as NaOH or potassium hydroxide (KOH) are instead used to delaminate TMDCs from SiO_2 growth substrates (Lin et al., 2012; Wang et al., 2014). PMMA is commonly removed in an acetone bath. This transfer method is efficient and reliable. Yet, the chemicals used in the process can contaminate the 2D layer and cause undesired doping or can physically damage the structure through corrosion; hence, alternative etchant-free methods were developed.

Water-based methods using capillary forces or bubble formation that drive the 2D layer/polymer stack separation are gaining increased interest. The capillary force-driven method employs water penetration between the hydrophobic 2D layer and the hydrophilic growth substrates. Once detached, the 2D layer/polymer stack floats to the water surface. The water is then pumped out, gradually lowering the stack onto the final substrate,

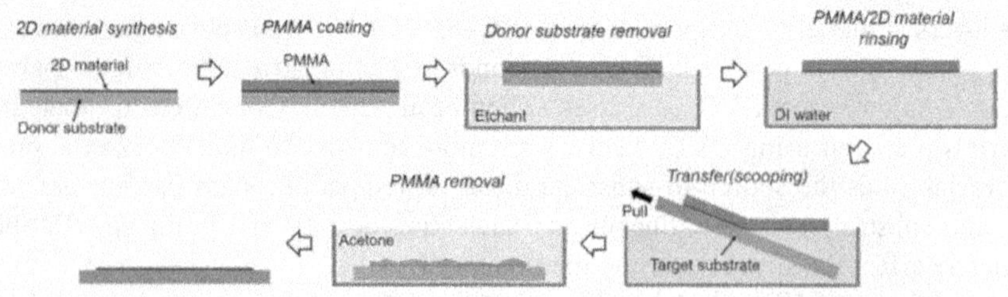

FIGURE 2.6 Schematic diagram of the PMMA-assisted 2D layer transfer method. (From Kim, C., Yoon, M.-A., Jang, B., Kim, J.-H., & Kim, K.-S., Tribology and Lubricants, 36(1), 1–10, 2020, https://doi.org/10.9725/KTS.2020.36.1.1. This figure is reproduced under the terms and conditions of the Creative Commons Attribution (CC BY) license (https://creativecommons.org/licenses/by/4.0/).)

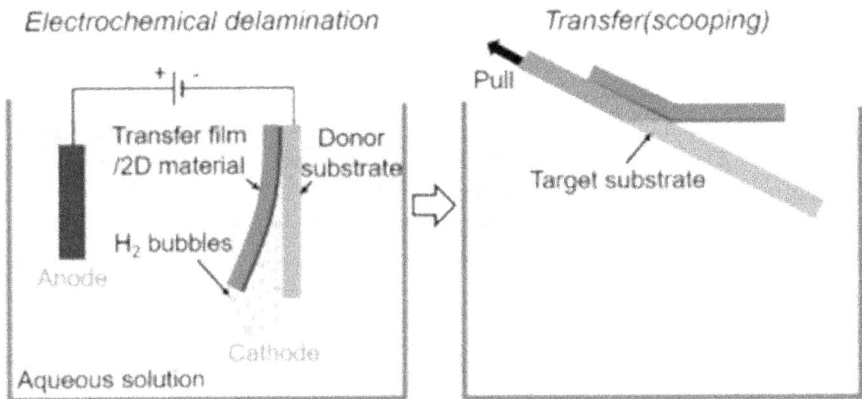

FIGURE 2.7 Schematic diagram of the electrochemical bubbling 2D layer transfer method using the development of H2 bubbles. (From Kim, C., Yoon, M.-A., Jang, B., Kim, J.-H., & Kim, K.-S., Tribology and Lubricants, 36(1), 1–10, 2020, https://doi.org/10.9725/KTS.2020.36.1.1. This figure is reproduced under the terms and conditions of the Creative Commons Attribution (CC BY) license (https://creativecommons.org/licenses/by/4.0/).)

and the polymer is removed using a solvent (Calado et al., 2012; Schneider et al., 2010). This method eliminates the use of chemicals. However, crack/wrinkle formation was observed due to water trapping and clusters of bubbles at the 2D layer/substrate interface when the growth substrate is not strongly hydrophobic (Calado et al., 2012). The bubbling transfer method utilizes the development of hydrogen gas at the polymer/2D layer/substrate interface as shown in Figure 2.7. This method is based on an electrochemical reaction where the metallic growth substrate acts as a cathode; therefore, it cannot be used with nonmetallic substrates (e.g., SiO_2 or sapphire) (Fan et al., 2020).

Recently, a similar approach was proposed based on ultrasonic bubbling transfer where a large number of micro-sized bubbles are generated by ultrasonication in a water bath. This method works well with insulating substrates commonly used for TMDCs (Ma et al., 2015). Both etchant-free methods allow the growth substrate to be reused for other deposition and transfer processes.

Polymer-free ultraclean transfer of 2D materials has gained attention in the 2D material community. A water-based and support layer-free (polymer-free) transfer method was recently developed. In this approach, a sacrificial water-soluble layer is deposited on the growth substrate before the development of a 2D layer. Once the process is complete, the 2D layer/soluble layer/growth substrate stack is washed with deionized (DI) water (Cho et al., 2018). The floating layer is then transferred onto the final substrate as already discussed. Perylene-3,4,9,10-tetracarboxylic acid tetrapotassium salt (PTAS) was successfully used as the sacrificial layer for the growth and transfer of MoS_2. It was also noted that PTAS can serve as seed promoters and support the nucleation of large-area, continuous, and uniform 2D planar films on a variety of substrates (Lee et al., 2013; Singh et al., 2020).

For the implementation of 2D materials at a larger commercial scale, the fabrication processes should be automated. Along with the ability to prepare large-area layers, this remains the major challenge as all transfer methods are laborious and complex.

2.3 FUNCTIONALITY OF 2D MATERIALS

2D materials are characterized by different, often improved, properties when compared to their bulk counterparts. This property enhancement is, in most cases, due to the discretization of electronic energy states by confining electrons in a material to a very small space and an increase in the overall reactivity and surface area as all constituting atoms are "exposed" to the surrounding environment.

Because of these unique properties, 2D materials are suited for a large variety of applications. As functional electronics, optoelectronics including flexible systems and battery electrode devices are expected to be the fastest-growing fields of application in the next decade, and this section provides an overview of mechanical, electrical, and optical properties of graphene, *h*-BN, TMDCs, and BP, which are among the most widely used 2D materials.

2.3.1 Mechanical Properties

2D materials, which are characterized by stronger in-plane covalent bonds with unique properties and weaker out-of-plane VdW bonds, can be easily exfoliated. Local strains can be generated by simply poking, bending, or folding the material like a piece of paper (Dai et al., 2019), which is not typically observed in bulk materials. To "scale" mechanical properties from 3D to 2D material systems and to reflect the planar configuration, it is necessary to normalize 3D parameters by dividing them by the thickness of the 2D material (Kim et al., 2019). Elastic (Young's) modulus and fracture strength will therefore have units of energy per area (J/m^2 or N/m) instead of per volume (J/m^3 or Pa).

Atomic force microscopy (AFM) nanoindentation has been well established in assessing materials' mechanical properties at the nanoscale. A novel setup was proposed in 2008 (Lee et al., 2008): The 2D material was suspended over circular wells and indented at constant speed by an AFM tip with nanoscale radius to record force–displacement curves. Despite the setup successfully measuring mechanical properties, the collected data may not properly reflect the properties of the overall layer as the load was focused only on a central point (Kim et al., 2019). To address the limitation of AFM nanoindentation, chip-based microelectromechanical systems (MEMS) tests have been developed. The MEMS are equipped with small actuators and detectors that enable the stretching of a sample under varying loading conditions allowing for uniform application of the force along the in-plane direction of a 2D membrane (Arshad et al., 2011; Ozkan et al., 2010).

To follow is an overview of experimentally measured mechanical properties of some common 2D materials. Lee et al. (2008) were the first to experimentally measure the elastic modulus ($E \sim 1$ TPa) and fracture strength ($\sigma_{max} \sim 130$ GPa) of monolayer graphene using AFM nanoindentation. Nonlinear elastic behavior and brittle fracture were observed. It was later found that the elastic modulus and fracture strength of graphene decrease with increasing numbers of layers (Wei et al., 2016; Zhang & Pan, 2012). This characteristic was related to interlayer slippage and subsequent energy dissipation during the testing loading–unloading cycle. In thicker graphene, nonuniform strain distribution is accelerated along the out-of-plane direction,

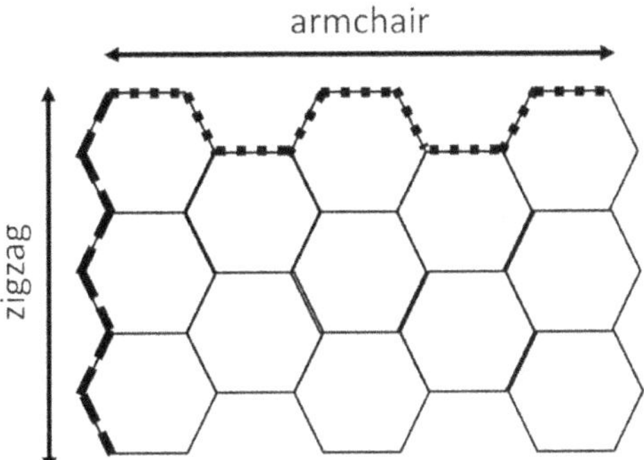

FIGURE 2.8 Armchair and zigzag directions within a 2D material layer with graphene-like hexagonal configuration.

resulting in decreased mechanical performance. It was reported that cracks tend to propagate along a preferential orientation and to then create zigzag edges (Fujihara et al., 2015). However, mechanically torn graphene progressed in a straight line with occasional changes in a direction toward either armchair or zigzag shapes (Kim et al., 2011). The armchair and zigzag directions within a 2D material layer with a graphene-like hexagonal configuration are illustrated in Figure 2.8.

The mechanical properties of *h*-BN, also known as "white graphene," were measured using the nanoindentation method and were found to be comparable to those of graphene with an elastic modulus of 0.865 TPa (Kim et al., 2015; Song et al., 2010). However, opposite to graphene, both the elastic modulus and fracture strength are not dependent on thickness variation (Falin et al., 2017). This phenomenon was explained by the fact that BN's orbitals have higher polarity than those of graphene, which ultimately causes the interlayer slipping energy to be increased (Kim et al., 2019).

Within the TMDC category, MoS_2 is the most popular 2D material. The mechanical properties of MoS_2 were characterized by AFM nanoindentation; elastic modulus and fracture strength were measured as ~270 and ~23 GPa, respectively, with a strain at failure that ranges between 6% and 11% (Bertolazzi et al., 2011). MoS_2 failure was identified as brittle in nature; however, it was found that with a 1% sulfur deficiency, the nature of failure can be shifted from brittle to plastic deformation (Ly et al., 2017). Enhanced mechanical properties were found for the tungsten-based TMDCs (WX_2) when compared to MoS_2. Also, for the same transition metal M, sulfides (MS_2) are the strongest, while tellurides (MTe_2) are the weakest due to the weakening of M–X (X = S, Se, Te) hybridization while going down the list from S to Te (Kim et al., 2019). In general, for TMDCs, the stress response was found to be stronger along the armchair direction illustrated in Figure 2.8. This phenomenon was linked to the strong hybridization occurring between the most external p orbitals of the chalcogens and the d orbitals of the transition metal, which causes a reallocation of the electronic charge to the shared region between the involved atoms (Li et al., 2013).

The mechanical properties of BP were evaluated using AFM nanoindentation. High anisotropy was observed along the crystalline directions under the applied stress (Tao et al., 2015), which was related to the puckered (nonplanar) structure of the layer

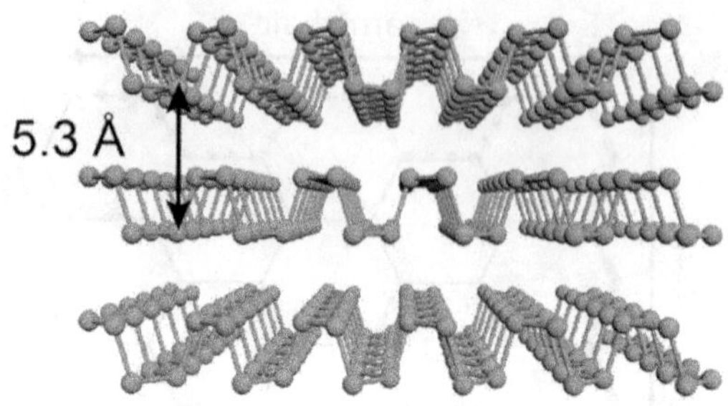

FIGURE 2.9 Nonplanar puckered structure of black phosphorous. (Adapted from Wang, D. et al., Front. Chem. 7, 2019.)

illustrated in Figure 2.9. Also, both the elastic modulus and the fracture strength tend to deteriorate when the layer thickness is decreased (Gallagher et al., 2016).

A degradation of the mechanical performance was observed in ambient air conditions due to the self-passivation process (Kim et al., 2019); therefore, BP is better suited for vacuum applications.

2.3.2 Electrical Properties

2D materials were found to show a wide range of electrical properties and transport characteristics. Due to their intrinsic crystal structures and stacking orientations, these sheets can behave as conductors (e.g., graphene), insulators (e.g., h-BN), semiconductors (e.g., MoS_2 and WS_2), or superconductors (e.g., $NbSe_2$ and NbS_2). Band gap engineering of 2D materials is an emerging field that offers a wide range of possibilities for tuning electronic properties. The structure of the energy band can be engineered through thickness control (number of layers) (Novoselov et al., 2016), elemental doping (Oliva-Leyva & Naumis, 2014; Tongay et al., 2011; Van Khai et al., 2012), and development of stacked heterostructures (Cai et al., 2018; Chen et al., 2018; Shi et al., 2018). In the last decade, research on the optimization of 2D material electronic properties has grown enormously due to their enhanced performance and lower energy requirement to power electronic systems such as solar cells, field effect transistors (FETs), and light-emitting diodes (LEDs) when compared to 3D counterparts.

Excellent electrical conductivity and adjustable work function make graphene relevant for FET applications. Nitrogen doping was found to be effective in improving the electrical conductivity of graphene (Deokar et al., 2022). The doped graphene was prepared by thermal annealing of reduced GO in ammonia gas. Variations in the carbon and oxygen content of the annealed product reduce the electrical resistance in doped graphene compared with GO and rGO (Pang et al., 2011; Van Khai et al., 2012).

The TMDCs based on periodic table groups VB and VIB metals (i.e., V, Nb, Ta, Cr, W, and Mo) are the most investigated (Fiori et al., 2014; Jariwala et al., 2014; Kappera et al., 2014; Wang et al., 2012; Wilson et al., 1975) because of the possibility to tune their electronic structure allowing for a range of behaviors including metallic, semimetallic, semiconducting, and superconducting. Properties tuned by doping were successfully achieved using nonmetal atoms such as H, B, C, N, O, and F. As a result, MoS_2, WS_2, and WSe_2 nanosheets developed total magnetic moment (H. Gao et al., 2020; Ma et al., 2011). Exposure of TDMCs to plasma oxygen led to the variation in the n-type and p-type conduction, allowing for the production of high-mobility FETs and planar monolayer p-n junctions utilized in semiconductor applications (Geim & Grigorieva, 2013; Hoffman et al., 2019). Nitrogen doping was found to make WS_2 electrochemically active. This makes such 2D material a suitable option for developing high-performance electro-catalysts (Sun et al., 2016).

The *h*-BN exhibits insulating properties and anisotropic resistivity (Pellegrino et al., 2011). Like graphene and TMDCs, doping can be used to alter their electrical characteristics. Semiconducting behavior was achieved through zinc doping in the range of 0%–4%, while beryllium implantation was used to achieve p-type conduction (Nose et al., 2006). One of the most interesting traits of *h*-BN is its flat atomic surface and graphene-like structure. These characteristics combined with a large electrical band gap make the material an ideal substrate. The electronic mobility of graphene on the *h*-BN substrate was found to be three times larger than that of the graphene without it, which makes this approach a viable strategy for enhancing the performance of large-area graphene electrical devices (Lee et al., 2012).

2.3.3 Optical Properties

Several researchers investigated the optical conductivity of graphene (Heersche et al., 2007; Liu & Hersam, 2019; Peres et al., 2007; Simsek, 2013) from the visible range to the near ultraviolet (UV) region. It was revealed that the optical conductivity of graphene increases with the increasing energy of the incident light (Liu & Hersam, 2019) and with the number of stacked layers (Heersche et al., 2007).

TMDCs are better suited for optical applications because of the large electronic density of states (DOS), which guarantees large optical adsorption. As graphene, the optical properties of the MoS_2 material are related to the number of layers but also to the interlayer distance, which allows one to control the spectral response in optical devices (Mak et al., 2010; Yu et al., 2017). The introduction of point defects in MoS_2 and WS_2 triggered new transitions in the optical range in line with photoconductivity measurements (Das et al., 2017). While in the visible range, disulfur vacancies activate the optical conductivity, molybdenum (Mo) and tungsten (W) vacancies activate it at low energies (Ribeiro et al., 2018).

More opportunities for efficient optoelectronic materials become available when VdW heterostructures are considered. The combination of materials with different work functions can lead to photoexcited electrons and holes accumulated in different layers. The intensity of this phenomenon can be controlled by tuning the distance between semiconductor layers (Fang et al., 2014; Rivera et al., 2015; Roy et al., 2013). Combinations of graphene, as a channel material, and TMDCs, as light-sensitive material where trapped charges are controlled by illumination, allow the creation of simple and efficient phototransistors (Roy et al., 2013). For multilayer BN/graphene/BN structures, it was found that the optical properties do not relate to thickness and/or stacking order; instead, they depend on light polarization. In particular, the frequency-dependent optical conductivity was found to exist only for the light polarized parallel to the plane (Farooq et al., 2015).

2.4 APPLICATIONS OF 2D MATERIALS

2D material-based FET is being widely investigated, particularly for sensors and nonvolatile memory (NVM) applications. rGO, BP, and TMDCs have successfully performed as charge channels for FET-based gas sensors. Multiple layers (2 to 4) of MoS_2 were deposited on a Si/SiO_2 wafer to manufacture a sensor for several gas detections, such as oxides of nitrogen. The system resulted in a detection limit of 0.8 ppm at room temperature (H. Li et al., 2012). More recently, a 2D heterostructure made of MoS_2 deposited on graphene by CVD was used for detecting low levels (0.2–1 ppm) of nitrogen dioxide (NO_2) gas at the temperature up to 200°C. The sensor showed a fast response/recovery time of less than one second with high reproducibility. Such advantages were linked to the synergistic effects of MoS_2 and graphene and the preferred exposure of active edge sites at the boundary of MoS_2 flakes (Hong et al., 2019).

NVMs are gradually replacing hard disk drives because of the growing need to access and transfer an ever-increasing amount of information in a short duration. As conventional silicon-based devices are quickly approaching their limit, NVMs are another promising application where 2D materials can serve as channel transistor. TMDCs, BP, and graphene are mainly involved in the production of resistive, ferroelectric, and flash memories (Bertolazzi et al., 2019; Kim et al., 2020; Ko et al., 2016; Lee et al., 2012, 2015). Graphene was found to act as a protective interfacial layer for decreasing power consumption in resistive NVMs (Ahn et al., 2018).

The large specific area, exceptional mechanical properties, tunable band gap, and good thermal and electrical conductivities made graphene, GO/rGO, TMDCs, and h-BN suitable for catalytic applications such as oxygen reduction or evolution reaction, photoinduced water splitting, and hydrogen evolution reaction. The catalytic performance of these 2D materials can be controlled through defect engineering and doping; defects or edges can act as active sites for catalytic reaction while doping changes the electronic states and the doped atoms can also serve as active sites (Chen et al., 2014; Deng et al., 2014; Wang et al., 2009; Yang et al., 2011).

2D materials such as graphene, GO, rGO, and *h*-BN were found to be very effective as anticorrosion coatings for a wide variety of materials and conditions including photocorrosion of semiconductors (Xi Chen et al., 2018; Khosravi et al., 2019; Weng et al., 2019) and biocorrosion of medical implants (Al-Saadi et al., 2017; Cui et al., 2017; Galbiati et al., 2017; Göncü et al., 2017; Mahvash et al., 2017; Parra et al., 2015; Zhang et al., 2017). The excellent anticorrosion properties were linked to the large surface area of graphene and graphene-like structure, which extend diffusion path length for the corrosive compounds. The high production prices have, however, limited this kind of application to specific fields in which corrosion results in great losses, such as aerospace, biomaterials, and advanced electronics.

The extended path for diffusion-based processes, present in the graphene structure, was also found to be beneficial for reducing the flammability of gases by delaying combustion (Chen et al., 2017; Shanmugam et al., 2020). It was shown that the time needed to burn pure cotton fabric and functional graphene-coated fabric was, respectively, 5 and 325 seconds (Chavali et al., 2020). This enables the use of graphene as reinforcement or coating material for increasing flame retardant properties.

Recently, graphene-based devices were developed for electromagnetic invisibility cloaking and adaptive thermal camouflage. Electromagnetic invisibility was achieved by modifying the refractive index and optical absorption of monolayer graphene through ion intercalation (Balci et al., 2015; Yang et al., 2013). Thermal camouflage was realized with adaptive thermal surfaces produced by the reversible intercalation of ionic liquid into multilayer graphene. This process resulted in the ability to control the surface thermal emission and absorption over the infrared (IR) spectrum. The fabricated device can disguise hot objects as cold and vice versa and blend itself with a varying thermal background in a few seconds when combined with a feedback mechanism (Phan et al., 2013; Salihoglu et al., 2018; Xiao et al., 2015; Zhao et al., 2019).

Currently, the library of 2D materials consists of more than 150 categories (Khan et al., 2020). Interest is quickly increasing, and several novel advanced 2D heterostructures, with selected compositions to target specific applications, are expected to be developed and introduced in the next few years.

REFERENCES

Adorinni, S., Cringoli, M. C., Perathoner, S., Fornasiero, P., & Marchesan, S. (2021). Green Approaches to carbon nanostructure-based biomaterials. *Applied Sciences*, *11*(6), 2490. https://doi.org/10.3390/APP11062490

Ahn, E. C., Wong, H. S. P., & Pop, E. (2018). Carbon nanomaterials for non-volatile memories. *Nature Reviews Materials*, *3*(3), 1–15. https://doi.org/10.1038/natrevmats.2018.9

Al-Saadi, S., Banerjee, P. C., Anisur, M. R., & Singh Raman, R. K. (2017). Hexagonal boron nitride impregnated silane composite coating for corrosion resistance of magnesium alloys for temporary bioimplant applications. *Metals*, *7*(12), 518. https://doi.org/10.3390/MET7120518

Anasori, B., Lukatskaya, M. R., & Gogotsi, Y. (2017). 2D metal carbides and nitrides (MXenes) for energy storage. *Nature Reviews Materials*, *2*(2), 1–17. https://doi.org/10.1038/natrevmats.2016.98

Arshad, S. N., Naraghi, M., & Chasiotis, I. (2011). Strong carbon nanofibers from electrospun polyacrylonitrile. *Carbon*, *49*(5), 1710–1719. https://doi.org/10.1016/J.CARBON.2010.12.056

Balci, O., Polat, E. O., Kakenov, N., & Kocabas, C. (2015). Graphene-enabled electrically switchable radar-absorbing surfaces. *Nature Communications*, *6*(1), 1–10. https://doi.org/10.1038/ncomms7628

Bertolazzi, S., Bondavalli, P., Roche, S., San, T., Choi, S. Y., Colombo, L., Bonaccorso, F., & Samorì, P. (2019). Nonvolatile memories based on graphene and related 2D materials. *Advanced Materials*, *31*(10), 1806663. https://doi.org/10.1002/ADMA.201806663

Bertolazzi, S., Brivio, J., & Kis, A. (2011). Stretching and breaking of ultrathin MoS 2. *ACS Nano*, *5*(12), 9703–9709. https://doi.org/10.1021/NN203879F/SUPPL_FILE/NN203879F_SI_001.PDF

Bian, R., Li, C., Liu, Q., Cao, G., Fu, Q., Meng, P., Zhou, J., Liu, F., & Liu, Z. (2022). Recent progress in the synthesis of novel two-dimensional van der Waals materials. *National Science Review*, *9*(5). https://doi.org/10.1093/NSR/NWAB164

Bonaccorso, F., Bartolotta, A., Coleman, J. N., & Backes, C. (2016). 2D-crystal-based functional inks. *Advanced Materials*, *28*(29), 6136–6166. https://doi.org/10.1002/ADMA.201506410

Cai, Z., Liu, B., Zou, X., & Cheng, H. M. (2018). Chemical vapor deposition growth and applications of two-dimensional materials and their heterostructures. *Chemical Reviews*, *118*(13), 6091–6133. https://doi.org/10.1021/

Cai, X., Luo, Y., Liu, B., & Cheng, H. M. (2018). Preparation of 2D material dispersions and their applications. *Chemical Society Reviews*, *47*(16), 6224–6266. https://doi.org/10.1039/C8CS00254A

ACS.CHEMREV.7B00536/ASSET/IMAGES/LARGE/CR-2017-00536M_0022.JPEG

Calado, V. E., Schneider, G. F., Theulings, A. M. M. G., Dekker, C., & Vandersypen, L. M. K. (2012). Formation and control of wrinkles in graphene by the wedging transfer method. *Applied Physics Letters*, *101*(10), 103116. https://doi.org/10.1063/1.4751982

Cartamil-Bueno, S. J., Cavalieri, M., Wang, R., Houri, S., Hofmann, S., & van der Zant, H. S. J. (2017). Mechanical characterization and cleaning of CVD single-layer h-BN resonators. *NPJ 2D Materials and Applications*, *1*(1), 1–7. https://doi.org/10.1038/s41699-017-0020-8

Chavali, K. S., Pethsangave, D. A., Patankar, K. C., Khose, R. V., Wadekar, P. H., Maiti, S., Adivarekar, R. V., & Some, S. (2020). Graphene-based intumescent flame retardant on cotton fabric. *Journal of Materials Science*, *55*(29), 14197–14211. https://doi.org/10.1007/S10853-020-04989-6

Chen, Xi, Chen, S., Liang, L., Hong, H., Zhang, Z., & Shen, B. (2018). Electrochemical behaviour of EPD synthesized graphene coating on titanium alloys for orthopedic implant application. *Procedia CIRP*, *71*, 322–328. https://doi.org/10.1016/J.PROCIR.2018.05.035

Chen, S., Duan, J., Jaroniec, M., & Qiao, S. Z. (2014). Nitrogen and oxygen dual-doped carbon hydrogel film as a substrate-free electrode for highly efficient oxygen evolution reaction. *Advanced Materials*, *26*(18), 2925–2930. https://doi.org/10.1002/ADMA.201305608

Chen, W., Liu, Y., Liu, P., Xu, C., Liu, Y., & Wang, Q. (2017). The preparation and application of a graphene-based hybrid flame retardant containing a long-chain phosphaphenanthrene. *Scientific Reports, 7*(1), 1–12. https://doi.org/10.1038/s41598-017-09459-9

Chen, Xin, Yang, H., Wu, B., Wang, L., Fu, Q., & Liu, Y. (2019). Epitaxial growth of h-BN on templates of various dimensionalities in h-BN-graphene material systems. *Advanced Materials*, *31*(12), 1805582. https://doi.org/10.1002/ADMA.201805582

Chen, P., Zhang, Z., Duan, X., & Duan, X. (2018). Chemical synthesis of two-dimensional atomic crystals, heterostructures and superlattices. *Chemical Society Reviews*, *47*(9), 3129–3151. https://doi.org/10.1039/C7CS00887B

Chhowalla, M., Liu, Z., & Zhang, H. (2015). Two-dimensional transition metal dichalcogenide (TMD) nanosheets. *Chemical Society Reviews*, *44*(9), 2584–2586. https://doi.org/10.1039/C5CS90037A

Cho, H.-Y., Nguyen, T. K., Ullah, F., Yun, J.-W., Nguyen, C. K., & Kim, Y. S. (2018). Salt-assisted clean transfer of continuous monolayer MoS2 film for hydrogen evolution reaction. *Physica B: Physics of Condensed Matter*, *532*, 84–89. https://doi.org/10.1016/J.PHYSB.2017.10.026

Choi, E. Y., Wray, C. A., Hu, C., & Choe, W. (2009). Highly tunable metal-organic frameworks with open metal centers. *CrystEngComm*, *11*(4), 553–555. https://doi.org/10.1039/B819707P

Corso, M., Auwärter, W., Muntwiler, M., Tamai, A., Greber, T., & Osterwalder, J. (2004). Boron nitride nanomesh. *Science*, *303*(5655), 217–220. https://doi.org/10.1126/SCIENCE.1091979

Cui, M., Ren, S., Chen, J., Liu, S., Zhang, G., Zhao, H., Wang, L., & Xue, Q. (2017). Anticorrosive performance of waterborne epoxy coatings containing water-dispersible hexagonal boron nitride (h-BN) nanosheets. *Applied Surface Science*, *397*, 77–86. https://doi.org/10.1016/J.APSUSC.2016.11.141

Dai, Z., Liu, L., Zhang, Z., Dai, Z., Liu, L., & Zhang, Z. (2019). Strain engineering of 2D materials: issues and opportunities at the interface. *Advanced Materials*, *31*(45), 1805417. https://doi.org/10.1002/ADMA.201805417

Das, O., Kim, N. K., Kalamkarov, A. L., Sarmah, A. K., & Bhattacharyya, D. (2017). Biochar to the rescue: balancing the fire performance and mechanical properties of polypropylene composites. *Polymer Degradation and Stability*, *144*, 485–496. https://doi.org/10.1016/J.POLYMDEGRADSTAB.2017.09.006

Deng, J., Ren, P., Deng, D., Yu, L., Yang, F., & Bao, X. (2014). Highly active and durable non-precious-metal catalysts encapsulated in carbon nanotubes for hydrogen evolution reaction. *Energy & Environmental Science*, *7*(6), 1919–1923. https://doi.org/10.1039/C4EE00370E

Deokar, G., Jin, J., Schwingenschlögl, U., & Costa, P. M. F. J. (2022). Chemical vapor deposition-grown nitrogen-doped graphene's synthesis, characterization and applications. N*pj 2D Materials and Applications 2022*, *6*(1), 1–17. https://doi.org/10.1038/s41699-022-00287-8

Dong, Z., Xu, H., Liang, F., Luo, C., Wang, C., Cao, Z. Y., Chen, X. J., Zhang, J., & Wu, X. (2019). Raman characterization on two-dimensional materials-based thermoelectricity. *Molecules*, *24*(1). https://doi.org/10.3390/MOLECULES24010088

Dresselhaus, M. S., & Dresselhaus, G., (2002). Intercalation compounds of graphite. *Advances in Physics*, *51*(1), 1–186. https://doi.org/10.1080/00018730110113644

Dreyer, D. R., Park, S., Bielawski, C. W., & Ruoff, R. S. (2009). The chemistry of graphene oxide. *Chemical Society Reviews*, *39*(1), 228–240. https://doi.org/10.1039/B917103G

Eda, G., Yamaguchi, H., Voiry, D., Fujita, T., Chen, M., & Chhowalla, M. (2011). Photoluminescence from chemically exfoliated MoS 2. *Nano Letters*, *11*(12), 5111–5116. https://doi.org/10.1021/NL201874W/SUPPL_FILE/NL201874W_SI_001.PDF

Falin, A., Cai, Q., Santos, E. J. G., Scullion, D., Qian, D., Zhang, R., Yang, Z., Huang, S., Watanabe, K., Taniguchi, T., Barnett, M. R., Chen, Y., Ruoff, R. S., & Li, L. H. (2017). Mechanical properties of atomically thin boron nitride and the role of interlayer interactions. *Nature Communications*, *8*(1), 1–9. https://doi.org/10.1038/ncomms15815

Fan, S., Vu, Q. A., Tran, M. D., Adhikari, S., Lee, Y. H., Fan, S., Vu, Q. A., Tran, M. D., Adhikari, S., & Lee, Y. H. (2020). Transfer assembly for two-dimensional van der Waals heterostructures. *2D Materials*, *7*(2). https://doi.org/10.1088/2053-1583/AB7629

Fang, H., Battaglia, C., Carraro, C., Nemsak, S., Ozdol, B., Kang, J. S., Bechtel, H. A., Desai, S. B., Kronast, F., Unal, A. A., Conti, G., Conlon, C., Palsson, G. K., Martin, M. C., Minor, A. M., Fadley, C. S., Yablonovitch, E., Maboudian, R., & Javey, A. (2014). Strong interlayer coupling in van der Waals heterostructures built from single-layer chalcogenides. *Proceedings of the National Academy of Sciences of the United States of America*, *111*(17), 6198–6202. https://doi.org/10.1073/PNAS.1405435111/SUPPL_FILE/PNAS.201405435SI.PDF

Farooq, M. U., Hashmi, A., & Hong, J. (2015). Thickness dependent optical properties of multilayer BN/Graphene/BN. *Surface Science*, *634*, 25–30. https://doi.org/10.1016/J.SUSC.2014.11.007

Fiori, G., Bonaccorso, F., Iannaccone, G., Palacios, T., Neumaier, D., Seabaugh, A., Banerjee, S. K., & Colombo, L. (2014). Electronics based on two-dimensional materials. *Nature Nanotechnology*, *9*(10), 768–779. https://doi.org/10.1038/nnano.2014.207

Fujihara, M., Inoue, R., Kurita, R., Taniuchi, T., Motoyui, Y., Shin, S., Komori, F., Maniwa, Y., Shinohara, H., & Miyata, Y. (2015). Selective formation of zigzag edges in graphene cracks. *ACS Nano*, *9*(9), 9027–9033. https://doi.org/10.1021/ACSNANO.5B03079/ASSET/IMAGES/NN-2015-03079Q_M001.GIF

Galbiati, M., Stoot, A. C., Mackenzie, D. M. A., Bøggild, P., & Camilli, L. (2017). Real-time oxide evolution of copper protected by graphene and boron nitride barriers. *Scientific Reports*, *7*(1), 1–7. https://doi.org/10.1038/srep39770

Gallagher, P., Lee, M., Amet, F., Maksymovych, P., Wang, J., Wang, S., Lu, X., Zhang, G., Watanabe, K., Taniguchi, T., & Goldhaber-Gordon, D. (2016). Switchable friction enabled by nanoscale self-assembly on graphene. *Nature Communications*, *7*(1), 1–7. https://doi.org/10.1038/ncomms10745

Gao, Y., Liu, Z., Sun, D. M., Huang, L., Ma, L. P., Yin, L. C., Ma, T., Zhang, Z., Ma, X. L., Peng, L. M., Cheng, H. M., & Ren, W. (2015). Large-area synthesis of high-quality and uniform monolayer WS2 on reusable Au foils. *Nature Communications*, *6*(1), 1–10. https://doi.org/10.1038/ncomms9569

Gao, H., Suh, J., Cao, M. C., Joe, A. Y., Mujid, F., Lee, K. H., Xie, S., Xie, S., Lee, J. U., Kang, K., Kim, P., Muller, D. A., & Park, J. (2020). Tuning electrical conductance of MoS2 monolayers through substitutional doping. *Nano Letters*, *20*(6), 4095–4101. https://doi.org/10.1021/ACS.NANOLETT.9B05247/ASSET/IMAGES/LARGE/NL9B05247_0005.JPEG

Geim, A. K., & Grigorieva, I. V. (2013). Van der Waals heterostructures. *Nature*, *499*(7459), 419–425. https://doi.org/10.1038/nature12385

Giesbers, A. J. M., Zeitler, U., Neubeck, S., Freitag, F., Novoselov, K. S., & Maan, J. C. (2008). Nanolithography and manipulation of graphene using an atomic force microscope. *Solid State Communications*, *147*(9–10), 366–369. https://doi.org/10.1016/J.SSC.2008.06.027

Göncü, Y., Geçgin, M., Bakan, F., & Ay, N. (2017). Electrophoretic deposition of hydroxyapatite-hexagonal boron nitride composite coatings on Ti substrate. *Materials Science and Engineering: C*, *79*, 343–353. https://doi.org/10.1016/J.MSEC.2017.05.023

Gorbachev, R. V., Riaz, I., Nair, R. R., Jalil, R., Britnell, L., Belle, B. D., Hill, E. W., Novoselov, K. S., Watanabe, K., Taniguchi, T., Geim, A. K., & Blake, P. (2011). Hunting for monolayer boron nitride: optical and Raman signatures. *Small (Weinheim an Der Bergstrasse, Germany)*, *7*(4), 465–468. https://doi.org/10.1002/SMLL.201001628

Gupta, A., Sakthivel, T., & Seal, S. (2015). Recent development in 2D materials beyond graphene. *Progress in Materials Science*, *73*, 44–126. https://doi.org/10.1016/J.PMATSCI.2015.02.002

Hao, S., Zhao, X., Cheng, Q., Xing, Y., Ma, W., Wang, X., Zhao, G., & Xu, X. (2020). A mini review of the preparation and photocatalytic properties of two-dimensional materials. *Frontiers in Chemistry*, *8*, 1177. https://doi.org/10.3389/FCHEM.2020.582146/BIBTEX

He, Q., Lin, Z., Ding, M., Yin, A., Halim, U., Wang, C., Liu, Y., Cheng, H. C., Huang, Y., & Duan, X. (2019). In situ probing molecular intercalation in two-dimensional layered semiconductors. *Nano Letters*, *19*(10), 6819–6826. https://doi.org/10.1021/ACS.NANOLETT.9B01898/ASSET/IMAGES/LARGE/NL9B01898_0004.JPEG

Heersche, H. B., Jarillo-Herrero, P., Oostinga, J. B., Vandersypen, L. M. K., & Morpurgo, A. F. (2007). Induced superconductivity in graphene. *Solid State Communications*, *143*(1–2), 72–76. https://doi.org/10.1016/J.SSC.2007.02.044

Hoffman, A. N., Stanford, M. G., Sales, M. G., Zhang, C., Ivanov, I. N., McDonnell, S. J., Mandrus, D. G., & Rack, P. D. (2019). Tuning the electrical properties of WSe2 via O2 plasma oxidation: towards lateral homojunctions. *2D Materials*, *6*(4), 045024. https://doi.org/10.1088/2053-1583/AB2FA7

Hong, H. S., Phuong, N. H., Huong, N. T., Nam, N. H., & Hue, N. T. (2019). Highly sensitive and low detection limit of resistive NO2 gas sensor based on a MoS2/graphene two-dimensional heterostructures. *Applied Surface Science*, *492*, 449–454. https://doi.org/10.1016/J.APSUSC.2019.06.230

Howard, C. A. (2021). Exfoliating large monolayers in liquids. *Nature Materials*, *20*(2), 130–131. https://doi.org/10.1038/S41563-020-00907-Y

Huang, J. K., Pu, J., Hsu, C. L., Chiu, M. H., Juang, Z. Y., Chang, Y. H., Chang, W. H., Iwasa, Y., Takenobu, T., & Li, L. J. (2014). Large-area synthesis of highly crystalline WSe(2) monolayers and device applications. *ACS Nano*, *8*(1), 923–930. https://doi.org/10.1021/NN405719X

Huang, Y., Sutter, E., Shi, N. N., Zheng, J., Yang, T., Englund, D., Gao, H. J., & Sutter, P. (2015). Reliable exfoliation of large-area high-quality flakes of graphene and other two-dimensional materials. *ACS Nano*, *9*(11), 10612–10620. https://doi.org/10.1021/ACSNANO.5B04258/ASSET/IMAGES/LARGE/NN-2015-04258Q_0008.JPEG

Ionescu, M. I., Sun, X., & Luan, B. (2015). Multilayer graphene synthesized using magnetron sputtering for planar supercapacitor application. *Canadian Journal of Chemistry*, *93*(2), 160–164. https://doi.org/10.1139/CJC-2014-0297/ASSET/IMAGES/LARGE/CJC-2014-0297F3.JPEG

Jariwala, D., Sangwan, V. K., Lauhon, L. J., Marks, T. J., & Hersam, M. C. (2014). Emerging device applications for semiconducting two-dimensional transition metal dichalcogenides. *ACS Nano*, *8*(2), 1102–1120. https://doi.org/10.1021/NN500064S/ASSET/IMAGES/LARGE/NN-2014-00064S_0010.JPEG

Ji, Q., Zhang, Y., Gao, T., Zhang, Y., Ma, D., Liu, M., Chen, Y., Qiao, X., Tan, P. H., Kan, M., Feng, J., Sun, Q., & Liu, Z. (2013). Epitaxial monolayer MoS2 on mica with novel photoluminescence. *Nano Letters*, *13*(8), 3870–3877. https://doi.org/10.1021/NL401938T/SUPPL_FILE/NL401938T_SI_001.PDF

Jiang, J. W., & Park, H. S. (2014). Mechanical properties of single-layer black phosphorus. *Journal of Physics D: Applied Physics*, *47*(38), 385304. https://doi.org/10.1088/0022-3727/47/38/385304

Kang, Z., Peng, Y., Qian, Y., Yuan, D., Addicoat, M. A., Heine, T., Hu, Z., Tee, L., Guo, Z., & Zhao, D. (2016). Mixed Matrix Membranes (MMMs) comprising exfoliated 2D covalent organic frameworks (COFs) for efficient CO2 separation. *Chemistry of Materials*, *28*(5), 1277–1285. https://doi.org/10.1021/ACS.CHEMMATER.5B02902/ASSET/IMAGES/LARGE/CM-2015-02902N_0008.JPEG

Kappera, R., Voiry, D., Yalcin, S. E., Branch, B., Gupta, G., Mohite, A. D., & Chhowalla, M. (2014). Phase-engineered low-resistance contacts for ultrathin MoS2 transistors. *Nature Materials*, *13*(12), 1128–1134. https://doi.org/10.1038/NMAT4080

Khan, K., Tareen, A. K., Aslam, M., Wang, R., Zhang, Y., Mahmood, A., Ouyang, Z., Zhang, H., & Guo, Z. (2020). Recent developments in emerging two-dimensional materials and their applications. *Journal of Materials Chemistry C*, *8*(2), 387–440. https://doi.org/10.1039/C9TC04187G

Khosravi, F., Nouri Khorasani, S., Rezvani Ghomi, E., Karimi Kichi, M., Zilouei, H., Farhadian, M., & Esmaeely Neisiany, R. (2019). A bilayer GO/nanofibrous biocomposite coating to enhance 316L stainless steel corrosion performance. *Materials Research Express*, *6*(8), 086470. https://doi.org/10.1088/2053-1591/AB26D5

Kim, K., Artyukhov, V. I., Regan, W., Liu, Y., Crommie, M. F., Yakobson, B. I., & Zettl, A. (2011). Ripping graphene: preferred directions. *Nano Letters*, *12*(1), 293–297. https://doi.org/10.1021/NL203547Z

Kim, J. H., Jeong, J. H., Kim, N., Joshi, R., & Lee, G. H. (2019). Mechanical properties of two-dimensional materials and their applications. *Journal of Physics D: Applied Physics*, *52*(8), 083001. https://doi.org/10.1088/1361-6463/AAF465

Kim, S. M., Hsu, A., Park, M. H., Chae, S. H., Yun, S. J., Lee, J. S., Cho, D. H., Fang, W., Lee, C., Palacios, T., Dresselhaus, M., Kim, K. K., Lee, Y. H., & Kong, J. (2015). Synthesis of large-area multilayer hexagonal boron nitride for high material performance. *Nature Communications*, *6*(1), 1–11. https://doi.org/10.1038/ncomms9662

Kim, T., Kang, D., Lee, Y., Hong, S., Gon Shin, H., Bae, H., Yi, Y., Kim, K., & Im, S. (2020). 2D TMD channel transistors with ZnO nanowire gate for extended nonvolatile memory applications. *Advanced Functional Materials*, *30*(40), 2004140. https://doi.org/10.1002/ADFM.202004140

Kim, C., Yoon, M.-A., Jang, B., Kim, J.-H., & Kim, K.-S. (2020). A review on transfer process of two-dimensional materials. *Tribology and Lubricants*, *36*(1), 1–10. https://doi.org/10.9725/KTS.2020.36.1.1

Ko, C., Lee, Y., Chen, Y., Suh, J., Fu, D., Suslu, A., Lee, S., Clarkson, J. D., Choe, H. S., Tongay, S., Ramesh, R., & Wu, J. (2016). Ferroelectrically gated atomically thin transition-metal dichalcogenides as nonvolatile memory. *Advanced Materials*, *28*(15), 2923–2930. https://doi.org/10.1002/ADMA.201504779

Lee, J. W., Dai, Z., Han, T. H., Choi, C., Chang, S. Y., Lee, S. J., De Marco, N., Zhao, H., Sun, P., Huang, Y., & Yang, Y. (2018). 2D perovskite stabilized phase-pure formamidinium perovskite solar cells. *Nature Communications*, *9*(1), 1–10. https://doi.org/10.1038/s41467-018-05454-4

Lee, Y. T., Kwon, H., Kim, J. S., Kim, H. H., Lee, Y. J., Lim, J. A., Song, Y. W., Yi, Y., Choi, W. K., Hwang, D. K., & Im, S. (2015). Nonvolatile ferroelectric memory circuit using black phosphorus nanosheet-based field-effect transistors with P(VDF-TrFE) polymer. *ACS Nano*, *9*(10), 10394–10401. https://doi.org/10.1021/ACSNANO.5B04592/ASSET/IMAGES/LARGE/NN-2015-04592M_0005.JPEG

Lee, H. C., Liu, W. W., Chai, S. P., Mohamed, A. R., Aziz, A., Khe, C. S., Hidayah, N. M. S., & Hashim, U. (2017). Review of the synthesis, transfer, characterization and growth mechanisms of single and multilayer graphene. *RSC Advances*, *7*(26), 15644–15693. https://doi.org/10.1039/C7RA00392G

Lee, H. S., Min, S. W., Park, M. K., Lee, Y. T., Jeon, P. J., Kim, J. H., Ryu, S., & Im, S. (2012). MoS2 nanosheets for top-gate nonvolatile memory transistor channel. *Small*, *8*(20), 3111–3115. https://doi.org/10.1002/SMLL.201200752

Lee, K. H., Shin, H. J., Lee, J., Lee, I. Y., Kim, G. H., Choi, J. Y., & Kim, S. W. (2012). Large-scale synthesis of high-quality hexagonal boron nitride nanosheets for large-area graphene electronics. *Nano Letters*, *12*(2), 714–718. https://doi.org/10.1021/NL203635V

Lee, C., Wei, X., Kysar, J. W., & Hone, J. (2008). Measurement of the elastic properties and intrinsic strength of monolayer graphene. *Science*, *321*(5887), 385–388. https://doi.org/10.1126/SCIENCE.1157996

Lee, Y. H., Yu, L., Wang, H., Fang, W., Ling, X., Shi, Y., Lin, C. Te, Huang, J. K., Chang, M. T., Chang, C. S., Dresselhaus, M., Palacios, T., Li, L. J., & Kong, J. (2013). Synthesis and transfer of single-layer transition metal disulfides on diverse surfaces. *Nano Letters*, *13*(4), 1852–1857. https://doi.org/10.1021/NL400687N

Lei, W., Mochalin, V. N., Liu, D., Qin, S., Gogotsi, Y., & Chen, Y. (2015). Boron nitride colloidal solutions, ultralight aerogels and freestanding membranes through one-step exfoliation and functionalization. *Nature Communications*, *6*(1), 1–8. https://doi.org/10.1038/ncomms9849

Li, Xuesong, Cai, W., An, J., Kim, S., Nah, J., Yang, D., Piner, R., Velamakanni, A., Jung, I., Tutuc, E., Banerjee, S. K., Colombo, L., & Ruoff, R. S. (2009). Large-area synthesis of high-quality and uniform graphene films on copper foils. *Science*, *324*(5932), 1312–1314. https://doi.org/10.1126/SCIENCE.1171245/SUPPL_FILE/LI.SOM.PDF

Li, J., Medhekar, N. V., & Shenoy, V. B. (2013). Bonding charge density and ultimate strength of monolayer transition metal dichalcogenides. *Journal of Physical Chemistry C*, *117*(30), 15842–15848. https://doi.org/10.1021/JP403986V/ASSET/IMAGES/LARGE/JP-2013-03986V_0006.JPEG

Li, S., Wang, S., Tang, D.-M., Zhao, W., Xu, H., Chu, L., Bando, Y., Golberg, D., & Eda, G. (2015). Halide-assisted atmospheric pressure growth of large WSe2 and WS2 monolayer crystals. *Applied Materials Today*, *1*(1), 60–66. https://doi.org/10.1016/j.apmt.2015.09.001

Li, H., Yin, Z., He, Q., Li, H., Huang, X., Lu, G., Fam, D. W. H., Tok, A. I. Y., Zhang, Q., & Zhang, H. (2012). Fabrication of single- and multilayer MoS2 film-based field-effect transistors for sensing NO at room temperature. *Small*, *8*(1), 63–67. https://doi.org/10.1002/SMLL.201101016

Li, Xiao, & Zhu, H. (2015). Two-dimensional MoS2: properties, preparation, and applications. *Journal of Materiomics*, *1*(1), 33–44. https://doi.org/10.1016/J.JMAT.2015.03.003

Li, Xuesong, Zhu, Y., Cai, W., Borysiak, M., Han, B., Chen, D., Piner, R. D., Colomba, L., & Ruoff, R. S. (2009). Transfer of large-area graphene films for high-performance transparent conductive electrodes. *Nano Letters*, *9*(12), 4359–4363. https://doi.org/10.1021/NL902623Y/ASSET/IMAGES/LARGE/NL-2009-02623Y_0005.JPEG

Lin, Z., Liu, Y., Halim, U., Ding, M., Liu, Y., Wang, Y., Jia, C., Chen, P., Duan, X., Wang, C., Song, F., Li, M., Wan, C., Huang, Y., & Duan, X. (2018). Solution-processable 2D semiconductors for high-performance large-area electronics. *Nature*, *562*(7726), 254–258. https://doi.org/10.1038/S41586-018-0574-4

Lin, Y. C., Zhang, W., Huang, J. K., Liu, K. K., Lee, Y. H., Liang, C. Te, Chu, C. W., & Li, L. J. (2012). Wafer-scale MoS2 thin layers prepared by MoO3 sulfurization. *Nanoscale*, *4*(20), 6637–6641. https://doi.org/10.1039/C2NR31833D

Liu, L., Feng, P., & Shen, X. (2003). Structural and electronic properties of h-BN. *Physical Review B*, *68*(10), 104102. https://doi.org/10.1103/PhysRevB.68.104102

Liu, X., & Hersam, M. C. (2018). Interface characterization and control of 2D materials and heterostructures. *Advanced Materials*, *30*(39), 1801586. https://doi.org/10.1002/ADMA.201801586

Liu, X., & Hersam, M. C. (2019). 2D materials for quantum information science. *Nature Reviews Materials*, *4*(10), 669–684. https://doi.org/10.1038/s41578-019-0136-x

Ly, T. H., Zhao, J., Cichocka, M. O., Li, L. J., & Lee, Y. H. (2017). Dynamical observations on the crack tip zone and stress corrosion of two-dimensional MoS2. *Nature Communications*, *8*(1), 1–7. https://doi.org/10.1038/ncomms14116

Ma, Y., Dai, Y., Guo, M., Niu, C., Lu, J., & Huang, B. (2011). Electronic and magnetic properties of perfect, vacancy-doped, and nonmetal adsorbed MoSe2, MoTe2 and WS2 monolayers. *Physical Chemistry Chemical Physics*, *13*(34), 15546–15553. https://doi.org/10.1039/C1CP21159E

Ma, L.-P., Ren, W., Cheng, H.-M., Ma, P. L., Ren, W. C., & Cheng, H.-M. (2019). Transfer methods of graphene from metal substrates: a review. *Small Methods*, *3*(7), 1900049. https://doi.org/10.1002/SMTD.201900049

Ma, D., Shi, J., Ji, Q., Chen, K., Yin, J., Lin, Y., Zhang, Y., Liu, M., Feng, Q., Song, X., Guo, X., Zhang, J., Zhang, Y., & Liu, Z. (2015). A universal etching-free transfer of MoS2 films for applications in photodetectors. *Nano Research*, *8*(11), 3662–3672. https://doi.org/10.1007/S12274-015-0866-Z

Mahvash, F., Eissa, S., Bordjiba, T., Tavares, A. C., Szkopek, T., & Siaj, M. (2017). Corrosion resistance of monolayer hexagonal boron nitride on copper. *Scientific Reports*, *7*(1), 1–5. https://doi.org/10.1038/srep42139

Mak, K. F., Lee, C., Hone, J., Shan, J., & Heinz, T. F. (2010). Atomically thin MoS2: a new direct-gap semiconductor. *Physical Review Letters*, *105*(13). https://doi.org/10.1103/PHYSREVLETT.105.136805

Morozov, S. V., Novoselov, K. S., Katsnelson, M. I., Schedin, F., Elias, D. C., Jaszczak, J. A., & Geim, A. K. (2008). Giant intrinsic carrier mobilities in graphene and its bilayer. *Physical Review Letters*, *100*(1), 016602. https://doi.org/10.1103/PHYSREVLETT.100.016602/FIGURES/4/MEDIUM

Najmaei, S., Liu, Z., Zhou, W., Zou, X., Shi, G., Lei, S., Yakobson, B. I., Idrobo, J. C., Ajayan, P. M., & Lou, J. (2013). Vapour phase growth and grain boundary structure of molybdenum disulphide atomic layers. *Nature Materials*, *12*(8), 754–759. https://doi.org/10.1038/nmat3673

Nicolosi, V., Chhowalla, M., Kanatzidis, M. G., Strano, M. S., & Coleman, J. N. (2013). Liquid exfoliation of layered materials. *Science*, *340*(6139). https://doi.org/10.1126/SCIENCE.1226419/ASSET/DF023776-42C8-403D-8B5B-5D426A69D270/ASSETS/GRAPHIC/340_1226419_F4.JPEG

Niu, L., Coleman, J. N., Zhang, H., Shin, H., Chhowalla, M., & Zheng, Z. (2016). Production of two-dimensional nanomaterials via liquid-based direct exfoliation. *Small*, *12*(3), 272–293. https://doi.org/10.1002/SMLL.201502207

Nose, K., Oba, H., & Yoshida, T. (2006). Electric conductivity of boron nitride thin films enhanced by in situ doping of zinc. *Applied Physics Letters*, *89*(11), 112124. https://doi.org/10.1063/1.2354009

Novoselov, K. S., Geim, A. K., Morozov, S. V., Jiang, D., Zhang, Y., Dubonos, S. V., Grigorieva, I. V., & Firsov, A. A. (2004). Electric field in atomically thin carbon films. *Science*, *306*(5696), 666–669. https://doi.org/10.1126/SCIENCE.1102896/SUPPL_FILE/NOVOSELOV.SOM.PDF

Novoselov, K. S., Jiang, D., Schedin, F., Booth, T. J., Khotkevich, V. V., Morozov, S. V., & Geim, A. K. (2005). Two-dimensional atomic crystals. *Proceedings of the National Academy of Sciences of the United States of America*, *102*(30), 10451–10453. https://doi.org/10.1073/PNAS.0502848102

Novoselov, K. S., Mishchenko, A., Carvalho, A., & Castro Neto, A. H. (2016). 2D materials and van der Waals heterostructures. *Science*, *353*(6298). https://doi.org/10.1126/SCIENCE.AAC9439/ASSET/4B54077E-F0A2-4BB2-BFA1-BC557DC78A9D/ASSETS/GRAPHIC/353_AAC9439_F6.JPEG

Oliva-Leyva, M., & Naumis, G. G. (2014). Anisotropic AC conductivity of strained graphene. *Journal of Physics: Condensed Matter*, *26*(12), 125302. https://doi.org/10.1088/0953-8984/26/12/125302

Ozkan, T., Naraghi, M., & Chasiotis, I. (2010). Mechanical properties of vapor grown carbon nanofibers. *Carbon*, *48*(1), 239–244. https://doi.org/10.1016/J.CARBON.2009.09.011

Pang, S., Hernandez, Y., Feng, X., Müllen, K., Pang, S., Hernandez, Y., Feng, X., & Müllen, K. (2011). Graphene as transparent electrode material for organic electronics. *Advanced Materials*, *23*(25), 2779–2795. https://doi.org/10.1002/ADMA.201100304

Parra, C., Montero-Silva, F., Henríquez, R., Flores, M., Garín, C., Ramírez, C., Moreno, M., Correa, J., Seeger, M., & Häberle, P. (2015). Suppressing bacterial interaction with copper surfaces through graphene and hexagonal-boron nitride coatings. *ACS Applied Materials and Interfaces*, *7*(12), 6430–6437. https://doi.org/10.1021/ACSAMI.5B01248

Pellegrino, F. M. D., Angilella, G. G. N., & Pucci, R. (2011). Linear response correlation functions in strained graphene. *Physical Review B: Condensed Matter and Materials Physics*, *84*(19), 195407. https://doi.org/10.1103/PHYSREVB.84.195407/FIGURES/1/MEDIUM

Peres, N. M. R., Guinea, F., & Castro Neto, A. H. (2006). Electronic properties of disordered two-dimensional carbon. *Physical Review B: Condensed Matter and Materials Physics*, *73*(12), 125411. https://doi.org/10.1103/PHYSREVB.73.125411/FIGURES/18/MEDIUM

Peres, N. M. R., Lopes Dos Santos, J. M. B., & Stauber, T. (2007). Phenomenological study of the electronic transport coefficients of graphene. *Physical Review B: Condensed Matter and Materials Physics*, *76*(7), 073412. https://doi.org/10.1103/PHYSREVB.76.073412/FIGURES/1/MEDIUM

Phan, L., Walkup IV, W. G., Ordinario, D. D., Karshalev, E., Jocson, J. M., Burke, A. M., & Gorodetsky, A. A. (2013). Reconfigurable infrared camouflage coatings from a cephalopod protein. *Advanced Materials*, *25*(39), 5621–5625. https://doi.org/10.1002/ADMA.201301472

Ramanathan, A., & Aqra, M. (2021). Review of the recent advances in nano-biosensors and technologies for healthcare applications. In B. Piro (Ed.), *CSAC2021: 1st International Electronic Conference on Chemical Sensors and Analytical Chemistry*, p. 76. MDPI: Basel. https://doi.org/10.3390/CSAC2021-10473

Reina, A., Jia, X., Ho, J., Nezich, D., Son, H., Bulovic, V., Dresselhaus, M. S., & Jing, K. (2009). Large area, few-layer graphene films on arbitrary substrates by chemical vapor deposition. *Nano Letters*, *9*(1), 30–35. https://doi.org/10.1021/NL801827V/SUPPL_FILE/NL801827V_SI_003.PDF

Ribeiro, H., Trigueiro, J. P. C., Owuor, P. S., Machado, L. D., Woellner, C. F., Pedrotti, J. J., Jaques, Y. M., Kosolwattana, S., Chipara, A., Silva, W. M., Silva, C. J. R., Galvão, D. S., Chopra, N., Odeh, I. N., Tiwary, C. S., Silva, G. G., & Ajayan, P. M. (2018). Hybrid 2D nanostructures for mechanical reinforcement and thermal conductivity enhancement in polymer composites. *Composites Science and Technology*, *159*, 103–110. https://doi.org/10.1016/J.COMPSCITECH.2018.01.032

Rivera, P., Schaibley, J. R., Jones, A. M., Ross, J. S., Wu, S., Aivazian, G., Klement, P., Seyler, K., Clark, G., Ghimire, N. J., Yan, J., Mandrus, D. G., Yao, W., & Xu, X. (2015). Observation of long-lived interlayer excitons in monolayer MoSe2-WSe2 heterostructures. *Nature Communications*, *6*. https://doi.org/10.1038/NCOMMS7242

Roy, K., Padmanabhan, M., Goswami, S., Sai, T. P., Ramalingam, G., Raghavan, S., & Ghosh, A. (2013). Graphene-MoS2 hybrid structures for multifunctional photoresponsive memory devices. *Nature Nanotechnology*, *8*(11), 826–830. https://doi.org/10.1038/NNANO.2013.206

Salihoglu, O., Uzlu, H. B., Yakar, O., Aas, S., Balci, O., Kakenov, N., Balci, S., Olcum, S., Süzer, S., & Kocabas, C. (2018). Graphene-based adaptive thermal camouflage. *Nano Letters*, *18*(7), 4541–4548. https://doi.org/10.1021/ACS.NANOLETT.8B01746/SUPPL_FILE/NL8B01746_SI_004.AVI

Sang, X., Li, X., Puretzky, A. A., Geohegan, D. B., Xiao, K., & Unocic, R. R. (2019). Atomic insight into thermolysis-driven growth of 2D MoS2. *Advanced Functional Materials*, *29*(52), 1902149. https://doi.org/10.1002/ADFM.201902149

Schneider, G. F., Calado, V. E., Zandbergen, H., Vandersypen, L. M. K., & Dekker, C. (2010). Wedging transfer of nanostructures. *Nano Letters*, *10*(5), 1912–1916. https://doi.org/10.1021/NL1008037/ASSET/IMAGES/LARGE/NL-2010-008037_0004.JPEG

Shanmugam, V., Das, O., Neisiany, R. E., Babu, K., Singh, S., Hedenqvist, M. S., Berto, F., & Ramakrishna, S. (2020). Polymer recycling in additive manufacturing: an opportunity for the circular economy. *Materials Circular Economy*, *2*(1), 1–11. https://doi.org/10.1007/S42824-020-00012-0

Shanmugam, V., Mensah, R. A., Babu, K., Gawusu, S., Chanda, A., Tu, Y., Neisiany, R. E., Försth, M., Sas, G., & Das, O. (2022). A review of the synthesis, properties, and applications of 2D materials. *Particle & Particle Systems Characterization*, *39*(6), 2200031. https://doi.org/10.1002/PPSC.202200031

Shaw, Z. L., Kuriakose, S., Cheeseman, S., Dickey, M. D., Genzer, J., Christofferson, A. J., Crawford, R. J., McConville, C. F., Chapman, J., Truong, V. K., Elbourne, A., & Walia, S. (2021). Antipathogenic properties and applications of low-dimensional materials. *Nature Communications*, *12*(1), 1–19. https://doi.org/10.1038/s41467-021-23278-7

Shenderova, O. A., Zhirnov, V. V., & Brenner, D. W. (2002). Carbon nanostructures. *Critical Reviews in Solid State and Materials Sciences, 27*(3–4), 227–356. https://doi.org/10.1080/10408430208500497

Shi, E., Gao, Y., Finkenauer, B. P., Akriti, A., Coffey, A. H., & Dou, L. (2018). Two-dimensional halide perovskite nanomaterials and heterostructures. *Chemical Society Reviews*, *47*(16), 6046–6072. https://doi.org/10.1039/C7CS00886D

Shi, Y., Li, H., & Li, L. J. (2015). Recent advances in controlled synthesis of two-dimensional transition metal dichalcogenides via vapour deposition techniques. *Chemical Society Reviews*, *44*(9), 2744–2756. https://doi.org/10.1039/C4CS00256C

Simsek, E. (2013). A closed-form approximate expression for the optical conductivity of graphene. *Optics Letters*, *38*(9), 1437–1439. https://doi.org/10.1364/OL.38.001437

Singh, A., Moun, M., Sharma, M., Barman, A., Kapoor, A. K., & Singh, R. (2020). NaCl-assisted substrate dependent 2D planar nucleated growth of MoS2. *Applied Surface Science*, *538*. https://doi.org/10.1016/j.apsusc.2020.148201

Smith, J. B., Hagaman, D., & Ji, H. F. (2016). Growth of 2D black phosphorus film from chemical vapor deposition. *Nanotechnology*, *27*(21). https://doi.org/10.1088/0957-4484/27/21/215602

Song, L., Ci, L., Lu, H., Sorokin, P. B., Jin, C., Ni, J., Kvashnin, A. G., Kvashnin, D. G., Lou, J., Yakobson, B. I., & Ajayan, P. M. (2010). Large scale growth and characterization of atomic hexagonal boron nitride layers. *Nano Letters*, *10*(8), 3209–3215. https://doi.org/10.1021/NL1022139/SUPPL_FILE/NL1022139_SI_001.PDF

Sun, J., Lu, C., Song, Y., Ji, Q., Song, X., Li, Q., Zhang, Y., Zhang, L., Kong, J., & Liu, Z. (2018). Recent progress in the tailored growth of two-dimensional hexagonal boron nitride via chemical vapour deposition. *Chemical Society Reviews*, *47*(12), 4242–4257. https://doi.org/10.1039/C8CS00167G

Sun, C., Zhang, J., Ma, J., Liu, P., Gao, D., Tao, K., & Xue, D. (2016). N-doped WS2 nanosheets: a high-performance electrocatalyst for the hydrogen evolution reaction. *Journal of Materials Chemistry A*, *4*(29), 11234–11238. https://doi.org/10.1039/C6TA04082A

Sutter, P., Lahiri, J., Albrecht, P., & Sutter, E. (2011). Chemical vapor deposition and etching of high-quality monolayer hexagonal boron nitride films. *ACS Nano*, *5*(9), 7303–7309. https://doi.org/10.1021/NN202141K

Tan, C., Cao, X., Wu, X. J., He, Q., Yang, J., Zhang, X., Chen, J., Zhao, W., Han, S., Nam, G. H., Sindoro, M., & Zhang, H. (2017). Recent advances in ultrathin two-dimensional nanomaterials. *Chemical Reviews*, *117*(9), 6225–6331. https://doi.org/10.1021/ACS.CHEMREV.6B00558

Tao, J., Shen, W., Wu, S., Liu, L., Feng, Z., Wang, C., Hu, C., Yao, P., Zhang, H., Pang, W., Duan, X., Liu, J., Zhou, C., & Zhang, D. (2015). Mechanical and electrical anisotropy of few-layer black phosphorus. *ACS Nano*, *9*(11), 11362–11370. https://doi.org/10.1021/ACSNANO.5B05151/ASSET/IMAGES/NN-2015-051517_M006.GIF

Tongay, S., Berke, K., Lemaitre, M., Nasrollahi, Z., Tanner, D. B., Hebard, A. F., & Appleton, B. R. (2011). Stable hole doping of graphene for low electrical resistance and high optical transparency. *Nanotechnology*, *22*(42). https://doi.org/10.1088/0957-4484/22/42/425701

Van Khai, T., Na, H. G., Kwak, D. S., Kwon, Y. J., Ham, H., Shim, K. B., & Kim, H. W. (2012). Significant enhancement of blue emission and electrical conductivity of N-doped graphene. *Journal of Materials Chemistry*, *22*(34), 17992–18003. https://doi.org/10.1039/C2JM33194B

Walker, P. (1981). *Chemistry and Physics of Carbon : A Series of Advances*, Vol. 17. Marcel Dekker: New York.

Wang, Xingli, Gong, Y., Shi, G., Chow, W. L., Keyshar, K., Ye, G., Vajtai, R., Lou, J., Liu, Z., Ringe, E., Tay, B. K., & Ajayan, P. M. (2014). Chemical vapor deposition growth of crystalline monolayer MoSe2. *ACS Nano*, *8*(5), 5125–5131. https://doi.org/10.1021/NN501175K/SUPPL_FILE/NN501175K_SI_001.PDF

Wang, Q. H., Kalantar-Zadeh, K., Kis, A., Coleman, J. N., & Strano, M. S. (2012). Electronics and optoelectronics of two-dimensional transition metal dichalcogenides. *Nature Nanotechnology*, *7*(11), 699–712. https://doi.org/10.1038/nnano.2012.193

Wang, Xinchen, Maeda, K., Thomas, A., Takanabe, K., Xin, G., Carlsson, J. M., Domen, K., & Antonietti, M. (2009). A metal-free polymeric photocatalyst for hydrogen production from water under visible light. *Nature Materials*, *8*(1), 76–80. https://doi.org/10.1038/NMAT2317

Wang, N., Mao, N., Wang, Z., Yang, X., Zhou, X., Liu, H., Qiao, S., Lei, X., Wang, J., Xu, H., Ling, X., Zhang, Q., Feng, Q., & Kong, J. (2021). Electrochemical delamination of ultralarge few-layer black phosphorus with a hydrogen-free intercalation mechanism. *Advanced Materials*, *33*(1), 2005815. https://doi.org/10.1002/ADMA.202005815

Wang, D., Yi, P., Wang, L., Zhang, L., Li, H., Lu, M., Xie, X., Huang, L., & Huang, W. (2019). Revisiting the growth of black phosphorus in Sn-I assisted reactions. *Frontiers in Chemistry*, *7*(Jan). https://doi.org/10.3389/FCHEM.2019.00021/PDF

Wei, X., Meng, Z., Ruiz, L., Xia, W., Lee, C., Kysar, J. W., Hone, J. C., Keten, S., & Espinosa, H. D. (2016). Recoverable slippage mechanism in multilayer graphene leads to repeatable energy dissipation. *ACS Nano*, *10*(2), 1820–1828. https://doi.org/10.1021/ACSNANO.5B04939

Wei, Q., & Peng, X. (2014). Superior mechanical flexibility of phosphorene and few-layer black phosphorus. *Applied Physics Letters*, *104*(25), 251915. https://doi.org/10.1063/1.4885215

Weng, B., Qi, M. Y., Han, C., Tang, Z. R., & Xu, Y. J. (2019). Photocorrosion inhibition of semiconductor-based photocatalysts: basic principle, current development, and future perspective. *ACS Catalysis*, *9*(5), 4642–4687. https://doi.org/10.1021/ACSCATAL.9B00313/ASSET/IMAGES/LARGE/CS-2019-00313E_0018.JPEG

Wilson, J. A., di Salvo, F. J., Mahajan, S., Wilson, J. A., di Salvo, F. J., & Mahajan, S. (1975). Charge-density waves and superlattices in the metallic layered transition metal dichalcogenides. *Advances in Physics*, *24*(2), 117–201. https://doi.org/10.1080/00018737500101391

Wu, S., Huang, C., Aivazian, G., Ross, J. S., Cobden, D. H., & Xu, X. (2013). Vapor-solid growth of high optical quality MoS2 monolayers with near-unity valley polarization. *ACS Nano*, *7*(3), 2768–2772. https://doi.org/10.1021/NN4002038/SUPPL_FILE/NN4002038_SI_001.PDF

Wu, G., Yi, M., Xiao, G., Chen, Z., Zhang, J., & Xu, C. (2019). A novel method for producing boron nitride nanosheets via synergistic exfoliation with pure shear ball milling and ultrasonication. *Ceramics International*, *45*(17), 23841–23848. https://doi.org/10.1016/J.CERAMINT.2019.08.058

Xiao, L., Ma, H., Liu, J., Zhao, W., Jia, Y., Zhao, Q., Liu, K., Wu, Y., Wei, Y., Fan, S., & Jiang, K. (2015). Fast adaptive thermal camouflage based on flexible VO2/graphene/CNT thin films. *Nano Letters*, *15*(12), 8365–8370. https://doi.org/10.1021/ACS.NANOLETT.5B04090

Xu, M., Liang, T., Shi, M., & Chen, H. (2013). Graphene-like two-dimensional materials. *Chemical Reviews*, *113*(5), 3766–3798. https://doi.org/10.1021/CR300263A/ASSET/IMAGES/MEDIUM/CR-2012-00263A_0025.GIF

Yan, J., Wei, T., Shao, B., Fan, Z., Qian, W., Zhang, M., & Wei, F. (2010). Preparation of a graphene nanosheet/polyaniline composite with high specific capacitance. *Carbon*, *48*(2), 487–493. https://doi.org/10.1016/J.CARBON.2009.09.066

Yang, Shubin, Feng, X., Wang, X., & Müllen, K. (2011). Graphene-based carbon nitride nanosheets as efficient metal-free electrocatalysts for oxygen reduction reactions. *Angewandte Chemie (International Ed. in English)*, *50*(23), 5339–5343. https://doi.org/10.1002/ANIE.201100170

Yang, J., Gurung, S., Bej, S., Zou, C., Sautter, J., Chen, P.-Y., Soric, J., Padooru, Y. R., Bernety, H. M., Yakovlev, A. B., & Aì, A. (2013). Nanostructured graphene metasurface for tunable terahertz cloaking. *New Journal of Physics*, *15*(12), 123029. https://doi.org/10.1088/1367-2630/15/12/123029

Yang, H., Wang, Y., Tiu, Z. C., Tan, S. J., Yuan, L., & Zhang, H. (2022). All-optical modulation technology based on 2D layered materials. *Micromachines*, *13*(1). https://doi.org/10.3390/MI13010092

Yang, Sheng, Zhang, P., Nia, A. S., & Feng, X., (2020). Emerging 2D materials produced via electrochemistry. *Advanced Materials*, *32*(10), 1907857. https://doi.org/10.1002/ADMA.201907857

Yin, Y., Han, J., Zhang, Y., Zhang, X., Xu, P., Yuan, Q., Samad, L., Wang, X., Wang, Y., Zhang, Z., Zhang, P., Cao, X., Song, B., & Jin, S. (2016). Contributions of phase, sulfur vacancies, and edges to the hydrogen evolution reaction catalytic activity of porous molybdenum disulfide nanosheets. *Journal of the American Chemical Society*, *138*(25), 7965–7972. https://doi.org/10.1021/JACS.6B03714/ASSET/IMAGES/LARGE/JA-2016-03714Y_0006.JPEG

Yu, W., Wang, Z., Zhao, X., Wang, J., Seng Herng, T., Ma, T., Zhu, Z., Ding, J., Eda, G., Pennycook, S. J., Ping Feng, Y., & Ping Loh, K. (2020). Domain engineering in ReS2 by coupling strain during electrochemical exfoliation. *Advanced Functional Materials*, *30*(31), 2003057. https://doi.org/10.1002/ADFM.202003057

Yu, S., Wu, X., Wang, Y., Guo, X., Tong, L., Yu, S. L., Wu, X. Q., Wang, Y. P., Guo, X., & Tong, L. M. (2017). 2D materials for optical modulation: challenges and opportunities. *Advanced Materials*, *29*(14), 1606128. https://doi.org/10.1002/ADMA.201606128

Zhan, Y., Liu, Z., Najmaei, S., Ajayan, P. M., & Lou, J. (2012). Large-area vapor-phase growth and characterization of MoS2 atomic layers on a SiO2 substrate. *Small*, *8*(7), 966–971. https://doi.org/10.1002/SMLL.201102654

Zhang, C., He, Y., Zhan, Y., Zhang, L., Shi, H., & Xu, Z. (2017). Poly(dopamine) assisted epoxy functionalization of hexagonal boron nitride for enhancement of epoxy resin anticorrosion performance. *Polymers for Advanced Technologies*, *28*(2), 214–221. https://doi.org/10.1002/PAT.3877

Zhang, Yupeng, & Pan, C. (2012). Measurements of mechanical properties and number of layers of graphene from nano-indentation. *Diamond and Related Materials*, *24*, 1–5. https://doi.org/10.1016/J.DIAMOND.2012.01.033

Zhang, Yu, Yao, Y., Sendeku, M. G., Yin, L., Zhan, X., Wang, F., Wang, Z., & He, J. (2019). Recent progress in CVD growth of 2D transition metal dichalcogenides and related heterostructures. *Advanced Materials*, *31*(41), 1901694. https://doi.org/10.1002/ADMA.201901694

Zhang, Yu, Zhang, Y., Ji, Q., Ju, J., Yuan, H., Shi, J., Gao, T., Ma, D., Liu, M., Chen, Y., Song, X., Hwang, H. Y., Cui, Y., & Liu, Z. (2013). Controlled growth of high-quality monolayer WS2 layers on sapphire and imaging its grain boundary. *ACS Nano*, *7*(10), 8963–8971. https://doi.org/10.1021/NN403454E/SUPPL_FILE/NN403454E_SI_001.PDF

Zhang, Yi, Zhang, L., & Zhou, C. (2013). Review of chemical vapor deposition of graphene and related applications. *Accounts of Chemical Research*, *46*(10), 2329–2339. https://doi.org/10.1021/AR300203N/ASSET/IMAGES/LARGE/AR-2012-00203N_0011.JPEG

Zhao, G., Han, S., Wang, A., Wu, Y., Zhao, M., Wang, Z., & Hao, X. (2015). "Chemical weathering" exfoliation of atom-thick transition metal dichalcogenides and their ultrafast saturable absorption properties. *Advanced Functional Materials*, *25*(33), 5292–5299. https://doi.org/10.1002/ADFM.201501972

Zhao, L., Zhang, R., Deng, C., Peng, Y., & Jiang, T. (2019). Tunable infrared emissivity in multilayer graphene by ionic liquid intercalation. *Nanomaterials, 9*(8), 1096. https://doi.org/10.3390/NANO9081096

Zheng, J., Zhang, H., Dong, S., Liu, Y., Tai Nai, C., Suk Shin, H., Young Jeong, H., Liu, B., & Ping Loh, K. (2014). High yield exfoliation of two-dimensional chalcogenides using sodium naphthalenide. *Nature Communications*, *5*(1), 1–7. https://doi.org/10.1038/ncomms3995

Zhou, J., Lin, J., Huang, X., Zhou, Y., Chen, Y., Xia, J., Wang, H., Xie, Y., Yu, H., Lei, J., Wu, D., Liu, F., Fu, Q., Zeng, Q., Hsu, C. H., Yang, C., Lu, L., Yu, T., Shen, Z., Lin, H., Yakobson, B. I., Liu, Q., Suenaga, K., Liu, G., Liu, Z. (2018). A library of atomically thin metal chalcogenides. *Nature*, *556*(7701), 355–359. https://doi.org/10.1038/S41586-018-0008-3

Zhu, Y., Murali, S., Cai, W., Li, X., Suk, J. W., Potts, J. R., & Ruoff, R. S. (2010). Graphene and graphene oxide: synthesis, properties, and applications. *Advanced Materials*, *22*(35), 3906–3924. https://doi.org/10.1002/ADMA.201001068

3 An Overview of Machine Learning

Dilanga Abeyrathna, Mahadevan Subramaniam, and Parvathi Chundi

3.1 INTRODUCTION

Though "*Learning*" is one of the basic tasks for humans and animals, defining its exact meaning is harder because of its wide range. The question "*what is learning?*" is philosophical; in this context, learning can be simply defined as, "modifications of a behavioral tendency by experience." [1]. There are several commonalities that can be seen between animal and machine learning tasks. In fact, many of the machine learning techniques are inspired by cognitive aspects of animals. ML has evolved as a subfield of artificial intelligence (AI), learning from the data collected historically or from experiments, and using it for future actions. In general, ML models consider the patterns of the input and adjusts internal structures to approximate the relationship between input and output. ML is also used to identify hidden patterns of data distributions to come up with meaningful relationships. The ability to learn unforeseen relationships from data without depending on explicitly programmed prior guidance is one of the main reasons why there are a plethora of ML-based applications. The very early definition for ML, "Field of study that gives computers the ability to learn without being explicitly programmed" [2] is still valid.

Here in this chapter, we provide an overview of multiple ML approaches including supervised, unsupervised, semi-supervised, self-supervised, and reinforcement learning. Though these learning categories use different learning techniques and generate different outputs to achieve the ML task, we use common terminology across all of them. We call the *target* to be learned as a "*concept*" and the *output* as "concept description" [3]. To learn the concepts, ML uses data in the form of a collection of "*instances*." Each instance is an individual and independent example of the concept, described by one or more *attributes*. Therefore, each instance is a collection of values, one for each of the attributes that describe the example. For example, a research publication dataset is a collection of research publication instances where a research publication instance is described by the following attributes: *title of the publication, date of publication, name of the publication venue*. Then, <*Study of ML, Mar 23, 2023, Journal of Machine Learning*> is one instance in the research publication dataset. Datasets are typically stored as tables where *rows* are instances and *columns* are attributes.

DOI: 10.1201/9781003132981-3

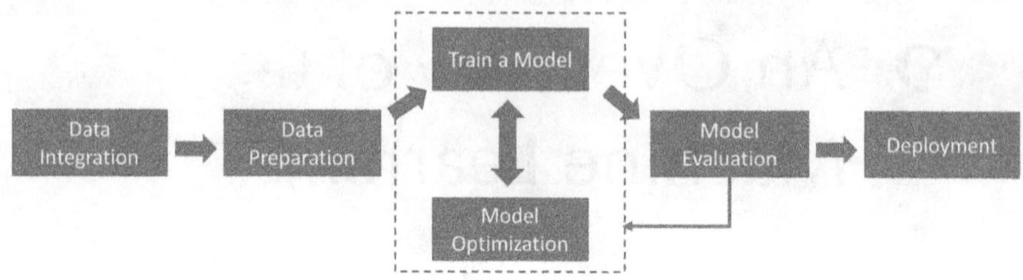

FIGURE 3.1 A typical ML pipeline.

3.1.1 The Processing Pipeline of an ML Task

The conventional machine learning pipeline comprises data integration, preprocessing, model building, and model evaluation. Figure 3.1 shows the flow chart of traditional machine learning processes. In this section, we discuss each step of the ML pipeline.

3.1.2 Data Integration

With the generation of large bulk of data each minute, in recent decades, the data needed for an ML task may be available across multiple sources in an organization. For example, research publication data may need to be integrated with material properties data to obtain an instance with all of attributes needed for learning a concept. Different data sources may store the data in different formats: structured storage such as tables or unstructured storage such as text files. Therefore, data from multiple data sources, which may be in multiple formats, should undergo a process, called *data integration*, to collect the data that can be analyzed by an ML task. Data integration is the process of integrating different sources of data into a single dataset to enable a unified view of the data. Data integration follows a set of standard steps as mentioned in Figure 3.2.

Figure 3.2 shows the *ETL* pipeline used for extracting the data from a raw data source. The *Extract* step uses the functions provided by the raw data sources to extract the raw data. The *Transform* followed by *Cleanse* steps then converts into a format needed by the application and cleans the raw data. The *Load* step stores the data in a database or data warehouse that can be eventually accessed by the application that executes the ML task. The ETL pipeline is applied to each data source to collect and integrate the data from multiple sources.

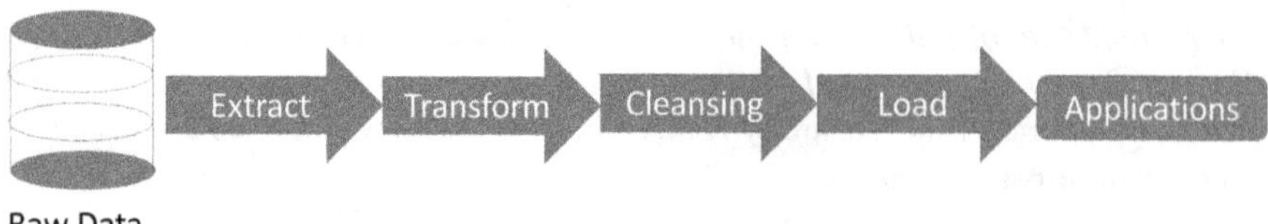

FIGURE 3.2 Data integration process.

To build a unified view of the data collected from multiple sources, diverse data from multiple sources must be integrated appropriately. For example, if one wants to integrate the material properties data and research publication data, one way to achieve this integration is to identify the names/symbols of the materials studied in each research publication and use the name/symbol to extract the properties of that material from the material properties database. The material name/symbol extracted from the research publication serves as the glue for integration of these two disparate data sources. Data integration is a nontrivial problem because the same data may be stored in different formats in different data sources. Take the simple example of "date" attribute. It can be recorded in DD/MM/YYYY format in one data source and in MM/DD/YYYY format in another data source. Data integration steps need to reconcile (using the *Transform* step) these formats so that a unified view can be constructed over the two data sources.

Currently, there are data integrations tools (e.g., *Meltano*) available to users that can automatically configure themselves so that they can process queries and extract data efficiently from multiple sources. The goal of these tools is to reduce the manual effort needed to integrate the data from these sources, and to obtain a unified view of the underlying data sources that are of high quality. Despite the availability of tools, data integration over multiple sources is a hard problem due to the disparate data formats, the uncertainty about how to resolve the differences between formats to build one standardized format, and therefore, it involves considerable manual effort.

3.1.3 Data Preparation

Real-world databases and data warehouses often contain inconsistent, incomplete, and inaccurate information due to many reasons: The manual recording of data may introduce these errors, the instruments that generated the data may be faulty, and some data was not known at the time of data entry and was left blank, etc. Before an ML task can be applied, the *input dataset* (i.e., the dataset to be analyzed by the ML task) must be "cleaned" by filling in missing values, removing outliers and inconsistencies, and smoothing noisy data. Otherwise, the "dirty" data can mislead the ML algorithm and lead to misleading concept descriptions.

If an instance in the input dataset has missing values for some of the attributes, the simplest action is to remove it from the input dataset. However, this may not work for small-size datasets. So, missing values in an instance are usually filled using a variety of ways: Substitute a missing value with the most commonly occurring value, mean, or median for that attribute, or use regression or inference-based methods to compute the missing values. It is important to make sure that the missing value computation methods do not introduce bias into the dataset. For this reason, regression- or inference-based methods are typically employed to find the most appropriate values to substitute for missing values in an instance.

The input dataset can contain *noise* such as inconsistent and/or outlier values that can mislead an ML algorithm. For numeric attributes, *data smoothing* techniques can be used to identify and remove noisy data from the input dataset. The numeric values from a column in the input dataset are sorted and the sorted value list binned.

Then, each value on a bin is replaced by the mean/median/maximum/minimum of the values in the bin which will replace outliers and inconsistent values. Regression analysis can also be used as a data smoothing technique where all values in the column that do not follow the function fitted to column values are replaced. Data smoothing techniques are also used for data discretizing (replace numeric values with nominal values) which is used for reducing noise in a column. For example, numeric values in a column that records salary can be replaced by levels of compensation: *high, medium,* and *low.*

Data preparation is an important step before building a model that learns a concept from the input dataset. Although much of it can be done using data smoothing and cleaning tools, it needs considerable manual effort as well.

3.1.4 Model Building

Once the dataset is prepared for the model building, the next step is to choose a machine learning algorithm which supports the best to solve the application. For example, if the dataset contains both attributes (predictors) and outputs for each instance, we can narrow the algorithm selection to supervised machine learning category. When the output of the input dataset is categorical, that is a finite set of values, the learning problem is called classification, whereas when the output is infinite, the learning problem is called regression (see Section 9.4.1). For supervised machine learning model building, the goal is to identify the relationship between predictors (X) with the output (Y). Here, f is a fixed but unknown function and ϵ is the error term of the model.

$$Y = f(X) + \epsilon$$

3.1.5 Model Evaluation

Model evaluation is another important step in the ML pipeline as it measures the progress of a model and understands how it works.

3.1.5.1 Training and Testing

The most obvious model evaluation would be calculating the error rate, that is the ratio between incorrect predictions to the total predictions. However, in practice the training inputs do not estimate accurate performance of a model, as, the training instances have been already seen by the model during the training process. The error rate calculated for the training set is called "resubstituting error," which is not a reliable measure of how model would perform on unseen data. Here, we consider partitioning the input dataset to extract independent datasets called test dataset, with the assumption that training, and test data follow similar distribution. The test set should not be used in any stage of model training but model evaluation stage. However, it is common to see a third partition called a validation set that is used to optimize the training process. After conducting the proper model evaluation, these partitions can be merged back to the training set to train the final model, in case of limited data availability. The basic partitioning mechanism is called the holdout method, that is extracting fixed portions for training and testing (and validation) and making sure not to use these for other purposes of model building.

In convention, 66% of the data is used as a training set, 20% of the remaining data is used for tests, and the rest as validation sets. The random holdout partitioning does not guarantee that the test set will be a representative set for the training set. For instance, the random selection would miss out complete data instances in training set that represent a class, so that model would not get a chance to train on those classes. The procedure called stratification comes as a savior in this matter, which helps in picking random samples from the input dataset to guarantee the class representation for both training and test set. The randomness of the holdout partitioning can be further mitigated using repeated holdout method, that is run multiple random partitioning to average overall error rate to achieve better model evaluation.

3.1.5.2 Cross-Validation

The repeated holdout method still does not guarantee the optimal representative partitioning experience. To make sure of the partitioning of the training testing datasets, a statistical technique called cross-validation can be used. The first step of the cross-validation is to determine a constant value (K) that represents the number of folds (partitions). Then, the process iterates such that in every iteration (1:K), one fold is considered as the test set and rest as training, until each fold gets a chance to be a test set. This approach is also known as K-fold cross-validation. This can be conducted both the variants random and stratified similar to the holdout method. This method generates K^2 number of models to evaluate the error rate which leads to obtain a better performance. However, the main downside of this method is computation intensiveness.

Leave-one-out cross-validation (LOOCV) is another variation of K-fold cross-validation where K=number of instances in the input dataset (n). Hence, each instance is used to evaluate the model, while the rest of the instances are used for model training. The main advantages of LOOCV are, (1) cross-validation does not depend on random selection (deterministic), (2) maximum number of instances (n–1) can be used for training in each iteration and works well with smaller datasets. However, the disadvantages attached to it are, (1) computationally costly and (2) prone to outliers in the dataset and lose the advantage of stratification technique. The bootstrap is another cross-validation-based estimator which utilizes the statistical technique of sampling with replacement [4].

3.1.5.3 Confusion Matrix

Confusion matrix (aka, contingency table) is one of the common ways of describing classification model performance and evaluation. As shown in Figure 3.3, confusion matrix summarizes correct and incorrect predictions, (1) True Negatives (TN), (2) False Positives (FP), (3) False Negatives (FN), and (4) True Positives (TP). A TN prediction for a test data item denotes that the model predicts the absence of a class value, and this coincides with the absence of the class value in the actual test data item. A FP prediction for a test data item denotes that the model predicts the presence of a class value, whereas the class value is absent in the actual data item. A FN prediction for a test data item denotes that the model predicts the absence of

		Predicted value		
		Positive	Negative	
Actual value	Positive	TP	FN (Type I error)	Sensitivity = $\frac{TP}{TP+FN}$
	Negative	FP (Type II error)	TN	Sensitivity = $\frac{TN}{TN+FP}$
		Precision $\frac{TP}{TP+FP}$	Negative Pred. Value $\frac{TP}{TN+FN}$	Accuracy $\frac{TP+TN}{TN+TN+FP+FN}$

FIGURE 3.3 Confusion matrix and associated metrics.

a class value, whereas the class value is present in the actual data item. A TP prediction for a test data item denotes that the model predicts the presence of a class value which coincides with the presence of the class value in the actual data item. A FP prediction is called Type I error. A FN prediction is called Type II error. The ultimate goal of learning a classification model is to achieve high TN and TP, mitigating Type I & Type II errors. The sum of the cells in the confusion table is equal to the total number of instances in the test data. Several evaluation metrics such as Recall, Precision, Specificity, Accuracy, and AUC-ROC curves can be used with these values. Figure 3.3 shows some of the metric equations with TN, TP, FN, and FP placeholders.

3.2 ML ALGORITHMS

Typically, machine learning approaches can be classified into two broad categories, Supervised and Unsupervised. Recently, several variations and combinations of these approaches including semi-supervised and self-supervised algorithms have been developed and applied successfully. In this section, we discuss some of the popular algorithms that present these two learning approaches and their variations. It is important to understand that selecting a proper algorithm that suits the data, and the application is important. When data is not enough or the model is not capable enough to detect the underlying patterns, it is identified as a model with underfitting. Similarly, when the model learns exactly the data provided in the training dataset yet performs poorly with unseen data, it creates a problem called overfitting. Using unforeseen data for testing the model is important to evaluate these model training complications. Cross-validation methods explained above are recommended to detect and mitigate these model training errors.

3.2.1 Bias and Variance

In the process of building and training a machine learning model, it is important to evaluate how well it generalizes on independent data (test data). In general, there

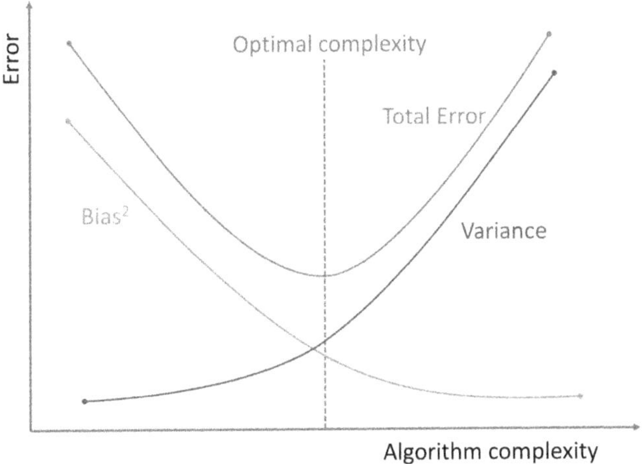

FIGURE 3.4 Bias–variance trade-off and model complexity.

is no single machine learning mode that can perform best on every task (No Free Lunch Theorem [5]). The better model suited for a given task is chosen between different models based on the performance. A given machine learning algorithm has to minimize the error of the test dataset. There are mainly two types of error categories namely, irreducible error and reducible error. The reducible error has two components, Bias and Variance.

In this context, the Bias factor refers to the model error introduced due to the inability to comprehend the underlying patterns in the data. If the prediction values and the actual values are positioned far from each other, it is an indication of high bias. Variance, on the other hand, refers to the error due to the overfitting of the training data. It provides an indication of how scattered the predicted values from the actual values are.

In the process of model training, the reducible error can be formulated as (**bias^2 + variance**). It is important to balance out these two factors to minimize the risk of a model prediction [6]. A model with high bias and low variance underfits the data and has poor accuracy, while a model with low bias and high variance overfits the data and has poor generalization performance. Hence the Bias–Variance Tradeoff is important to find the optimal model complexity (see Figure 3.4) to achieve the sweet spot where a machine model performs with the minimum error introduced by the bias and the variance.

3.3 UNSUPERVISED LEARNING

Unsupervised learning is a type of machine learning in which an algorithm is trained on a dataset without any labels or predefined categories. The main task of Unsupervised learning is to extract information and underlying patterns from the data whose classes are not known. In this section, two main techniques used in Unsupervised learning are discussed, specifically cluster analysis and principal component analysis (PCA).

3.3.1 Cluster Analysis

One of the fundamental approaches in unsupervised learning is cluster analysis [7]. The main idea of clustering is to determine subgroups in a dataset using a similarity measure so that the within-group data points are more similar to each other than to those in other groups. These groups are called clusters. As clustering refers to a vast variety of techniques focusing on partitioning coherent sets in a dataset, there have been various categorizations. One of the broader categorizations is hard clustering vs soft clustering. Hard clustering, also known as crisp clustering, is a method that assigns each data point to a single cluster. Hence, a single data point can only be designated to a single cluster. On the other hand, soft clustering, also known as fuzzy clustering, permits a single data point to be allocated to multiple clusters. Another broader categorization would be with respect to the clustering technique used. Here, we mainly discuss three main categories, (1) Hierarchical, (2) Partition, and (3) Density clustering.

3.3.1.1 Hierarchical Clustering

In Hierarchical clustering algorithms, the subgroups are created based on the hierarchy of nested groups either merging (Agglomerative clustering) or splitting (Divisive clustering) based on some similarity measure. Agglomerative clustering starts with each data point as a separate cluster and merges the clusters until the stopping criteria trigger. Divisive clustering, on the other hand, starts with all data points in one cluster and recursively splits them based on their dissimilarity. Dendrograms are commonly used to visualize the hierarchical clustering results (Figure 3.5).

Hierarchical clustering is often preferred over other clustering techniques as it does not require specifying the number of clusters in advance, as the number of clusters is determined by the hierarchy of the dendrogram. The ability to handle nonconvex

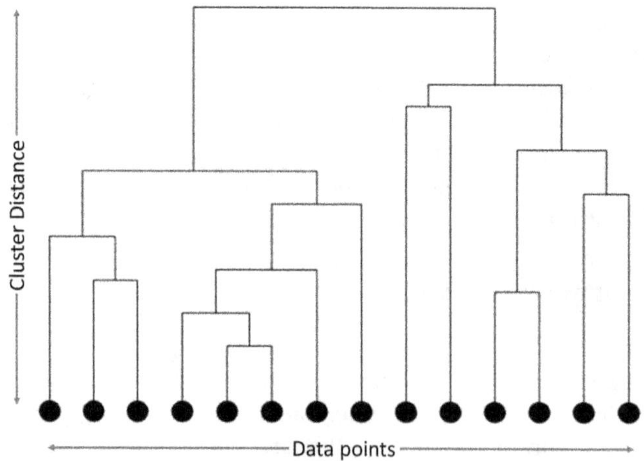

FIGURE 3.5 Dendrogram of a hierarchical clustering.

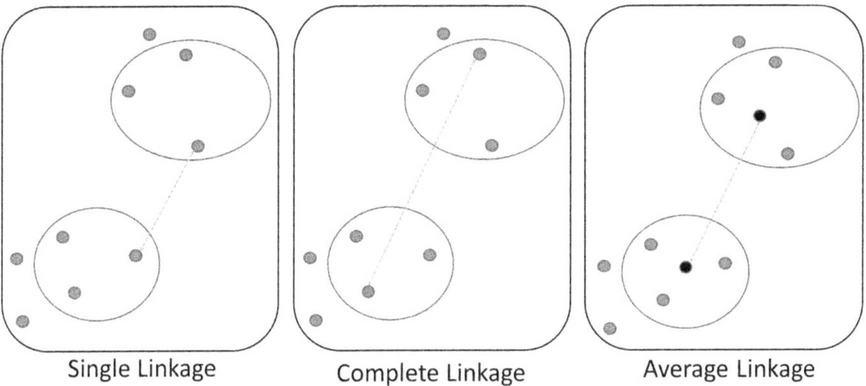

FIGURE 3.6 Hierarchical clustering linkage techniques.

clusters is an important strength of this clustering technique. However, hierarchical clustering can be computationally expensive, especially for large datasets, and it is sensitive to the choice of similarity measure and linkage method (Single, Complete, and Average (Figure 3.6)).

3.3.1.2 Partition-Based Clustering

Partition-based clustering algorithms divide the data into a prespecified number of clusters, and the algorithms belonging to this category are hard clustering methods. The *k*-means clustering [8,9] approach is one of the most popular partition-based clustering algorithms. *K*-means clustering algorithm iteratively partitions the data into *k*, and predefines the number of clusters minimizing the sum of squared distances between the data points and their assigned cluster centers. Partition-based methods are mostly simple, efficient, and scallions well to large datasets. The limitations would be, determining the *k* (number of clusters) parameter in advance, whenever a point is close to the center of another cluster; it gives poor results due to the overlapping of data points.

3.3.1.3 Density-Based Clustering

In density-based clustering [8,10], the subgroups are formed based on the density of the data space. The most popular density-based clustering algorithm is DBSCAN (Density-Based Spatial Clustering of Applications with Noise). DBSCAN algorithm starts with an unvisited data point and finds the number of data points within a predefined distance (ε). If the points within the ε-neighborhood are greater than or equal to the redefined parameter (*minPoints*), the point becomes a core point and a cluster is formed. This algorithm continues until all the data points have been visited. The main advantage of these algorithms is the clusters are discovered based on the density of the data points so that there is no requirement for a predefined value for a number of clusters. Still, the parameters ε, and minPoints have to be determined in advance. Similarly, the inability to handle data with varying densities in the same cluster is another downside of this clustering approach.

3.3.1.4 Cluster Evaluation

Compared to the evaluation techniques used in supervised learning approaches, there is no direct way of evaluating the quality of an unsupervised learning task due to the absence of the ground truth. To estimate the quality of clustering, or to compare and validate which clustering approach works well for a given dataset, several measurements have been proposed in the literature [11]. The assessments can be performed internally, aka "internal cluster evaluation", where the clustering results are evaluated based on the data used for the clustering process. Examples of such evaluations are the Silhouette coefficient [12], Dunn index [13], and Davies–Bouldin index [11]. On the other hand, in "external evaluation" the quality of the clusters is evaluated using external data. Some of the measures of the quality of processed clusters that use external evaluation are the Jaccard index, F-measure, Mutual Information, and Confusion matrix.

3.3.2 Principal Component Analysis (PCA)

Another approach in unsupervised learning is dimensionality reduction. Dimensionality reduction is the process of reducing the number of features or variables in a dataset while preserving as much of the relevant information as possible. When there are large sets of correlated variables in high dimensional space, the usage of data efficiently could be degraded. For example, when the datasets are huge due to multiple variables and/or multiple observations per variable. Some of the disadvantages related to high dimensional correlated data are,

- High requirements of storage and computational power,
- Multicollinearity issues,
- Inability to visualize the data, and
- High quantity of noise.

PCA [14] is a popular method for dimensionality reduction, which identifies the most important directions of variation in the data and projects the data in these directions. In other words, PCA projects the data into a lower-dimensional subspace so that the original highly correlated variables are transformed into a set of linearly uncorrelated variables. These variables are identified as principal components. PCA performs transformations on the vector space to extract the principal components. Then these reduced dimension data be used to analyze data such as pattern recognition, outlier detection, and trends. Some of the main applications that utilize PCA are image compression, blind source separation, visualizing multidimensional data, reducing the number of dimensions in healthcare data, and finding patterns in high-dimensional datasets [15,16].

3.3.2.1 Limitations of PCA

One of the main downsides of PCA is the assumption of linear relationships, such that principal components are determined by using a linear combination of the original features. This assumption may not be true for all datasets and that may lead to suboptimal results. An example of a possible alternative are Autoencoders, which consider nonlinear feature relationships while reducing the feature dimensions. The results of PCA depend on the scaling of the variables. If the scaling of the data has not been carefully addressed, the resulting principal components may be biased toward features with larger variance. However, some research studies introduced a scale-invariant form of PCA to mitigate this limitation. Not only scale, but PCA is also sensitive to outliers in the data. Other than that, lack of interpretability, limited applicability with non-Gaussian data, information loss due to dimensional reduction, and computational complexity can be mentioned as limitations of PCA [17].

3.4 SUPERVISED LEARNING

Supervised ML approaches are perhaps the most popular and very widely used in automating applications across several domains including materials engineering, microbiology and medicine. Some of the popular examples for supervised learning would be recommendation systems, speech recognition, image recognition, weather forecasting, and many more. The basic idea underlying this approach is to learn an unknown function from a sample specifying the known mapping of that function. Supervised learning approaches essentially attempt to fit an unknown function from known input-to-output mapping data sample. The known input-to-output mapping data sample is also commonly referred to as *ground truth or labeled data* with the mapping inputs referred to as *instances* and their corresponding outputs referred to as the *labels*. The unknown inputs are referred to as *unlabeled data*. Supervised learning algorithms involve building an ML model by training them on labeled, which can then be used to infer the outputs on unseen, unlabeled data. The types of labels used in supervised learning approaches depend on the ML task at hand. Categorical values, including Boolean values, are commonly used as labels in many applications. For instance, labels formed by categorical values, cells, microbial byproducts, and nonoccluded surfaces could be used in an ML task that detects objects in biofilm images. The images consisting these objects could be captioned with these values to produce labelled data. On the other hand, labels could be more complex and may consist of sub-images of these objects for an ML task that segments these objects in biofilm images. Labels could also be text strings or numeric values drawn from an infinite and continuous set of values such as real numbers. Supervised learning tasks can be broadly categorized into classification and regression. When the label values are finite, we call it a classification task, whereas when the label values are infinite, we call it a regression task.

3.4.1 Regression

Regression comprises a variety of approaches for studying correlations between outputs (or labels in ML) and one or more inputs (features in ML). Regression approaches are employed in ML primarily for predictive tasks. In ML, usually the outputs involve variables taking values from continuous domains. Regression algorithms may be categorized depending on the class function (also model) that is used to fit the output variables to the inputs and the methods used to find this function. Regression problem is usually formulated in terms of the following components—a set of weights/parameters (w_0, w_2,... ,w_k), input variables (xj_1, xj_2,...,xj_k), in the jth data instance with the output variable, y_j. along with the error term, e_j. Usually, each weight and the input variable instance are either scalar or vector (row of values), whereas the output variable instances are scalars. The regression problem can be formulated as discovering function *f*, relating the input and output variables based on *n* observed data instances.

$$y_j = f\left(w_0, \ldots w_k, \ldots, x_{j1}, \ldots, x_{jk}\right) + e_j \;\; j = 1, \ldots, n.$$

3.4.1.1 Linear Regression

In linear regression, the function *f* is a linear combination of the weights and the input variables. *Simple linear regression* relates one output and to *n* instances of input variable as follows

$$y_j = w_0 + w_1 x_{j1} + e_j, \; j = 1, \ldots, n$$

whereas *multiple linear regression* relates one output variable to n instances of *k* input variables

$$y_j = \left(w_0 + w_1 x_{j1} + \ldots + w_k x_{jk}\right) + e_j j = 1, \ldots, n.$$

The output variable y_j is a scalar in *univariate linear regression,* where it is a vector (multiple outputs) in *multivariate linear regression.* A regression is linear as long as the weights are linear; that is, the following regression is still linear despite the quadratic input variable term.

In a simple linear regression, the model is obtained from using the *n* instances of input x_{j1},

$$\widehat{y_j} = \widehat{w_0} + \widehat{w_1} x_{j1},$$

where $\widehat{w_0}$ and $\widehat{w_1}$ are the estimated weights and $\widehat{y_j}$ is the value of the output variable predicted by the model. The difference between the value of the output variable predicted by the model and its true value is called *residual*, $e_j = y_j - \widehat{y_j}$. Several methods are available for model estimation. The most common method, the least squares, estimates the weights and predicts the output value by minimizing the sum of squared residuals,

$$SR = \sum_{j=1}^{n} e_j^{\,2}$$

and obtains,

$$\widehat{w_1} = \frac{\sum \left(x_i - \overline{x}\right)\left(y_i - \overline{y}\right)}{\sum \left(x_i - \overline{x}\right)^2}, \widehat{w_0} = \overline{y} - \widehat{w_1}\,\overline{x},$$

where $\overline{x}$ and $\overline{y}$ are the means of the input and output values, respectively.

3.4.2 Classification

Classification predictive modeling is the task of mapping input x to discrete output variable y. The output is often identified as categories, classes, or labels. There are many classification techniques to conduct classification tasks is imperative. Each classifier would have its own pros and cons; hence, identifying the better-performing or more suitable classifier for a given classification task is important. Based on the number of categories assigned to each instance, classification problems can be divided into three categories—binary, multiclass, and multilabel classification. In a binary classification problem, each instance is associated with a single label whose values are positive/negative (or true/false) to denote class membership. In a multiclass classification problem, each instance is associated with a single label drawn from a finite set containing more than two labels. Both classification tasks aim to predict a single class outcome for a given data instance, hence also known as single label classification techniques. Multilabel classification on the other hand is an extension of the traditional single label classification problem where each example is associated with a set of categories.

3.4.2.1 Logistic Regression

Logistic regression is another variant of regression task that has been used widely in the field of AI and ML. As mentioned above, linear regression models predict continuous dependent variables given a set of independent variables, whereas logistic regression is a classification algorithm that predicts categorical values. Logistic regression models use logistic function (also known as the logit function), which maps $p(x)$ as a sigmoid function of x

$$p(x) = \frac{e^{w_0 + w_1 x}}{1 + e^{w_0 + w_1 x}}$$

The plot of the logistic regression equation follows an S-curve, ranging the outcomes from 0 to 1, regardless of the value range of x. Hence, the simple prediction can be done by considering $p(x) < 0$ for some values of x and $p(x) > 1$ for the rest of the x values. The equation then can be manipulated to obtain the odds.

$$e^{w_0 + w_1 x} = \frac{p(x)}{1 + p(x)}$$

$p(x)/[1 + p(x)]$ takes any value ranging from 0 to infinity and determines the probability of a given condition. By taking the logarithm of the above equation, log odds can be calculated.

$$y = \log\left(\frac{p(x)}{1 + p(x)}\right) = w_0 + w_1 x$$

The advantages of logistic regression models are mathematically less complex, less processing time for large volumes of data, and flexibility to handle a large range of values. *Multiple Logistic Regression* is another variant of Logistic regression that predicts binary responses for multiple predictors.

$$p(x) = \frac{e^{w_0 + w_1 x_1 + w_2 x_2 + \cdots + w_k x_k}}{1 + e^{w_0 + w_1 x_1 + w_2 x_2 + \cdots + w_k x_k}}$$

3.4.2.2 Decision Trees

Decision tree learning represents one of the simplest models of machine learning. Decision trees represent functions that take a vector of attributes as its input and return a single output denoting a decision. The inputs and output can be continuous or discrete valued variables. In case of Boolean valued decisions, the input assignment that leads to false output is called a negative example and one that leads to true output is called a positive example. A decision tree learning algorithm learns a decision tree from a given set of training examples. Decision trees can be exponentially large for certain functions (e.g., majority). Further, the search space for decision trees is also excessively big. For instance, there are 2^{2^n} of Boolean decision trees for n input variables. One of the limitations of decision trees is overfitting to their training set; that is, the function that they learn is close to perfect on training examples but does not generalize to newer examples.

3.4.2.3 Artificial Neural Networks (ANNs)

Another classification technique is ANNs inspired by the human brain's structure and function. These networks consist of many interconnected perceptrons which mimic the functionality of human neurons. The information propagates front and/or back in layer form. Each layer consists of a set of neurons performing a dedicated functionality. These layers can mainly be categorized into three categories, input layer, output layer, and hidden layers (see Figure 3.7). The input layer receives the input data (usually in the form of a multidimensional vector), and the output layer delivers the final output. All the layers in between the input and output layers are identified as hidden layers. The input layer passes the input data to the hidden layers, and the hidden layers then make the decisions to maintain the weighting scheme. These weights are updated on the go using optimization techniques such as Stochastic Gradient Descent (SGD). Having deeper stacks of hidden layers are identified as deep neural networks, which is discussed in the later part of this chapter.

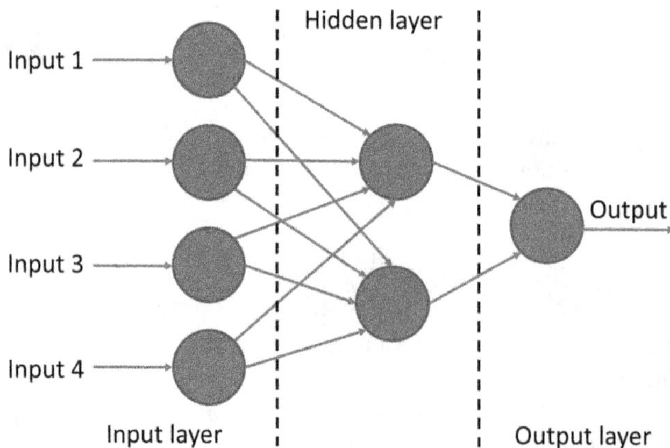

FIGURE 3.7 An example of ANN.

3.4.3 Supervised Learning Variants: Self-Supervised Learning

Supervised learning methods provide powerful feature learning given better quality and quantity of data. Larger models such as deep neural networks are extremely data hungry. However, the success of such systems hinges on a large amount of labeled data, which is not always available and often prohibitively expensive to acquire. The annotation bottleneck has motivated a wave of research in self-supervised representation learning methods that have been widely studied and advanced rapidly in recent years [18–20]. Self-supervised learning process transforms an unsupervised problem into a supervised problem when a dataset contains a huge quantity of unlabeled data. In conventional self-supervised model building, there are two main training stages, (1) *pretext task* which determines the invariance of the representations, and (2) *downstream task* which consumes the learned representations effectively [21].

3.4.3.1 Pretext Task

During the pretext task stage, the model learns to extract intermediate representations of the input data. A large quantity of unlabeled data is used to extract the underlying patterns and structures within the data. A large number of the pretext tasks for self-supervised learning have been studied in recent research studies.

3.4.3.2 Downstream Task

Downstream task can also be defined as the knowledge transfer process of the representations learned during the pretext task. This defines the model's purpose. These downstream tasks can be of various types such as image classification, object detection, semantic segmentation, machine translation, sentiment analysis, and so on. The goal of self-supervised representation learning is to learn the underlying structure and features of the input data without any explicit supervision and then utilize these learned representations in various downstream tasks to improve their performance.

3.4.3.3 Types of Self-Supervised Learning

According to the literature [22], self-supervised techniques can be categorized into contrastive learning and noncontrastive learning. The main difference between them lies in how they utilize unlabeled data to generate useful representations.

3.4.3.3.1 Contrastive Learning

The main idea of contrastive learning is to attract positive samples while repelling negative samples. The positive sample could be semantically related instances, whereas negative samples are semantically dissimilar. Here, one sample from the input dataset is considered as an anchor, and its own augmented version is treated as a positive sample, while the rest of the input data is treated as a negative sample. For instance, transformation-based pretext tasks such as resizing, flipping, and blurring can be used to generate a stochastically distorted perspective of the anchor while keeping the semantics of positive samples the same. Even though contrastive algorithms prevent complete collapse through negative examples, they are still prone to representation (dimensional) collapse [23]. Some of the popular vison-based models are MoCo [24], and SimCLR [25].

3.4.3.3.2 Non-Contrastive Learning

Compared to contrastive learning techniques, noncontrastive learning does not involve creating positive and negative pairs of samples. Instead, it directly trains the model to predict the properties of the input data that are relevant to the downstream task. Some examples of noncontrastive learning models are Bootstrap Your Own Latent (BYOL) [26] and SimSiam [27]. These noncontrastive learning techniques do not suffer dimensional collapse [28].

3.4.3.4 Challenges in Self-Supervised Learning

Self-supervised learning is a promising approach that learns representations from unlabeled data and then tunes these to the task at hand by using limited amounts of labeled data. While it addresses the labeling bottleneck of machine learning, it comes with a set of its own challenges as described below.

- **Pretext Task:** As the pretext task is one of the major steps in the self-supervised learning process selection of a more suitable one is important. However, there is no straightforward approach to determining the most suitable pretext task to extract better representations out of unlabeled data for a given downstream task.
- **Model Performance:** Performance metrics such as accuracy show convincing results given a large amount of unlabeled data. Self-supervised tasks with a moderate to a small amount of unlabeled data could generate inaccurate pseudo-labels.
- **Computational Complexity:** Most of the proposed self-supervised approaches in recent literature require a tremendous amount of computational power and time. Specifically, due to multistaged architectures, such as generating pseudo-labels, learning representations, and downstream tasks, the required computational power and time are considerably higher compared to their supervised counterparts.
- **Dataset Bias:** Any learning task can be affected by dataset bias due to reasons such as dataset imbalance and long tail distribution. Such situations can be mitigated if the dataset is preprocessed and annotated properly. However, in self-supervised learning applications, the representations learned from a large portion of the biased unlabeled datasets could perform poorly in the inference stage. Identifying these dataset bias factors from the data is challenging and requires more attention to resolve such biases.

3.5 DEEP LEARNING

Deep learning is conventionally categorized as a subfield of machine learning. Deep learning was introduced by Hinton et al. [29] in 2006 which was based on the concept of Artificial Neural Networks (ANN). It is prevailing in a wide range of domains including health care, visual recognition, text analytics, cybersecurity, and many more [30]. Deep learning models typically follow the same processing pipeline as conventional machine learning modeling. However, the major difference of deep learning modeling is that the feature extraction is automated rather than the manual extraction used in conventional machine learning. Most of the conventional machine learning models tend to show decreased performance increments given more data. Deep learning models on the other hand perform better when data grows exponentially [30].

3.5.1 Convolutional Neural Networks (CNNs)

CNNs have attracted enormous interest in Deep learning applications due to their performance over a vast variety of domains. CNNs learn and extract features from a given data input automatically [31]. The main idea behind CNN architectures is to extract low-level features such as textures, edges, and corners, and combine them to extract high-level features of parts of objects to identify the complete object as an output. Figure 3.8 shows an example of CNN architecture taking an image input and delivering an output classification prediction.

3.5.1.1 Basic Building Blocks of CNN Architecture

A typical CNN architecture is built using the following components.

- **Convolutional Layers:** The main building block of a CNN is *convolution layers* which perform convolutional filtering of the input data and produces a set of feature maps that represent different learned features. Each filter extracts a different type of feature from the input data, and the output feature maps can be used as input to subsequent layers in the network. In simpler words, the convolution operation in CNNs can be described as a multiplication of an array of input data with an array of two-dimensional weights, called a filter or a kernel. Here, the filter has to be smaller than the input data to perform the dot product and sum up to a single value, referred to as the scalar product. Figure 3.9 shows a convolutional operation with 3×3 input with a 2×2 filter, which results in 2×2 output.
- **Pooling Layers:** The pooling layer is another important component of CNN architectures. It is important to consider reducing the dimensionality of the feature maps in order to reduce the computational complexity and enhance generalizability. Pooling layers operate on each feature map independently and perform a down-sampling operation by taking the maximum (Max-Pooling) or average (Average-Pooling) value of each nonoverlapping region of the feature map. The output of a pooling layer is a smaller feature

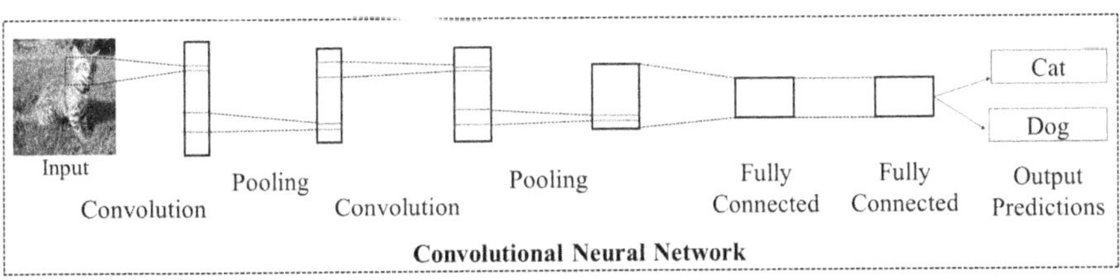

FIGURE 3.8 An example of CNN architecture.

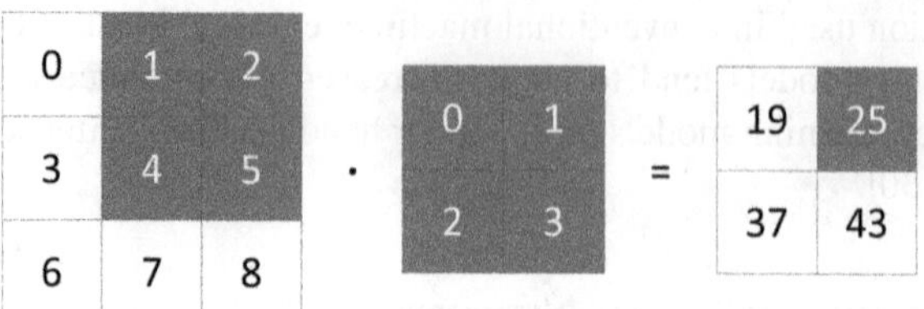

FIGURE 3.9 Convolution.

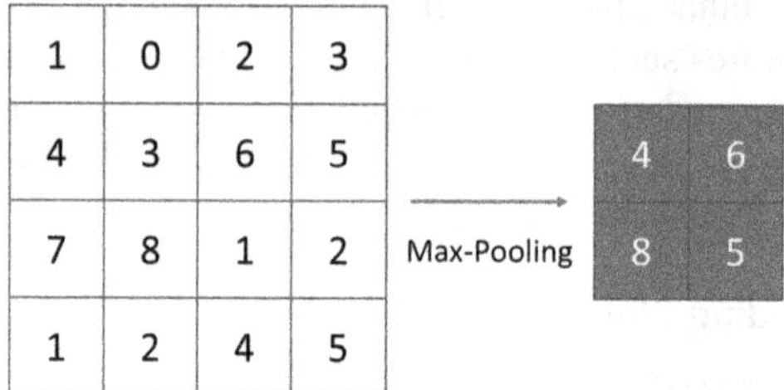

FIGURE 3.10 Max-Pooling.

map with reduced spatial dimensions. Max-Pooling is the most commonly used type, which selects the maximum value within each pooling window. Average pooling, on the other hand, calculates the average value within each pooling window. An example Max-Pooling operation is shown in Figure 3.10, 2×2 filter on a 4×4 input with stride 2. Besides these two most common pooling techniques, Global Average Pooling [32], which takes the average value of each feature map across all locations, is also widely used in recent studies [33,34].

3.5.1.2 Activation Functions

When learning feature representations from a given input, extracting only linear relationships is not enough. Activation functions help to extract nonlinear relationships and patterns through the output of a neuron or a group of neurons. Activation functions determine enabling a neuron as well as controlling the output range of a neuron. In CNNs, the activation functions are triggered after the convolutional and pooling layers to introduce nonlinearity into the model. The following are some of the popular activation functions that are commonly used in CNNs.

- **Sigmoid:** This activation function maps input values to values between 0 and 1 to determine active and inactive states. The sigmoid function is mostly used in binary classification problems. Even though this function is easy to implement and compute, it mostly ends up with a vanishing gradients effect, which slows down the training process and shows poor performance.

$$\sigma(z)=\frac{1}{1+e^{-z}}$$

- **Tanh (Hyperbolic Tangent):** Tanh activation function outputs values range from −1 to 1 for a given input value. This activation function is closely similar to the Sigmoid activation function, but the output range is wider. Tanh activation functions are preferable when strong gradients and big learning steps are required. However, similar to the Sigmoid activation functions, Tanh too leads to the vanishing gradient problem.

$$\sigma(z) = \frac{e^z - e^{-z}}{e^z + e^{-z}}$$

- **ReLU (Rectified Linear Unit):** ReLu is one of the most used activation functions which has a simple function to achieve nonlinearity. The function returns the input if it is positive, and zero otherwise. The main advantage of this activation function is that it is computationally efficient and easy to implement. However, one major drawback of ReLu function is that sometimes some of the neurons constantly output zero (also known as dead neurons). One of the ReLu variants called Leaky ReLU allows a small negative slope for negative input values to avoid dead neurons.

$$\text{relu}\ (z) = \max(0,\ z)$$

- **Softmax:** When it comes to multiclass classification tasks, the Softmax activation function shows better performances compared to other activation functions. In order to determine a final outcome from a set of values output from neurons, determining the corresponding probabilities is important. The softmax activation function converts the outcomes from each neuron into a class-specific probability distribution. These probabilities then be used to approximate the most likely class for a given input.

$$\sigma(z)_i = \frac{e^{z_i}}{\sum_{j=1}^{K} e^{z_j}}$$

Each activation function has its own strengths and weaknesses; hence, the choice of activation function can have a significant impact on the performance of a CNN. Figure 3.11 shows the Sigmoid and the Relu activation functions which gained larger attention in CNN model building.

3.5.1.3 Fully Connected Layers (FCLs)

FCLs typically follow a series of convolutional and pooling layers that extract features from the input (image). FCLs are also referred to as dense layers at every node in a layer connected to every node in the proceeding layer. The main task of an FCL is to convert the last convolutional layer output into a single-dimensioned array (also called flattening). The number of nodes in the FCL is determined by the number of classes in the classification problem.

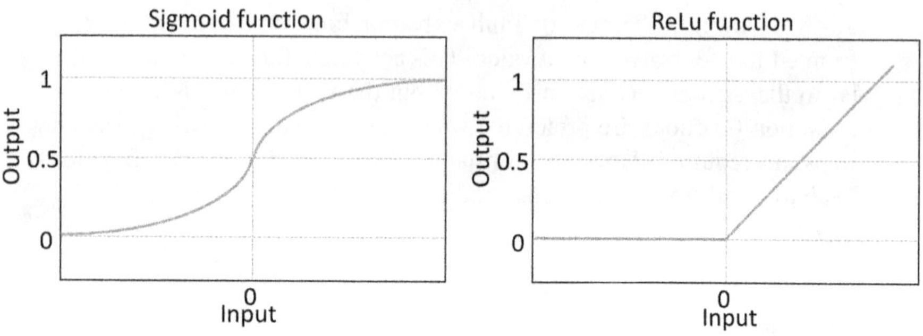

FIGURE 3.11 Sigmoid activation function and ReLu activation function.

3.5.1.4 Advanced CNN Architectures for Classification

One of the main applications of using CNN is image classification. CNNs use a series of feature extractions using different layers and perform classification tasks as an output. Some of the commonly used CNN models are discussed below.

- **LeNet5:** LeNet CNN architecture can be identified as one of the earliest compared to the other popular architectures. This was the starting point of convolution and pooling layers to be used in CNN architectures. The architecture of LeNet consists of seven layers, including two convolutional layers, two pooling layers, and three fully connected layers. The layers are arranged in a sequential manner so that the output of each layer is an input to the next layer. The first layer generates six feature maps with a size of 5×5, and the second subsamples the features maps from the first layer into the factor of 2×2. Then, the third layer generates 16 feature maps of size $\times 5$ and performs subsampling of the output by a factor of 2×2 using a pooling layer.
- **VGGNet (Visual Geometry Group):** VGG deep CNN architecture was introduced in 2014 by the Visual Geometry Group at the University of Oxford [35]. The main goal of this architecture is to show the deeper the network the better the learned feature representation. The VGG architecture consists of a series of convolutional layers with 3×3 filters and a stride of 1, followed by a Max-Pooling layer with a 2×2 filter and a stride of 2. This pattern is repeated several times, increasing the number of filters in each layer as the spatial resolution decreases. Finally, the output of the convolutional layers is flattened and fed into a series of fully connected layers for classification. VGG also introduced the concept of transfer learning, where the weights of a pretrained network can be used as a starting point for a new network on a different task.
- **ResNet (Residual Network):** Resnet Deep Convolutional neural network was introduced in 2015 by a Microsoft Research team [36]. One of the main goals of developing such CNN was to address the problem of vanishing gradients in deep neural networks. During the learning process, the gradients tend to vanish when a CNN has a large number of layers. However, training a deep model to capture better feature representations requires deeper

architectures. The key feature of ResNet architecture is residual connections. These connections support the network to learn residual functions and skip layers to pass the current output to later layers. This creates a shortcut for the gradient to propagate through the network and helps to avoid the problem of vanishing gradients.

In summary, CNNs are able to perform hierarchical feature extraction and can be used for a wide range of computer vision tasks, including image classification, object detection, and semantic segmentation.

3.5.1.5 Advanced CNN Architectures for Object Detection

The image classification describes or annotates as a whole, while the object detection task aims to determine the exact location of an object in an image. Hence, the object detection task has two aims, (1) classification of the objects and (2) localize the object. The architectures built to address object detection can be mainly categorized into two groups, namely single-staged and two-staged. In general, the two-staged models first generate possible region proposals and then process them further to construct the final bounding boxes to localize the objects. However, the single-staged approaches attempt the object detection task directly from the input image. Hence, single-staged models perform the detection tasks faster but less accurately, whereas two-staged models perform the detection task slower than the single-staged approaches yet with better accuracy.

- **Region-based CNN (R-CNN):** The family of R-CNN architectures is highly popular in object detection applications which comes under two-staged architectures. The initial R-CNN [37] used selective search techniques to generate region proposals in the first stage. Then, these region proposals were cropped out and then classified using a classification model. However, the selective search approach tends to propose a large amount of object region proposals which leads to higher computational cost. As an extension to the R-CNN architecture, Fast R-CNN [38] was proposed with Region of Interest (ROI)-based pooling to obtain fixed-size feature output. Further, Faster R-CNN [39] architecture was proposed to overcome the selective search bottleneck, by extracting the region proposals from internal feature maps using the outcomes of intermediate activation functions. This proposal bounding boxes represent the location of the object.
- **You Only Look Once (YOLO):** The family of YOLO [40] architectures can be identified as commonly used single-staged object detection architectures. In YOLO architectures, the object detection task is converted to a classification task to complete the task in a single pass. In the YOLO network, the image is divided into grids, and for each grid, there is a set of classifiers to determine the region of interests belonging to the grid. The predefined anchor boxes are aligned in identified grids, and with the help of a regression network, the final object can be localized. Currently, there are many extended and advanced variations available with different object detection capabilities [41–43].

3.5.1.6 Advanced CNN Architectures for Segmentation

Similar to object detection applications, object segmentation techniques also have gained drastic advancement with CNN architectures. Even though the detection tasks with R-CNN and YOLO architectures show convincing improvements, the localization is mostly bounded by a box. The localization precision is higher when pixel-level classification is performed for objects. Here in this section, two CNN-based segmentation architectures are discussed.

- **Fully Convolutional Networks (FCN) [33]:** As discussed in the CNN-based classification architectures, the input-to-output process is carried out by the feature extractors followed by a fully connected classifier. However, these fully connected layers cannot preserve the spatial information of the objects to identify the exact localization of an object [44]. Eliminating these fully connected layers and introducing full-size average pooling layers can be used to preserve the spatial information of a set of two-dimensional activation maps. This process is also known as Global Average Pooling [32] as all the weights corresponding to each class are summed from all the layers.
- **U-Net:** The U-Net architecture [45] which was proposed in 2015 mainly targets biomedical image segmentation tasks. This architecture type belongs to the encoder–decoder category, where the encoder is a convolutional neural network, and the decoder is a deconvolutional neural network. The decoder contains a mirror sequence of the encoder CNN. One of the reasons for the better segmentation performance of this architecture is its capability to achieve the level of abstraction through skip connections. These skip connections are connected from the encoder blocks to their mirrored counterparts in the decoder (Figure 3.12).
- **Mask R-CNN:** Mask R-CNN [46] is an extension of Faster R-CNN object detection architecture. Other than the Region Proposal Network (RPN), classification and bounding box regression network of Faster R-CNN network, Mask R-CNN consists of an additional branch of mask predictions for

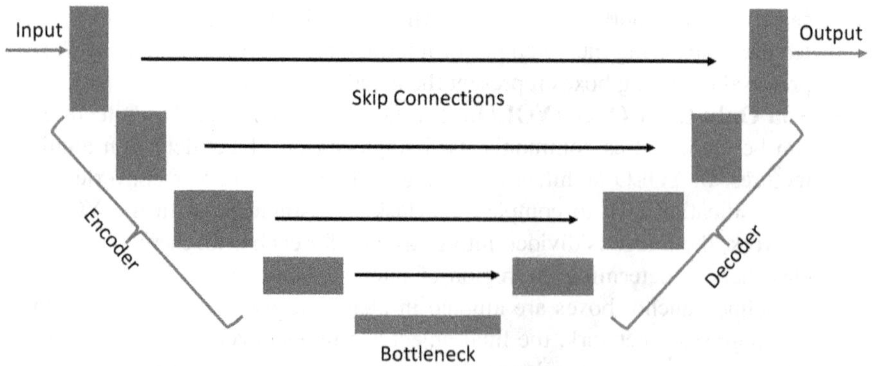

FIGURE 3.12 U-Net encoder decoder structure with skip connections.

each object proposal. The complete pipeline can be trained end-to-end, for instance segmentation tasks.

3.6 RECURRENT NEURAL NETWORKS (RNN)

Similar to CNNs, RNNs too have become popular and commonly used in deep learning applications. RNNs are mostly used with sequential data such as natural language texts, Speech, Audio, Video, Physical processes, and real-time embedded system outputs. Sequential processing demands two important qualities: the ability to maintain the length variability of input/output and comprehending the order of a sequence to learn and predict.

Figure 3.13 shows simple RNN architecture and its unfolded representation. RNN models learn by looping the output of the previous state of it to itself as an input. This looping structure supports RNNs to capture long-term dependencies in the data. However, the vanilla RNN models suffer from the vanishing gradient problem. Long Short-Term Memory (LSTM) networks [47], and Gated Recurrent Unit (GRU) networks are examples of extensions of RNN architectures to address vanishing gradient issues and other performance enhancements.

- **Long Short-Term Memory (LSTM):** LSTMs were introduced in 1997 by Hochreiter and Schmidhuber [47], and it has been widely used in several sequel data processing applications. LSTM uses gated units to address the problem of vanishing gradient. A single memory cell in an LSTM unit contains three gates namely, Forget Gate, Input Gate, and Output Gate. The Forget Gate controls the information which should be memorized and forgotten based on its usefulness. The Input Gate is responsible for controlling which information should be input to the cell state, whereas The Output gate determines and controls the outputs.
- **Gated Recurrent Unit (GRU):** Gated Recurrent Unit (GRU) architecture was introduced in 2014 by Cho et al. [48]. This architecture too employs gates to control the flow of information and mitigate the vanishing gradient problem. Specifically, GRU has two gates, namely the update gate and the reset gate. GRU has similar characteristics as LSTM, yet processes data faster due to the fact that it has less number of gates. The update gate GRU captures the dependencies from large sequences of data adaptively without discarding the information gathered from the previous states.

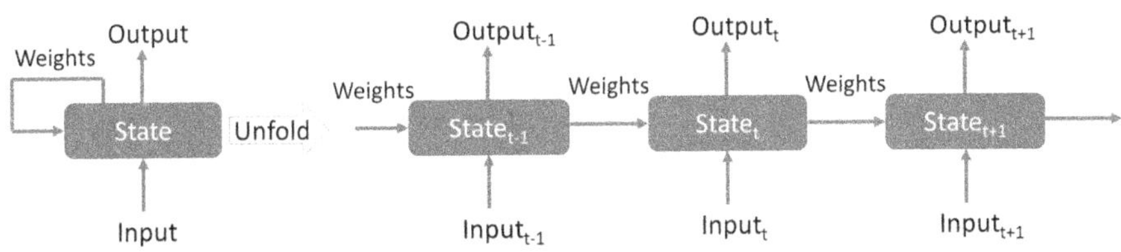

FIGURE 3.13 Typical RNN architecture.

REFERENCES

[1] Guilford, J. P. (1967). *The Nature of Human Intelligence*. McGraw-Hill: New York.

[2] Samuel, A. L. (1959). Some studies in machine learning using the game of checkers. *IBM Journal of Research and Development*, 3(3), 210–229.

[3] Witten, I. H., Frank, E., Hall, M. A., Pal, C. J., & DATA, M. (2005, June). *Data Mining: Practical Machine Learning Tools and Techniques* (Vol. 2, No. 4). Morgan Kaufmann Publishers Inc.: San Franisco.

[4] A. Ramezan, C., A. Warner, T., & E. Maxwell, A. (2019). Evaluation of sampling and cross-validation tuning strategies for regional-scale machine learning classification. *Remote Sensing*, 11(2), 185.

[5] Wolpert, D. H., & Macready, W. G. (1997). No free lunch theorems for optimization. *IEEE Transactions on Evolutionary Computation*, 1(1), 67–82.

[6] Wasserman, L. (2004). *All of Statistics: A Concise Course in Statistical Inference* (Vol. 26). Springer: New York.

[7] Hastie, T., Tibshirani, R., Friedman, J. H., & Friedman, J. H. (2009). *The Elements of Statistical Learning: Data Mining, Inference, and Prediction* (Vol. 2, pp. 1–758). Springer: New York.

[8] Jain, A. K., Murty, M. N., & Flynn, P. J. (1999). Data clustering: A review. *ACM Computing Surveys (CSUR)*, 31(3), 264–323.

[9] Hastie, T., Tibshirani, R., Friedman, J. H., & Friedman, J. H. (2009). *The Elements of Statistical Learning: Data Mining, Inference, and Prediction* (Vol. 2, pp. 1–758). Springer: New York.

[10] Ester, M., Kriegel, H. P., Sander, J., & Xu, X. (1996). A density-based algorithm for discovering clusters in large spatial databases with noise. In Proceedings of the Second International Conference on KDD (Vol. 96, No. 34, pp. 226–231). AAAI Press: Portland.

[11] Davies, D. L., & Bouldin, D. W. (1979). A cluster separation measure. In IEEE Transactions on Pattern Analysis and Machine Intelligence (Vol. PAMI-1, no. 2, pp. 224–227). IEEE.

[12] Rousseeuw, P. J. (1987). Silhouettes: A graphical aid to the interpretation and validation of cluster analysis. *Journal of Computational and Applied Mathematics*, 20, 53–65.

[13] Dunn, J. C. (1974). Well-separated clusters and optimal fuzzy partitions. *Journal of Cybernetics*, 4(1), 95–104.

[14] Jolliffe, I. (2005). Principal component analysis. *Encyclopedia of Statistics in Behavioral Science*. doi:10.1002/0470013192.bsa501

[15] Wold, S., Esbensen, K., & Geladi, P. (1987). Principal component analysis. *Chemometrics and Intelligent Laboratory Systems*, 2(1–3), 37–52.

[16] Abdi, H., & Williams, L. J. (2010). Principal component analysis. *Wiley Interdisciplinary Reviews: Computational Statistics*, 2(4), 433–459.

[17] Vidal, R., Ma, Y., &Sastry, S. S. (2016). *Generalized Principal Component Analysis* (pp. 25–62). Springer: New York.

[18] Cai, L., Xu, X., Liew, J. H., & Foo, C. S. (2021). Revisiting superpixels for active learning in semantic segmentation with realistic annotation costs. In Proceedings of the IEEE/CVF Conference on Computer Vision and Pattern Recognition (pp. 10988–10997). IEEE: Nashville, TN.

[19] Chen, Y., & Joo, J. (2021). Understanding and mitigating annotation bias in facial expression recognition. In Proceedings of the IEEE/CVF International Conference on Computer Vision (pp. 14980–14991). IEEE: Montreal.

[20] Hu, W., Liu, B., Gomes, J., Zitnik, M., Liang, P., Pande, V., & Leskovec, J. (2019). Strategies for pre-training graph neural networks. arXiv preprint arXiv:1905.12265.

[21] Jaiswal, A., Babu, A. R., Zadeh, M. Z., Banerjee, D., & Makedon, F. (2020). A survey on contrastive self-supervised learning. *Technologies*, 9(1), 2.
[22] Rani, V., Nabi, S. T., Kumar, M., Mittal, A., & Kumar, K. (2023). Self-supervised learning: A succinct review. *Archives of Computational Methods in Engineering*, 30(4), 1–15.
[23] Hua, T., Wang, W., Xue, Z., Ren, S., Wang, Y., & Zhao, H. (2021). On feature decorrelation in self-supervised learning. In Proceedings of the IEEE/CVF International Conference on Computer Vision (pp. 9598–9608). IEEE: Montreal.
[24] He, K., Fan, H., Wu, Y., Xie, S., & Girshick, R. (2020). Momentum contrast for unsupervised visual representation learning. In Proceedings of the IEEE/CVF Conference on Computer Vision and Pattern Recognition, , June 13–19, 2020 (pp. 9729–9738). IEEE: Seattle, WA.
[25] Chen, T., Kornblith, S., Norouzi, M., & Hinton, G. (2020). A simple framework for contrastive learning of visual representations. In Proceedings of the 37th International Conference on Machine Learning (pp. 1597–1607). PMLR: Vienna.
[26] Grill, J. B., Strub, F., Altché, F., Tallec, C., Richemond, P., Buchatskaya, E., Doersch, C., Pires, B. A., Guo, Z., Azar, M. G., Piot, B., Kavukcuoglu, K., Munos, R., & Valko, M. (2020). Bootstrap your own latent-a new approach to self-supervised learning. *Advances in Neural Information Processing Systems*, 33, 21271–21284.
[27] Chen, X., & He, K. (2021). Exploring simple siamese representation learning. In Proceedings of the IEEE/CVF Conference on Computer Vision and Pattern Recognition (pp. 15750–15758). IEEE: Nashville.
[28] Bardes, A., Ponce, J., & LeCun, Y. (2022). Vicregl: Self-supervised learning of local visual features. arXiv preprint arXiv:2210.01571.
[29] Russell, S. J. (2010). *Artificial Intelligence a Modern Approach*. Pearson Education, Inc.: London.
[30] Sarker, I. H. (2021). Deep learning: A comprehensive overview on techniques, taxonomy, applications and research directions. *SN Computer Science*, 2(6), 420.
[31] Albawi, S., Mohammed, T. A., & Al-Zawi, S. (2017, August). Understanding of a convolutional neural network. *In 2017 International Conference on Engineering and Technology (ICET)* (pp. 1–6). IEEE.
[32] Szegedy, C., Liu, W., Jia, Y., Sermanet, P., Reed, S., Anguelov, D., Erhan, D., Vanhoucke, V., & Rabinovich, A. (2015). Going deeper with convolutions. In Proceedings of the IEEE Conference on Computer Vision and Pattern Recognition, June 7–12, 2015 (pp. 1–9). IEEE: Boston, MA.
[33] Long, J., Shelhamer, E., & Darrell, T. (2015). Fully convolutional networks for semantic segmentation. In Proceedings of the IEEE Conference on Computer Vision and Pattern Recognition, June 7–12, 2015 (pp. 3431–3440). IEEE: Boston, MA.
[34] Sun, W., & Wang, R. (2018). Fully convolutional networks for semantic segmentation of very high resolution remotely sensed images combined with DSM. *IEEE Geoscience and Remote Sensing Letters*, 15(3), 474–478.
[35] Simonyan, K., & Zisserman, A. (2014). Very deep convolutional networks for large-scale image recognition. arXiv preprint arXiv:1409.1556.
[36] He, K., Zhang, X., Ren, S., & Sun, J. (2016). Deep residual learning for image recognition. In Proceedings of the IEEE Conference on Computer Vision and Pattern Recognition, June 27–30, 2016 (pp. 770–778). IEEE: Las Vegas, NV.
[37] Girshick, R., Donahue, J., Darrell, T., & Malik, J. (2014). Rich feature hierarchies for accurate object detection and semantic segmentation. In Proceedings of the IEEE Conference on Computer Vision and Pattern Recognition, June 23–28, 2014 (pp. 580–587). IEEE: Columbus, OH.
[38] Girshick, R. (2015). Fast R-CNN. In Proceedings of the IEEE International Conference on Computer Vision, December 7–13, 2015 (pp. 1440–1448). IEEE: Santiago.

[39] Ren, S., He, K., Girshick, R., & Sun, J. (2015). Faster R-CNN: Towards real-time object detection with region proposal networks. In Advances in Neural Information Processing Systems, 28 (pp. 91–99). Curran Associates: Montreal.

[40] Redmon, J., Divvala, S., Girshick, R., & Farhadi, A. (2016). You only look once: Unified, real-time object detection. In Proceedings of the IEEE Conference on Computer Vision and Pattern Recognition, June 27–30, 2016 (pp. 779–788). IEEE: Las Vegas, NV.

[41] Redmon, J., & Farhadi, A. (2017). YOLO9000: Better, faster, stronger. In Proceedings of the IEEE Conference on Computer Vision and Pattern Recognition, July 21–26, 2017 (pp. 7263–7271). IEEE: Honolulu, HI.

[42] Redmon, J., & Farhadi, A. (2018). Yolov3: An incremental improvement. arXiv preprint arXiv:1804.02767.

[43] Wang, C. Y., Bochkovskiy, A., & Liao, H. Y. M. (2022). YOLOv7: Trainable bag-of-freebies sets new state-of-the-art for real-time object detectors. arXiv preprint arXiv:2207.02696.

[44] Sun, W., & Wang, R. (2018). Fully convolutional networks for semantic segmentation of very high resolution remotely sensed images combined with DSM. *IEEE Geoscience and Remote Sensing Letters*, 15(3), 474–478.

[45] Ronneberger, O., Fischer, P., & Brox, T. (2015). U-net: Convolutional networks for biomedical image segmentation. In Medical Image Computing and Computer-Assisted Intervention-MICCAI 2015: 18th International Conference, October 5–9, 2015, Proceedings, Part III 18 (pp. 234–241). Springer International Publishing: Munich.

[46] He, K., Gkioxari, G., Dollár, P., & Girshick, R. (2017). Mask R-CNN. In Proceedings of the IEEE International Conference on Computer Vision, October 22–29, 2017 (pp. 2961–2969). IEEE: Venice.

[47] Hochreiter, S., & Schmidhuber, J. (1997). Long short-term memory. *Neural Computation*, 9(8), 1735–1780.

[48] Cho, K., Van Merriënboer, B., Gulcehre, C., Bahdanau, D., Bougares, F., Schwenk, H., & Bengio, Y. (2014). Learning phrase representations using RNN encoder-decoder for statistical machine translation. arXiv preprint arXiv:1406.1078.

4 Discovery of 2D Materials with Machine Learning

Md Mahmudul Hasan, Rabbi Sikder, Bharat K. Jasthi, Etienne Z. Gnimpieba, and Venkataramana Gadhamshetty

4.1 INTRODUCTION: HIGH-THROUGHPUT SCREENING

The outstanding properties of 2D materials from classes of semi-metal (graphene), insulator (hexagonal boron nitride) and metallic carbides, nitride, and carbonitrides (MXenes) have been utilized in different industries. Such properties include high carrier mobilities, superconductivity, mechanical flexibility, as well as high optical absorption compared to their bulk counterparts. These properties collectively render them as excellent candidates in application such as barrier coatings [1–4], energy devices [5–7], catalyst [8–10], biosensors [11–13], spintronics [14–16], and supercapacitors [17–19]. Following the 2010 Nobel Prize for isolating graphene and demonstrating its remarkable properties, there has been a significant interest in discovering other promising 2D materials. Such a discovery, especially for a specific scientific or industrial application, entails significant financial resources, extensive time (10–20 years) [20], along with a failure risk (e.g., inability to exfoliate stable 2D materials). The discovery process typically entails six different key steps including discovery, development, optimization, system design, certification, and manufacturing (Figure 4.1). It is unlikely that these different stages will be overseen by the same scientific teams and same places. Even if this is the case, any communication lapse among them can slow down the entire discovery process.

Computational modeling and experiments are the two key methodologies used in 2D materials research. Both offer innate benefits and limitations. While experiments are relatively general easy way of exploring new materials, they require extensive infrastructure (human and hardware), time, and money. On the other hand, modeling efforts solely rely upon the theories and computational power. Molecular dynamic simulation [21], density functional theory (DFT) [22], Monte Carlo techniques [23], and phase-field methods [24,25] can be used to run tests virtually and in a shorter period of time [26]. Their major limitations arise from their overreliance on: (1) intrinsic microstructural properties of the materials, (2) sophisticated computing equipment, (3) and data from previous studies that many not be relevant for new systems. Thus, computational efforts are often integrated with experimental efforts to study 2D materials properties and correlate them with synthesis and process functions. In certain cases, both experiments and computations fail to achieve a desirable function. For example, it is difficult to examine transition temperature of

DOI: 10.1201/9781003132981-4

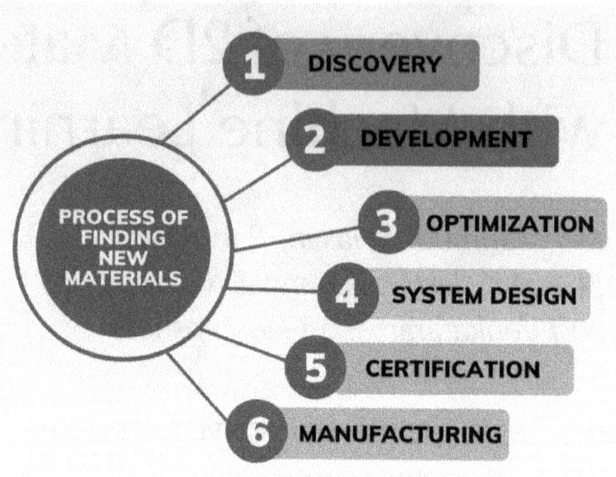

FIGURE 4.1 The process of finding new 2D materials.

glass experimentally owing to their changes across a large temperature range [27]. The transition temperature also cannot be simulated accurately as it is influenced by multiple variables (e.g., pressure, structure, and fundamental constitutive traits) [28] that cannot be implemented easily using computer tools. Artificial intelligence (AI) can overcome flaws of these two methods, where computational and experimental approaches are combined to develop analytical tools for predicting functional properties of previously unexplored 2D materials. High-throughput first-principles calculations can be used to study the vast 2D materials space. The resulting databases can be leveraged by AI methods to develop computational tools for predicting 2D materials that display the highest probability of existence and offer desired (predicted) properties. The predicted results are then verified with experimental results.

Following the establishment of the Materials Genome Initiative (MGI) in 2011 [29], many other 2D materials property databases have been created that meet findability, accessibility, interoperability, and reusability (FAIR) principles. Examples include the inorganic crystal structure database (ICSD) [30], Computational 2D Materials Database (C2DB) [31], 2D Materials Encyclopedia (2dmatpedia) [32], Open Quantum Materials Database (OQMD) [33], Materials Project [34], Cambridge Structural Databases [35], and Harvard Clean Energy Project (HCEP) [36]. AI-based machine learning (ML) tools can be developed to leverage these data sets, generate hypotheses about the optimum experimental circumstances and parameters, learn and adapt without any explicit instructions, find, analyze and draw insights from patterns observed from the data [37,38]. Such ML approaches have been used recently to unearth new 2D materials and correlate their fundamental properties with desirable functions. For example, such ML approaches have been used to develop new catalysts [39], battery materials [40], and light-emitting diodes (LEDs) [41] based on the 2D materials. ML tools can be paired with computer modeling processes to develop efficient and reliable solutions for 2D materials discovery. For instance, Sorkun et al. [42] leveraged ab initio theoretical predictions and data-driven approaches for virtually screening 2D materials from a large compositional space for energy conversion and storage applications. They downselected nearly 316,500 stable 2D materials and identified promising candidates for energy conversion and storage applications. Such data-driven approaches can also leverage data from even failed experiments [43].

This chapter will introduce effective ML techniques for discovering new 2D materials and predicting their properties. It will be organized under the following key topics: (1) ML approaches for 2D materials research, (2) prediction of 2D material properties, (3) application of ML approaches for discovering novel 2D materials, (4) ML for other purposes, and (5) countermeasures for common problems, and finally, (5) conclusions.

4.2 ML APPROACHES FOR 2D MATERIALS RESEARCH

ML tools can be used in different fields related to high-dimensional data, such as classification, regression. They can extract insight and knowledge from massive databases, learn from different computations from previous studies, and predict reliable decisions. Although ML techniques became common in diverse fields including image recognition [44], natural language processing [45], speech recognition [46], and banking [47], their application in the materials science research became prominent only in the past decade. The ML techniques were first applied in materials science in 1990s [48,49]. However, they were limited to studies focusing on identifying and projecting the physio-chemical behavior of fiber/matrix boundaries in composite materials [49]. This application allowed scientists to realize the utility of ML in other topics including materials discovery and properties prediction.

4.2.1 Three ML Approaches for 2D Materials Research

Three major forms of ML approaches are supervised, unsupervised, and reinforcement learning. While supervised ML requires defined or labelled forms of input and output training datasets [50], unsupervised ML use raw data or unclassified datasets [51]. As the name implies, reinforcement learning uses positive reinforcement to encourage desirable actions. Negative reinforcement can also be used to discourage undesirable ones. This model observes the surroundings and plans and executes appropriate activities to improve its performance over time. The final choice of an ML approach depends upon the quality and quantity of 2D materials datasets and the research problem under consideration. Three key classes of ML algorithms [52] include (1) Regression (2) Classification and clustering (deep learning, DL), and (3) Probability estimation (see Figure 4.2 for subclasses under these key categories).

4.2.1.1 Construction of an ML Model

An ML system is typically built after identifying a specific problem goal and empirical function that denotes it. The sample is the subcategory of the data [53] obtained after the raw data is processed using data cleaning and feature engineering steps. Cleaning step identifies incorrect, incomplete, and irrelevant data. Such incorrect data is then revised and cleaned [54]. Feature engineering uses the information to

COMMONLY USED MACHINE LEARNING ALGORITHMS

REGRESSION	CLASSIFICATION AND CLUSTERING	PROBABILITY ESTIMATION
Support Vector Machine (SVM) Regression	Decision Tree (DT)	Naive Bayes
Artificial Neural Network (ANN)	Artificial Neural Network (ANN)	Expectation Maximization (EM)
Kernel Ridge Regression (KRR)	Support Vector Machine (SVM) Classification	
Multiple Linear Regression (MLR)	Density-Based Spatial Clustering of Applications with Noise (DBSCAN)	

FIGURE 4.2 Three classes of ML algorithms used in 2D materials research.

generate features which in turn are used to operate the ML algorithms. Feature engineering step encompasses tasks including selection of features, extraction of features, feature learning, and feature construction. This step critically influences the quality and accuracy of the model. In spite of the recent technological advances, feature engineering remains a tough and expensive job that requires trained experts, substantial time, and resources. An ML model construction includes two fundamental parts: (1) ML algorithm and (2) model optimization algorithm [55]. The model represents a system based on complex statistical and mathematical ideas, as well as the algorithm that was learned from the sample. Eventually, the overall process can give refined optimized output.

4.2.1.1.1 Three Stages to Build an ML Model

The process of building an ML method can be broken down into three distinct stages [55]: (1) generating the training data to create samples, (2) building the ML model using those clean data (samples), and (3) model evaluation and optimization (e.g., cross-validation, hyper-parameter optimization) [56]. The overall construction process of an ML system is illustrated in Figure 4.3.

The first step, training data, is the process of collecting raw data from computational simulations and experimental data. These raw datasets are then cleaned into usable form by data cleaning and feature engineering technique. The overall datasets are divided into three categories: (1) Training dataset (to train the ML model), (2) Validation dataset (validate the model), and (3) Test dataset (actual testing of the final model). For example, in case of developing protective coatings based on 2D materials, the experimental and simulation data is based on the EIS, Tafel, LPR, and corrosion data.

Data cleaning process is then used to tailor and modify the raw data into a more suitable form. The next step involves feature engineering that refers to the process of

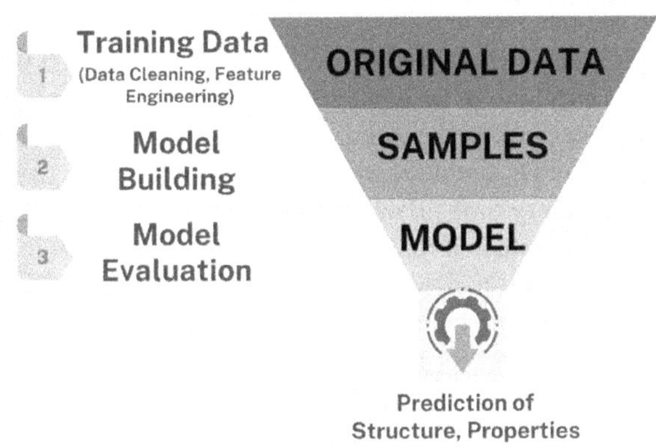

FIGURE 4.3 Three stages to build an ML model.

identifying the consistent, nonredundant, and essential features to use in model construction. Hence, it is important to identify only key conditional attributes that affect the obtained datasets. For example, in Li-ion battery applications, although various external and internal factors may influence the battery performance, only key factors (e.g., cell potential, gravimetric capacity, volumetric expansion) are considered while performing the experiment [57]. That is why it is critical to use feature selection properly to find out the attributes that affect the most [58]. The second step, model building, uses different linear or nonlinear functions to link input data to output data. Model evaluation is the last step. This step aims to calculate the generalization accuracy of a model on the data that could come across in the future.

Methods used for evaluation are listed in Table 4.1. These methods need test data to evaluate the model's performance, and it is recommended not to use the data used for building the model as test data. Because that might result in overfitting [59].

TABLE 4.1
Analogy of Different Assessment Techniques

Method	Condition	Advantages	Limitations
Bootstrapping	Small data volume	Effective separation of training and testing data	Original dataset is different from training data
Hold-out	Enough data volume	Less complex computational data	Training data volume is smaller than original dataset
Cross-validation or LOOCV	Enough data volume/ Small data volume and training and testing data can be separated	Change in volume of training data as no effect	High-level computational complexity

Overfitting is a phenomenon when a model remembers its training data set and always predicts the accurate label for any point that's in the training set [60]. Holdout is the evaluation method where different data is used for evaluation and the whole dataset is categorized randomly into three different sets, namely (1) training set, (2) validation set, and (3) test set. In fact, the training set is indeed a fraction of that same data that was initially utilized to build the model. On the other hand, the validation set is essentially the subcategory of the data utilized to assess the model's efficiency. This helps to test the model build and refine the parameters to finalize the best version of the model in the end. Test set, also known as unseen data, is the subset of the data used to evaluate how well the model would perform in the future. Holdout is a very handy process for it is known for its speed, flexibility, and simplicity. However, difference in training and test data set might cause high variance [61]. Cross-validation and bootstrapping are also used for model evaluation.

4.2.1.2 Data Collection and Representation

Although there are only 92 naturally occurring elements in the modern periodic table, there can be unimaginable variations of materials with different combinations. It is remarkable to see how ML and HTS (high-throughput screening) flourish by offering somewhat an appropriate solution within a short amount of time and with minimal means. For ML models to be successful, they must have extremely high-quality and exact data [62]. Following list is the most significant and trustworthy open-source research-related experimental, theoretical, and computer modeling databases relevant to 2D materials development.

- Inorganic Crystal Structural Database (ICSD) (https://icsd.products.fiz-karlsruhe.de/)
- Computational 2D Materials Database (C2DB) for structural, thermodynamic, elastic, electronic, magnetic, and optical properties (https://cmr.fysik.dtu.dk/c2db/c2db.html)
- Joint Automated Repository for Various Integrated Simulations (JARVIS) for Database of DFT-, MD-, and ML-based calculations (https://jarvis.nist.gov/)
- Crystallographic open database (COD) for crystallographic data (http://www.crystallography.net/cod/)
- Open Quantum Materials Database (OQMD) for thermodynamic and structural properties calculated from DFT (http://oqmd.org/)
- Materials Project (MP) for computational data under the Materials Genome Initiative (MGI) (https://materialsproject.org/)
- 2D materials encyclopedia (http://www.2dmatpedia.org/)
- PubChem (https://pubchem.ncbi.nlm.nih.gov/)

Feature engineering is a critical step within data representation process [63]. For making the raw data more suitable for an algorithm, it needs to be converted and this procedure is called feature engineering. In this process, data is represented in such a way that has meaning for an ML algorithm. Any material's data can be expressed structurally or in elemental form. For elemental form, data such as charge number, atomization energy, etc., can be used and structural form data such as bond order parameters, Fourier series, etc., are incorporated. Often, a combination of elemental and structural data is required for better comprehension. That is why the selection of molecular descriptors is critical to have an efficient HTS and ML model to solve any problems. Some common descriptors are radial distribution functions (RDF), principal component analysis (PCA), adjacency matrix, and coulombic matrix. The RDF descriptor is a perfect example of a crystal structure descriptor [64]. While considering the Fourier series, the atomistic RDF (FR) is the descriptor for the space of chemical component that is built on the distance among molecules. Moreover, it fulfills major portion of the prerequisites of being a descriptor. This is also a reliable source of projection when considering molecules' energy surfaces. On the other hand, PCA is used to reduce dimensionality [65]. The calculation of this tool is derived from the eigenvalue of a matrix and its corresponding eigenvectors. The base of this approach is linear algebra and so the mathematical foundation is very sound which eventually results in the straightforward interpretation of results. However, the assumption of linearity can be attributed to the shortcoming of this method. As most materials science-related cases are nonlinear, this method is seldom used in advanced machine learning methods. Additionally, adjacency matrix is very handy while considering chargeless particles, whereas the combination of both charge and structural information is incorporated [66] for coulombic matrix. In fact, if the chemistry of the atomic species is included, an adaptation of adjacency matrix, Coulombic matrix representation can be obtained.

4.2.1.3 Selection and Evaluation Procedure of Model

There are different and effective machine learning algorithms ranging from as simple as linear regression curves to some intricate neural networks. Due to scarcity of enough data, not all algorithms are employed to find and predict new materials. As ML in materials science is a new direction and this area itself is still emerging and a lot of new directions are being generated every day. From those, the most useful ones are listed in Figure 4.2. Here, Naïve Bayes can work as validating any theory [67], decision tree can show the routes of materials synthesis [52], artificial neural network can predict reaction product [68], and support vector machine can establish the structure–property relationship [69]. As stated earlier, for employing ML in materials science, the setback is the scarcity of high-quality datasets [70]rather than algorithms. However, thanks to DFT, in near future this scarcity of theoretical data will no longer exist. But performing ML analysis using big set of data is still not a cost-efficient technique due to the fundamental parts of ML such as feature engineering [71]. By substituting first-principles calculations with machine learning, it can help save money and effort throughout this situation.

In DFT, the Kohn–Sham (KS) equation can be utilized to measure the entire energy of any molecules. The KS equation exists in a KS system, a hypothetical system that is made of particles which do not interact with each other still can produce same density like any other system that as particles interacting with each other [72,73].

The KS equation can be written as:

$$E(n) = T_s(n) + U_H(n) + E_{xc}(n) \tag{4.1}$$

here

T_s = noninteracting electrons' kinetic energy (K.E.)
U_H = Hartree potential energy
V = external potential
n = electronic density
E_{xc} = exchange-correlation term showing owing to energy estimations in the Kohn–Sham approach.

Calculating the value of overall energy through a computational approach is always intensive. Through ML approaches, the outcome of DFT simulations can be predicted without even performing it. For example, Brockherde et al. [74] circumvented the KS equation while doing DFT computation by studying the energy functional applying the kernel ridge regression (KRR) program. Or each accessible experimental test result could be fed into machine learning programs and from there pattern of structure–property can be comprehended. ML models were at first used in the chemical and pharmaceutical industries.

One of the ML approaches used in pharmaceuticals [75] medicines is quantitative structure–activity relationship (QSAR) modeling, which has aided to assess energy, expense, and improved pharmacology and pharmacological activity. Artificial neural network (ANN), decision trees (DT), random forest (RF), and support vector machines (SVM) are the examples of QSAR approach, which have been incorporated in finding new medicines [76]. The same mindsets can help in the revolution of novel electrode and catalyst materials discovery and their property prediction. Some of the commonly used ML models are as follows:

4.2.1.3.1 Regressors

Linear models such as linear regression and Bayesian ridge regression, neural networks, RF, and KRR are the regressors ML models.

4.2.1.3.1.1 Kernel Regression This is one of the common and popular models. Similarities between two sets of data are measured as input in Kernel-based methods [77]. Furthermore, its outcome could be understood mostly as a linear set of kernel functions for the given data. Gaussian fit is a popular fitting method in which several Gaussian curves are used to fit data in a model. It has been established that the linear model outperforms the Kernel ridge regression in terms of efficiency. However, the latter one is more flexible. These models have done their job by predicting formation energies [78], potential energy surfaces [79], electronic density of states [80], etc.

4.2.1.3.2 Neural Networks

This is a modeling technique designed after the brain [81]. The input of this system is called feature and the output is prediction. Between input and output, there might be a single or several layers which consist of several functions which help to make the output or prediction. Neural networks have been considered cutting-edge machine learning modeling systems that are regarded as one of the most common [82] algorithms. However, ANN requires a cornucopia of quality data and turns out to be a very exorbitant technique in terms of computational expenses. Contrarily, it has demonstrated the ability to just be reliable and cost-effective when dealing with smaller datasets. The model learns from different features of molecule, such as charge, the interatomic distance, etc. It can compare data with the data that has been already learned and can differentiate any irregularities. It can also capture underlying patterns among the input data [83]. Although neural networks' primary design constrained the model's ability only to deal with massive amounts of data, recent improvements have allowed them to function effectively with modest amounts of data as well. As an example, after almost 10,000 datasets of training, Grossman's model of a generalized crystal graph convolutional neural network (CGCNN) effectively projected several key features of perovskite crystal structure. [84] Surprisingly, this prediction has very high accuracy when compared to DFT prediction as compared to experimental results.

4.2.1.3.3 *Transfer Learning*

It is widely known that insufficient data is the major setback for applying ML models in materials research. Transfer learning is a new machine learning technique that addresses this limitation by moving data between learning activities. Hutchinson et al. [85] tested several architectures to forecast the bandgap of crystalline compounds. By using this procedure, he became capable of predicting bandgaps. He basically took the differences amongst the replies of the several systems and then used them to teach other systems. By learning the difference between computational and experimental data, this model predicted the bandgap only using a very small amount of data.

4.2.1.3.4 *Natural Language Processing*

This is the ML technique that is incorporated with human language. For example, how data from keyboard input can return results such as Google, Bing, etc., google translation, speech recognition, auto text correction, etc. So, NLP can process any data in textual form even materials science-based text or literature. Kim et al. [86] used the NLP technique to find out key parameters from more than 12,000 pieces of literature for hydrothermal synthesis of titania nanotubes.

4.2.1.3.5 *Machine Learning Toolkits*

AMP (Atomistic Machine-Learning Package) is an open-source ML language framework. It is compatible with many of these DFTs to constructing learning prospects, as well as GPAW (projector-augmented wave), VASP (Vienna Ab initio Simulation Package), and other atomistic simulation tools. The following is a list of some packages and libraries that can be used for modeling:

- **Keras**: Open-source neural network Python library (http://keras.io/)
- **Atomistic Machine-learning Package (AMP)**: ML for atomistic calculations (https://amp.readthedocs.io/en/latest/)
- **Classification And REgression Training (CaReT)**: for classification and regression models (https://github.com/topepo/caret)
- **MAterials Simulation Toolkit for Machine Learning (MAST-ML)**: Open-source Python package (https://github.com/uw-cmg/MAST-ML)
- **COMmon Bayesian Optimization Library (COMBO)**: Python library for ML techniques (https://github.com/tsudalab/combo)

4.2.1.4 Model Optimization

Validation datasets are being employed to justify the performance of the ML model built. Prediction mistakes occur when an outcome is predicted incorrectly based on a set of inputs. This error can be classified into two categories:

1. Variance errors: if multiple teaching datasets are applied, different flaws are caused by differences in algorithm outcomes [87]. In a flawless ideal case, there would not be any variance even if varying datasets are used; however, ML algorithms like KNN, RF are prone to high variance errors.
2. Bias errors: indicate the difference between the predicted output and actual practical output of test data. An excessive bias error implies the mapping function being too approximate which was created to function the overall process in less time with a low computational cost, whereas a moderate or low bias error suggests the mapping function of having fewer approximations which will lead to higher cost.

These errors can result in critical problems such as underfitting, overfitting, etc. That is why these must be balanced. Typically, underfitting is there when the mapping function does not include any significant data. It represents low variance and high bias. Then again, overfitting is indicative of high variance and low bias and it has superfluous data [88]. So, to avoid under- and overfitting, an optimized balance between these two types of errors should be achieved in the model. Models are assessed using a variety of measures such as:

i. Mean absolute relative error,
ii. Coefficient of determination, R^2,
iii. Learning rate,
iv. Loss function,
v. Mean absolute error,
vi. ROC curve

4.2.2 A Summary of the Use of Machine Learning in 2D Materials Research

Machine learning has been applied in materials research greatly in recent years because of the efficiency of the process in time and money and accuracy in prediction. Figure 4.4 exhibits the application of machine learning throughout the realm of materials science, which can be divided into three groups, for instance, (1) prediction of materials properties, (2) discovery of new materials, and (3) different other purposes such as process optimization, battery monitoring, etc. Similar approaches are applicable to 2D materials too. Using regression analysis methods, 2D materials property prediction, both micro and macroscopic, can be obtained. On the other hand, probabilistic model such as Markov chain [89] is used to screen combinations of components and structures [90], and out of the few good options, candidate having relatively superior performance is selected finally by using DFT-centered verification.

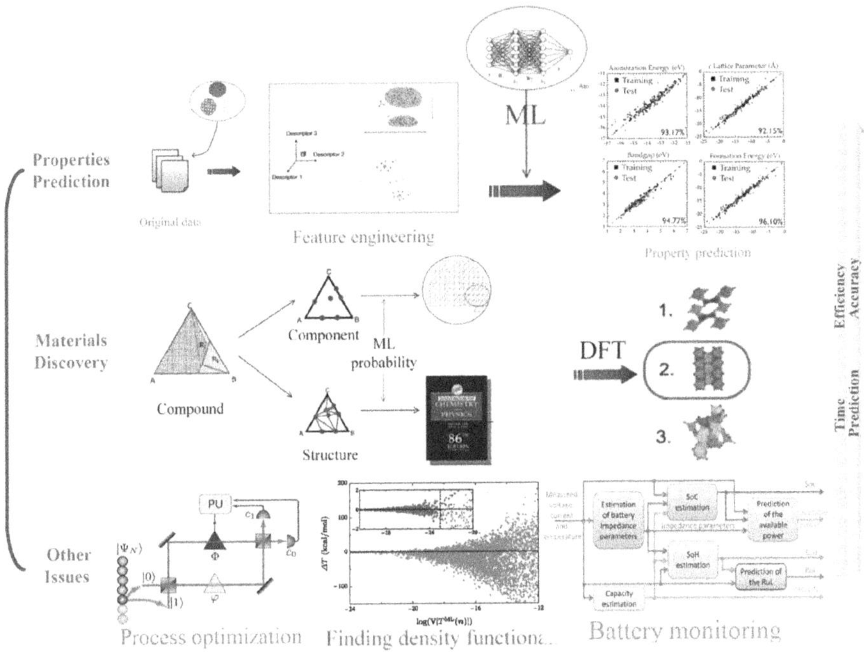

FIGURE 4.4 A summary of the use of machine learning in materials research. (Reprinted with the permission from Liu, Y., et al., Materials discovery and design using machine learning. *Journal of Materiomics*, 2017. **3**(3): pp. 159–177.)

4.3 PREDICTION OF 2D MATERIAL PROPERTIES USING MACHINE LEARNING

2D material property can be predicted via machine learning models. The conventional property of materials can be obtained either by experimental procedures or by computational simulation. Both of these processes require complex operation and experimental setup. So, it is difficult to obtain data on the property of the materials and some may remain even unknown. Another point to consider is that these experiments are carried out at the end of the selection process; therefore, if the conditions are unfavorable, then all the investments made so far would be wasted. Moreover, sometimes properties of materials cannot be studied even through a colossal amount of effort in experimental or computational efforts. And so intelligent prediction systems must be developed which can determine material characteristics efficiently and precisely at a cheap expense and promptly. In this case, machine learning can be applied because it examines the creation and analysis of computer programs that can extract insights and patterns from data. By analyzing and figuring out existing relations amongst various characteristics of materials as well as the other factors associated with those properties through the extraction of insights and information from existing practical data, machine learning can help predict the property of materials. Figure 4.5 demonstrates the basic framework of how these models can efficiently predict material property.

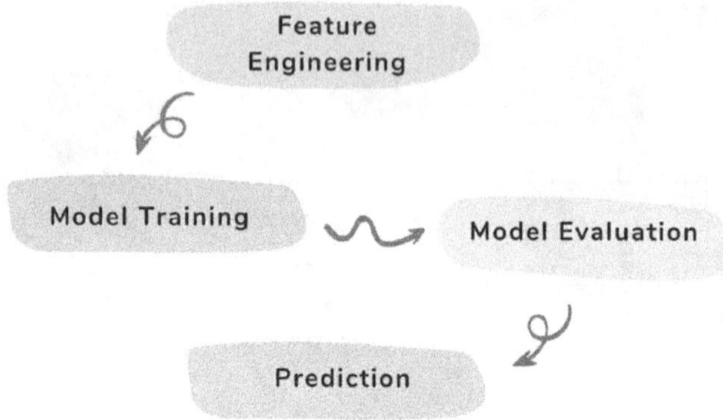

FIGURE 4.5 Fundamental framework for the application of ML in 2D materials property prediction.

Firstly, for the determination of the conditional attributes associated with property prediction, feature engineering is conducted. After that, the system is trained to define the association amongst these conditional factors and the decision parameters. In the end, the model projects the properties of the materials as output.

The trained model can predict materials properties in two classes, such as macroscopic performance prediction and microscopic property prediction. While investigating the macroscopic functionalities of materials, researchers emphasize on physical as well as mechanical properties. The structure–activity correlation of the material's characteristics and microstructure is such an example [90]. Neural networks, support vector machine (SVM), and optimization techniques were employed to investigate the macroscopic performance of materials. These machine learning algorithms have an excellent track record for addressing regression and classification tasks.

Artificial neural networks are by far the most extensively employed algorithms for evaluating parameters which are typically undefined as well as rely on even a huge set of input data. ANNs are nonlinear statistical analysis techniques that are based on biological neural networks with the capability to learn by themselves and adjust [91]. Backpropagation ANNs (BP-ANNs) and radial basis function ANNs (RBF-ANNs) are also useful neural network techniques. BP-ANNs produce acceptable precise estimation with a high degree of adaptability [92]. But the convergence rate of this procedure is slow and at times faces local minima problems. On the other hand, RBF-ANNs can overcome the problem of local minima through integrating both the ANN and the radial basis function (RBF) concept. RBF-ANNs have superiority in the convergence rate also. The main convenience of ANNs is that they can learn from observed data and little prior comprehension of the target material is required [20]. However, in order to predict the attributes, it is essential to have a large, diversified dataset for training purposes [20].

Dunn et al. [93] developed machine learning tools, such as Automatminer and Matbench. Automatminer develops a machine learning (ML) system that has the capability to generate projections of materials as output by taking materials' structural and compositional data as input. The Matbench has 13 supervised machine learning tasks from 10 datasets. Samples in each task may vary from 312 to 132,752. Every single assignment has its datasets, also with input being compositional or structural component and the return being the specimen's targeted characteristic. Rajan et al. applied kernel ridge regression (KRR), support vector regression (SVR), Gaussian process regression (GPR), and decision tree (DT) boosting processes the prediction of the G0W0 bandgaps of 2D material. LASSO159 was applied for producing as well as choosing characteristics. Finally, the feature space was enhanced for every single approach.

Table 4.2 represents the use of ML in 2D materials properties prediction using various conditional attributes. The ML models used are also represented in the same table. As an example, Tawfik et al. [94] predicted the interlayer distance of 1431 bilayers using four machine learning models, namely relevance vector machine (RVM), feedforward neural network (FNN), and decision tree (DT). They used LASSO algorithm to summarize the optimum number of descriptors which is 35. The BR1 descriptor was employed to teach the machine learning methods. The bandgap was also predicted by the ML models except for this time the number of descriptors was 11. Using density functional theory (DFT), the bandgap was also calculated and compared to the results obtained by the ML models. In total, the bandgap was predicted for 210 bilayers applying the models educated by the BR1 representation.

4.4 APPLICATION MACHINE LEARNING APPROACHES TO DISCOVER NOVEL 2D MATERIALS

After discussing the application of ML in materials property prediction in the previous section, here we demonstrate the application of ML in 2D materials discovery. Discovering material having good performance has always been a key topic in materials science. Screenings for new materials either computationally or experimentally require structure transformation and element replacement. Nevertheless, structural, and compositional search space is usually not flexible rather constrained [95]. These screening methods require a huge amount of time in experimentation or computation without the assurance of being successful and typically result in efforts being pointless. Considering the fact of these conventional methods being imprudent in case of money and time, machine learning model combined with computational simulation is adapted for the discovery of novel 2D materials.

Figure 4.6 depicts a typical machine learning procedure for finding new 2D materials. This overall method is divided into two stages: The first is referred to as a learning system, and the second is referred to as a predictive model. Data screening, feature engineering, model selection, learning, as well as validation are all covered by the learning stage. The model obtained from the learning system is used by the prediction system to give output as the projection of components as well as structure. The following method is then used to choose new materials: DFT computations are utilized to examine the feasibility of the candidates, and the prediction system provides suggestions for candidate structure via structural and compositional approval. For finding new materials, numerous machine learning techniques are currently being used and the sole purpose of these attempts is to find materials with good performance. These are classified into two types: forecasting of crystalline and composition structure. Table 4.3 demonstrates a list of different 2D materials, and their properties predicted by different machine learning approaches.

TABLE 4.2
Use of ML in 2D Materials Properties Prediction

Conditional Attributes	Compounds	ML Methods	Predicted Properties	Reference
Exfoliation energy Crystal structure Bond strength	Graphene hexagonal boron nitride MoS_2	ANN, Brute Force Method	Stability, Heat of formation, Bandgap, Valence and Conduction band, Magnetic state	Sorkun et al. [42]
Ionic radii Oxidation state electronegativities	Perovskites Cubic perovskites	SVR, GRNN, MLR, ANN, RF	Lattice constants	Majid et al. [95]
Ionicity Atomic radius Electron affinity	AB, A2B2, AB2-noX, AB2X	LASSO (Least absolute shrinkage and selection operator), Ridge Regression	Bandgap, exciton binding energy	Liang et al. [96]
Atomic properties Structural properties	AB2, A2BB’	DFT, XGBoost, SISSO	Thermodynamic stability: formation energy, energy above the convex hull	Schleder et al. [97]

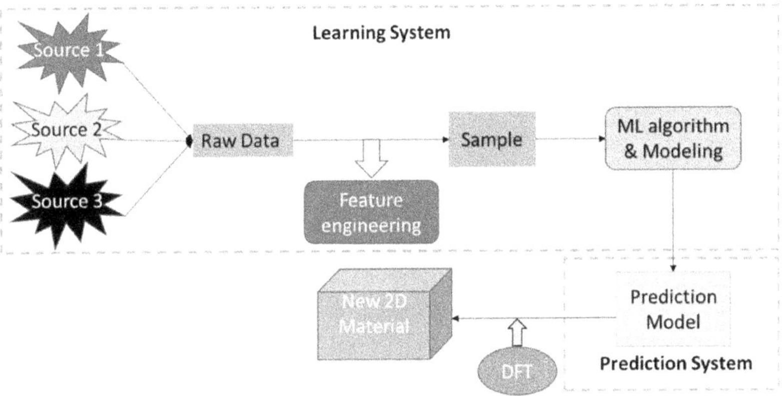

FIGURE 4.6 Typical machine learning procedure for finding new 2D materials.

TABLE 4.3
List of 2D Materials and Properties Predicted by Different ML Methods

Class of 2D Material	2D Material Pedicted	Atomic Structure	ML Method
Metal	$CrCuTe_2$ $CrCuSe_2$ $CrCuS_2$	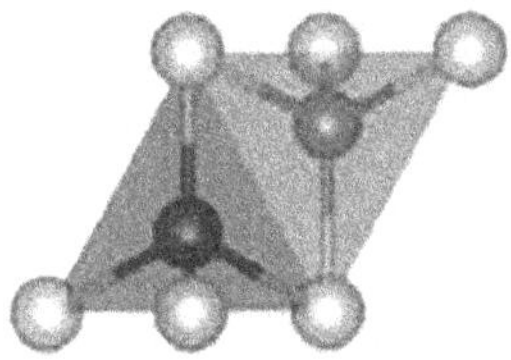 $CrCuTe_2$ [99]	Gradient Boosting Classifier (GBC) [99]
Half-metal	$MnCl_3$ $MnBr_3$ V_2CoO_6 V_2C_{15} V_3I_8 CrO_2	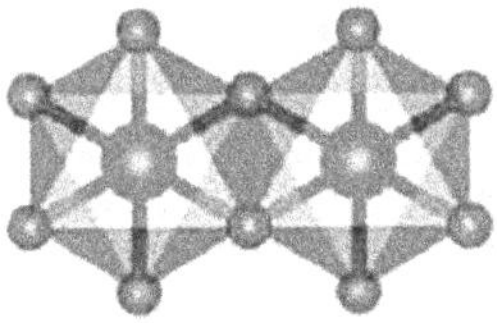 CrO_2 [99]	MatGAN (Generative Adversarial Network) [100] Gradient Boosting Classifier (GBC) [99]
Insulator	Bi2Se3	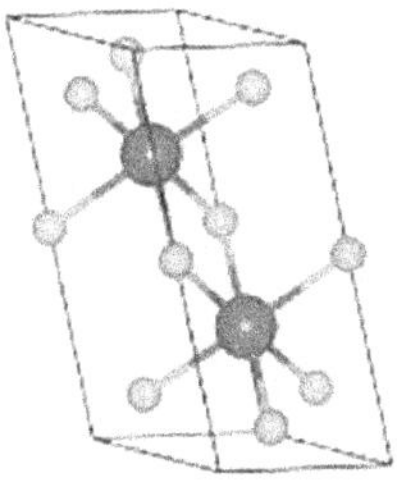 Bi_2Se_3 [101]	Gradient Boosting Classifier (GBC) [102]

(Continued)

TABLE 4.3 (*Continued*)
List of 2D Materials and Properties Predicted by Different ML Methods

Class of 2D Material	2D Material Pedicted	Atomic Structure	ML Method
Semiconductor	$CsSnCl_2Br$ $CsSnBr_2I$ $CsSnBr_2Cl$ $CrWBr_6$ CrOF $CrSiTe_3$	$CrWBr_6$ [99]	Regression Model & Gradient Boosting Classifier [103] Gradient Boosting Classifier (GBC) [99]

4.4.1 Predictions of Crystal Structure

Predictions of crystal structure remain one of the key factors in finding new materials and development. However, it remains one of the crucial problems that form the base for any materials design. By predicting the crystal structure, experiments related to structure can be avoided and that can save time and resources. Prediction of crystal structures based on both chemical reaction and first-principles crystal structure prediction is difficult due to the inherent complex mechanism. Following a chemical reaction, for example, it necessitates detailed and explicit knowledge regarding the entire reaction's potential energy surface (PES). Then again, prediction of crystal structure using first principles requires the consideration of a colossal amount of component arrangements by utilizing high-level computational quantum chemistry techniques [104]. However, machine learning uses different algorithms to analyze an enormous amount of experimental data and extract insights and empirical rules, and this technique has been used widely at this time.

Cluster resolution was utilized to choose features [105], which were then used for the inputs into partial least-squares discriminant analysis (PLS-DA) and SVMs by Oliynyk et al. [106]. To keep the problem simple, just the seven most prevalent samples were chosen. A total of 706 compounds were classified into three groups: (1) 235 compounds for feature, (2) 235 compounds for optimization of SVMs and PLS-DA, and (3) 236 for validation. Support vector machines' performance was comparatively superior compared to that of PLS-DA with an accuracy of 93.2% for SVMs and 77.1% for PLS-DA. SVMs were applied on a dataset having around 1505 compounds adapted from Pearson's crystal database. Then, it was tested to see what would happen provided that the number of the feature was reduced from a high to a low number. For example, the resulting sensitivity was 97.3%, accuracy 96.9%, and specificity 93.9% after changing the features from 1000 to 110.

In the overall designing process, the prediction of crystal structure is just the first step. The next step is to combine stability determination with property design. A set of 60,000 potential perovskites were studies by Balachandran et al. [107] via several machine learning techniques. First, the compounds were classified into perovskites or nonperovskites by SVM machine learning method. Then, the curie temperature of those perovskites was predicted. A candidate is added to the training set after it was synthesized experimentally, and the cycle continued. Six perovskites were discovered out of the ten synthesized compounds, and 898 K was found to be the greatest Curie temperature.

To investigate the relationship between electronegativity and particle measurement with crystalline structure, Ceder et al. [108] used principal component regression and Bayesian probability. This research shed light on the physical explanation which then governs crystallographic estimation. Moreover, Fischer et al. [109] ventured into the area of extracting knowledge from computational or experimental data and built a model that can predict structure based on information gathered from experimental data. This model is known as Data Mining Structure Predictor (DMSP). DMSP collects as well as analyzes empirical information in order to guide quantum approaches for the study of balanced crystalline formations. DBSCAN and OPTICS are programs developed by Phillips and Voth [110] which can find novel kinds of structure from a large number of datasets. Liu et al. [111] aimed to overcome issues regarding multi-objective design requirements and suggested a standardized machine learning structure. This novel framework includes arbitrary data production, feature engineering, and also several classification algorithms, and its overall function is projecting microstructures of Fe-Ga alloys. Obtained result of this method was promising as it outperformed traditional computational techniques. The average time required for this process was decreased by 80% and an efficiency that cannot be obtained by other methods.

Hautier et al. [112] built a probabilistic model by integrating theoretical and experimental data to predict new compositions and their possible crystal structures. Ab initio computations then validate these predictions. Based on the theory, machine learning provided the probability density of distinct structures existing side by side in a system. This approach was applied and Hautier et al. went through 2211 A-B-O system compounds (where A and B were from 67 different elements) in the inorganic crystal structure database (ICSD) [113] where no ternary oxide was found. There were a total of 1261 compounds with 5546 crystalline structures as a result of this. DFT was used to compute the energies of all these molecules. To assess the stability, energy of decomposing pathways was estimated, yielding 355 new compounds upon the convex hull.

Sendek and colleagues [114] applied logistic regression (LR) model and screened Li-ion conductor materials. Materials project database was used for the screening, and 317 candidates were finalized from a total of 12,831 candidates. Then LR was applied to create a classification model for more refined selection which resulted in 21 truly suitable candidates in the end, having an overall reduction of 99.8%.

4.4.2 Prediction of Components

Another method for discovering novel materials is component prediction. In a nutshell, it is required to figure out whether certain chemical entities are likely to result in compounds or not. When compared to the projection of crystalline systems, component prediction turns out to be a relatively common practice of machine learning. The search space for components is quite narrow for empirical or semi-empirical approaches which involve numerous authentication computations as well as experimentations, which might stymie considerable growth of novel material discovery. Currently, the overall prediction system of components utilizing the service of machine learning can be distinguished as two categories: (1) element combination recommendations after a group of components intended for a specified system and (2) demonstration of ionic substitutions designed for novel compound finding. While regression models can be applied to predict crystalline structure with no prior knowledge, component prediction is done using a Bayesian statistical model to solve for a posteriori probability. The distinction between classical and Bayesian statistical models is whether or not previous information is incorporated [115]. The utilization of trial data as well as the compilation, extracting, and handling of previous information are all very important in a Bayesian statistical model. Because they perform well in posterior probability estimates, these models are utilized to forecast material components.

Hautier et al. [112] used a Bayesian statistical analysis to obtain insights of 183 popular oxides compounds in the ICSD library to accurately forecast 209 novel ternary oxides. When compared to the previous (exhaustive) method, the expense of analysis was dropped about 30 times compared to that of before. KRR was used by Faber et al. [116] to compute the formation energy of two million elpasolites (stoichiometry ABC_2D_6) crystals containing main group elements up to bismuth. For a training set of 104 compositions, errors of roughly 0.1 eV/atom remained. 78 phase diagrams were created through energies and data from the materials project, along with 90 new-found stoichiometries that were projected on the convex hull.

Ward and colleagues [117] used typical RFs to estimate formation energies using Voronoi tessellations and atomic characteristics as inputs. The descriptors outperformed Coulomb matrices [118] and partial RDFs [64] on a training set of about 30,000. Remarkably, information after the Voronoi tessellation had little effect on the 30,000-material training set's findings. This is because the dataset contains extremely limited materials having same composition yet distinct formation. When the number of training sets was boosted to 400,000 materials from the open quantum materials database [33], overall inaccuracy of the composition-only system increased by 37% compared to the model that included structure information.

Li et al. [119] analyzed a dataset of roughly 2150 $A_{1-x}A'_xB_{1-y}B'_yO_3$ perovskites, which are employed as cathodes in elevated temperature fuel cells, using several regression and classification approaches. All techniques employed elemental characteristics as a feature. The top classifiers were highly randomized trees (having the best regression performance) and KRR. The first one had an average error of 17 meV/atom. Because of typical elemental composition space being so constrained, it is difficult to compare the faults in this work to others.

Instead of only ternary oxides, Meredig and colleagues [98] employed an exact similar technique for the prediction of components of ternary compounds. When compared to a normal first-principles analysis, they were capable of predicting 4500 potential ternary compounds exhibiting thermodynamic feasibility, and also the execution time was cut approximately six times. Convolutional neural networks and transfer learning were used by Zheng et al. [120] to predict stable full-Heusler compounds AB_2C. The idea behind a transfer learning [121] model is that at first a model is trained for a certain type of topic and then employing elements, insights, and knowledge learned during the initial phase for different training purposes, lowering quantity of data needed. Convolutional neural networks for image identification were applied by employing an image of the periodic table representation. The system was originally taught to forecast the formation energy of about 65,000 full-Heusler compounds from a certain database (OQMD) [33], with a mean absolute error of 7 meV/atom (for 60,000 data points) and 14 meV/atom (for a training set of 5000 compositions).

As previously said, strong machine learning algorithms can be constructed to find novel materials by predicting crystal structure and component structure. However, when using machine learning approaches for finding new materials and their property prediction, there are still certain challenges in the data collection stage. Due to the high expense of library synthesis, it is difficult to obtain large and high-quality datasets, and hence, it is considered a major challenge.

4.5 MACHINE LEARNING FOR MISCELLANEOUS FUNCTIONS

In addition to the application of ML in predicting materials properties and developing novel materials, it is widely used in various fields of materials science and engineering. ML has been used in research that requires mass simulations and experimental studies which is challenging to solve via typical research methodology.

ML has been used to determine density function. Snyder et al. [122] recently used a machine learning model for solving a newly designed density functional issue. The precision attained while estimating the kinetic energy (KE) of the structure, having mean absolute errors less than 1 kcal/mol is surprisingly amazing. This is by the far contrastingly great result compared to other approximations as it took even less than a hundred densities to train. ML has also been applied to fields like battery monitoring [123–125], optimization of overall process [126–129], corrosion prediction [130–132], etc. Although ML has been gradually applied in the field of corrosion, corrosion prediction, and application of protective coatings to prevent corrosion, the corrosion industry could have profited far more from the revolutionary progress in the field of ML. Wen et al. [133] utilized SVM model and input variables such as temperature, salinity, dissolved oxygen (DO), pH, and oxidation–reduction potential to predict corrosion rate of 3C steel in marine sea water. Another study reports the application of 29,100 electrochemical data (corrosion current (I_{corr}), corrosion potential (E_{corr}), Tafel, Bode, Nyquist) to train regression, DT, and gradient boosting ML models to predict corrosion behavior of high-strength nickel-based superalloy (Inconel 718). Application of ML as a datamining tool has predicted the corrosion behavior of another nickel-based alloy (alloy-22) [134]. In this study, temperature, exposure time, surface area, and weight loss were taken as input variables and then the neural network-based ML model predicted the corrosion rate and weight loss of the nickel alloy. Corrosion defect depth of oil and gas pipelines have been predicted in another study which utilizes a collection of ML techniques (PCA, GBM, RF, NN), and several input variables such as temperature, pressure, pH, ion concentration [135].

However, very few studies have been performed to predict the corrosion resistance of 2D materials as protective coating. Allen et al. [136] utilized EIS data to predict the corrosion resistance of 2D materials (graphene) in corrosive environment. They utilized deep learning-based model called variational autoencoder (VAE) to generate 1000 synthetic datasets from the original 49 experimental datasets. Input variables such as capacitance (double-layer (C_{dl}), coating (C_c)), resistance (solution (R_{soln}), charge-transfer (R_{ct}), polarization (R_{po})), and open circuit potential (OCP) from these datasets were then used to train ML model (Deep Neural Network

TABLE 4.4
ML for Miscellaneous Application

Input Variables	ML Methods	Model Solution	Reference
Current, Resistance, Voltage	NN, SVM	State of charge (SoC) Estimation of Lithium Ion Battery (LIB)	Meng et al. [138]
Degree of Deformation, Temperature	Fuzzy Neural Network (FNN)	Mechanical property (strength, deformation) Prediction of Titanium Alloy	Han et al. [139]
Aging Temperature and Aging Time	SVM	Prediction of Hardness, Electrical Conductivity of Aluminum Alloys	Fang et al. [140]
Amplitude of Wave, Age	Naive-Bayes	Damage Detection in Composite Materials	Addin et al. [141]

(DNN), Extreme Gradient Boosting (XGBoost)) to predict the corrosion resistance of 2D materials coating against microbial induced corrosion (MIC). The structural features of sulfate-reducing bacterial (SRB) biofilm, contributor to MIC to arbitrary substrates, were analyzed using neural network based deep learning models [137]. Although this study does not directly contribute to the field of corrosion resistance prediction of 2D materials protective coating, similar idea can help in that regard.

Application of ML in various field of science has been explored to solve relevant problems. Table 4.4 demonstrates instances where ML techniques such as neural network, SVM, Bayesian Network have been applied to tackle issues related to battery monitoring, process optimization, etc. Application of ML model helps achieve the ability to predict battery life (LIB) and mechanical properties (hardness, strength) of alloys and detect damage in composite materials. These examples illustrate the extensive reliance of science and engineering on machine learning.

Although the application of ML to predict new 2D materials with superior corrosion resistance property has been hardly explored so far, application of ML in the prediction of 2D material's various properties (electronic, structural) [99,142,143] has been explored successfully. With the availability of hundreds of thousands of structural, electronic, mechanical, and chemical data in several databases, Big Data technique can utilize these datasets to train and predict new 2D materials with superior corrosion resistance applications against both abiotic and biotic corrosion.

4.6 ASSESSMENT OF COMMON CHALLENGES AND THEIR PREVENTION METHODS

4.6.1 The Problems with Model Building

Data is required in machine learning model building and for the overall evaluation technique. That is why a large group of datasets are required for the whole process. And from that source datasets, a subgroup of the source data is chosen for research in a certain way. The expression sample denotes the fundamental dataset in machine learning, which often includes both training and test data. The three types of sample creation challenges that exist currently are the origin of its sample data, issues regarding creating feature vectors and assessing the size of the sample.

Computer modeling and experimental data are the source of sample data that are accumulated by several institutions, laboratories, organizations, or schools. However, these data collecting processes hardly have a centralized administrative structure. Although the development in the field of materials data infrastructures has addressed this difficulty, the applicability of machine learning is still limited because each database is independent, and the data format is not uniform.

Feature vectors are crucial because they compare the validity of model prediction. The feature vectors should, in theory, provide a basic physical foundation for extracting fundamental chemical and structural patterns, allowing for quick forecasts of original material chemistries. Electron density, Coulomb matrix, structure, and Composition are some of the most often utilized feature vectors in material science. There is no universal eigenvector that would be efficient for certain purposes in material science because each eigenvector is designed for just a distinct purpose. It is evident that different representations of features may yield different forecasting results [144].

In machine learning, determining sample mass is considered an important component throughout the sample construction process which remains associated with dimension reduction. The size of the sample size has influences if the sample datasets contain implicit knowledge about the sample's intrinsic characteristics, which is highly dependent on the particular topic and the machine learning technology used. Provided that simple techniques with only a few attributes as well as easy implementation, namely the SVM technique [145] may function well even if the size of the sample is not that large, advanced algorithms, including neural networks [48], which may attain highly accurate results on standard sample datasets irrespective of the size.

4.6.2 Usability

The level of difficulty in employing different machine learning algorithms for tackling pragmatic issues is denoted as usability. Machine learning's intricacy in materials research presents itself in two ways. (1) Machine learning is a difficult method that demands expert knowledge and guidance to complete. When employing machine learning for material property predictions, for instance, the analysis of correlation must be used to improve the prediction model accuracy. In some studies, it is evident that reducing the high dimensionality of a given problem having high dimensionality helps in the accuracy prediction [146]. (2) Determining parameters is likewise a challenging task. Since these parameters and kernel functions are all so important to machine learning methods, determining these is a crucial step in that process. In materials research, the parameters of machine learning systems are mostly defined by manual modification or experience. Moreover, to optimize such parameters, several optimization procedures are used. As a result, figuring out how to make machine learning approaches more usable is an issue that needs to be tackled immediately.

4.6.3 Learning Efficiency

Machine learning speed is proportional to its pragmatic use. While promptness is constantly sought while training as well as testing machine learning models, but in reality, achieving it all at the same time is difficult. The KNN approach, for example, has a fast-learning rate but very slow testing rate; contrarily, neural networks models take little time to train yet very long time to test. Due to the tiny nature of these samples, learning efficiency is not a big issue in materials science machine learning applications right now. But this problem will remain in these models in materials research when materials genome project spread and materials science will adopt the idea of big-data techniques having sample datasets with massive volume. Thus, this learning efficiency issue regarding how machine learning can be improved will also become a question which needs to be resolved as a matter of urgency.

4.7 CONCLUSIONS

Machine learning exists as a sophisticated and important method through which computer systems can gain insight. This is a part of artificial intelligence and is the trendiest type of analytical method. It is used for a variety of objectives in materials research, namely new material discovery, material property prediction, and many other reasons on a macroscopic to microscopic level. This technique in materials research has wide range of applications, covering electrolyte materials, different types of oxides, functional and metallic materials. Machine learning can be utilized to produce precise and effective methods for materials research, as evidenced by a wide range of relevant works. These works related to discovering new 2D materials with attractive functionalities, predicting 2D materials property using ML are gaining more and more attention as the range of application of 2D materials has widened significantly.

2D materials have captured traction in the field of electronics (sensors, spintronics), photonics and optoelectronics, and power and energy (batteries, supercapacitors) applications due to their attractive properties (bandgap, spin-orbit coupling, magnetic properties, barrier properties). In this chapter, we have already discussed that many studies have already predicted hypothetical 2D materials using various ML techniques. These predictions would narrow down the options for potential materials and help to reduce the cost of time and resource. Overall, the integration of ML in 2D materials research will accelerate the overall discovery process and change the field for the better. Moreover, varieties of topics are incorporated with machine learning and they turn out to be great according to various studies done on related topics. In the meantime, machine learning algorithms are deserving of more exploration.

In materials design and discovery, machine learning is mostly used to tackle probability estimation, regression, clustering, and classification problems. Furthermore, machine learning does well when addressing issues regarding sorting, correlation, and other similar tasks. As a result, these approaches would be used for tackling additional difficulties in materials research, which would probably lead to even more advancements. Usually, only a single machine learning method, such as ANN, SVM, DT, etc., is used for a certain problem. Sometimes, the results of these methods are compared for a particular topic and the best option is selected. As a result, each model's application range is severely limited. And so, developing a unifying context to apply this process to different strategies for resolving problems would considerably increase applications of machine learning approaches. This would enhance the machine learning model's efficacy as well as the generalization ability considerably. Big data is a trendy issue right now, and it is gaining a lot of traction in a variety of industries.

In materials scientific study, as well as many other disciplines, the subject of how to reserve, organize, analyze an enormous volume of data is one of the difficult issues to overcome. As a result, exploring machine learning model's applications in different scientific fields especially materials research alongside the big data technique is expected to be a critical study course for days to come. Deep learning showed excellence at processing massive amounts of data and has paved the way for significant advances in image processing, speech recognition, and other domains. As a result, in materials research, deep learning technologies involving sophisticated big data study ought to be explored.

REFERENCES

1. Othman, N.H., et al., Graphene-based polymer nanocomposites as barrier coatings for corrosion protection. *Progress in Organic Coatings*, 2019. **135**: pp. 82–99.
2. Kousalya, A.S., et al., Graphene: an effective oxidation barrier coating for liquid and two-phase cooling systems. *Corrosion Science*, 2013. **69**: pp. 5–10.
3. Galbiati, M., et al., Real-time oxide evolution of copper protected by graphene and boron nitride barriers. *Scientific Reports*, 2017. **7**(1): pp. 1–7.
4. Joseph, A., et al., 2D MoS2-hBN hybrid coatings for enhanced corrosion resistance of solid lubricant coatings. *Surface and Coatings Technology*, 2022. **443**: p. 128612.
5. Yadav, P., T.D. Raju, and S. Badhulika, Self-poled hBN-PVDF nanofiber mat-based low-cost, ultrahigh-performance piezoelectric nanogenerator for biomechanical energy harvesting. *ACS Applied Electronic Materials*, 2020. **2**(7): pp. 1970–1980.
6. Sup Choi, M., et al., Controlled charge trapping by molybdenum disulphide and graphene in ultrathin heterostructured memory devices. *Nature Communications*, 2013. **4**(1): pp. 1–7.
7. Gong, Y., et al., Two-dimensional hexagonal boron nitride for building next-generation energy-efficient devices. *ACS Energy Letters*, 2021. **6**(3): pp. 985–996.
8. Ikram, M., et al., 2D chemically exfoliated hexagonal boron nitride (hBN) nanosheets doped with Ni: synthesis, properties and catalytic application for the treatment of industrial wastewater. *Applied Nanoscience*, 2020. **10**(9): pp. 3525–3528.
9. Fan, F.R., et al., Emerging beyond-graphene elemental 2D materials for energy and catalysis applications. *Chemical Society Reviews*, 2021. **50**: pp. 10983–11031.
10. Yam, K.M., et al., Graphene-based heterogeneous catalysis: role of graphene. *Catalysts*, 2020. **10**(1): p. 53.
11. Zhang, L., et al., Graphene-based optical nanosensors for detection of heavy metal ions. *TrAC Trends in Analytical Chemistry*, 2018. **102**: pp. 280–289.
12. Akbar, F., et al., Graphene synthesis, characterization and its applications in nanophotonics, nanoelectronics, and nanosensing. *Journal of Materials Science: Materials in Electronics*, 2015. **26**(7): pp. 4347–4379.
13. Sorkin, V. and Y.W. Zhang, Graphene-based pressure nano-sensors. *Journal of Molecular Modeling*, 2011. **17**(11): pp. 2825–2830.
14. Han, W., et al., Graphene spintronics. *Nature Nanotechnology*, 2014. **9**(10): pp. 794–807.
15. Choudhuri, I., P. Bhauriyal, and B. Pathak, Recent advances in graphene-like 2D materials for spintronics applications. *Chemistry of Materials*, 2019. **31**(20): pp. 8260–8285.
16. Seneor, P., et al., Spintronics with graphene. *MRS Bulletin*, 2012. **37**(12): pp. 1245–1254.

17. Tan, Y.B. and J.-M. Lee, Graphene for supercapacitor applications. *Journal of Materials Chemistry A*, 2013. **1**(47): pp. 14814–14843.
18. Gunday, S.T., et al., Nanocomposites composed of sulfonated polysulfone/hexagonal boron nitride/ionic liquid for supercapacitor applications. *Journal of Energy Storage*, 2019. **21**: pp. 672–679.
19. Zhu, Y., et al., Modifications of MXene layers for supercapacitors. *Nano Energy*, 2020. **73**: p. 104734.
20. Liu, Y., et al., Materials discovery and design using machine learning. *Journal of Materiomics*, 2017. **3**(3): pp. 159–177.
21. Rapaport, D.C., *The Art of Molecular Dynamics Simulation*. 2004: Cambridge University Press: Cambridge. https://doi.org/10.1017/CBO9780511816581
22. Neugebauer, J. and T. Hickel, Density functional theory in materials science. *Wiley Interdisciplinary Reviews: Computational Molecular Science*, 2013. **3**(5): pp. 438–448.
23. Baumgärtner, A., et al., *The Monte Carlo Method in Condensed Matter Physics*. Vol. 71. 2012: Springer Science & Business Media: Berlin/Heidelberg.
24. Steinbach, I., Phase-field models in materials science. *Modelling and Simulation in Materials Science and Engineering*, 2009. **17**(7): p. 073001.
25. Olson, G.B., Designing a new material world. *Science*, 2000. **288**(5468): pp. 993–998.
26. Hautier, G., A. Jain, and S.P. Ong, From the computer to the laboratory: materials discovery and design using first-principles calculations. *Journal of Materials Science*, 2012. **47**(21): pp. 7317–7340.
27. Camacho-Zuñiga, C. and F.A. Ruiz-Treviño, A new group contribution scheme to estimate the glass transition temperature for polymers and diluents. *Industrial & Engineering Chemistry Research*, 2003. **42**(7): pp. 1530–1534.
28. Yu, X.-L., B. Yi, and X.-Y. Wang, Prediction of the glass transition temperatures for polymers with artificial neural network. *Journal of Theoretical and Computational Chemistry*, 2008. **7**(05): pp. 953–963.
29. de Pablo, J.J., et al., The materials genome initiative, the interplay of experiment, theory and computation. *Current Opinion in Solid State and Materials Science*, 2014. **18**(2): pp. 99–117.
30. Allmann, R. and R. Hinek, The introduction of structure types into the Inorganic Crystal Structure Database ICSD. *Acta Crystallographica Section A: Foundations of Crystallography*, 2007. **63**(5): pp. 412–417.
31. Haastrup, S., et al., The computational 2D materials database: high-throughput modeling and discovery of atomically thin crystals. *2D Materials*, 2018. **5**(4): p. 042002.
32. Zhou, J., et al., 2DMatPedia, an open computational database of two-dimensional materials from top-down and bottom-up approaches. *Scientific Data*, 2019. **6**(1): pp. 1–10.
33. Saal, J.E., et al., Materials design and discovery with high-throughput density functional theory: the open quantum materials database (OQMD). *JOM*, 2013. **65**(11): pp. 1501–1509.
34. Jain, A., et al., Commentary: the materials project: a materials genome approach to accelerating materials innovation. *APL Materials*, 2013. **1**(1): p. 011002.
35. Groom, C.R., et al., The Cambridge structural database. *Acta Crystallographica Section B: Structural Science, Crystal Engineering and Materials*, 2016. **72**(2): pp. 171–179.
36. Hachmann, J., et al., Lead candidates for high-performance organic photovoltaics from high-throughput quantum chemistry: the Harvard Clean Energy Project. *Energy & Environmental Science*, 2014. **7**(2): pp. 698–704.
37. Murphy, K.P., *Machine Learning: A Probabilistic Perspective*. 2012: MIT Press: Cambridge.
38. Bishop, C.M., *Pattern Recognition and Machine Learning*. 2006: Springer: New York.
39. Jennings, P.C., et al., Genetic algorithms for computational materials discovery accelerated by machine learning. *npj Computational Materials*, 2019. **5**(1): p. 46.

40. Wong, A.J.Y., et al., Battery materials discovery and smart grid management using machine learning. *Batteries & Supercaps*, 2022: p. e202200309.
41. Gómez-Bombarelli, R., et al., Design of efficient molecular organic light-emitting diodes by a high-throughput virtual screening and experimental approach. *Nature Materials*, 2016. **15**(10): pp. 1120–1127.
42. Sorkun, M.C., et al., An artificial intelligence-aided virtual screening recipe for two-dimensional materials discovery. *npj Computational Materials*, 2020. **6**(1): p. 106.
43. Raccuglia, P., et al., Machine-learning-assisted materials discovery using failed experiments. *Nature*, 2016. **533**(7601): pp. 73–76.
44. Wu, M. and L. Chen. Image recognition based on deep learning. In *2015 Chinese Automation Congress (CAC)*, Wuhan, November 27–29 2015, (pp. 542–546). 2015. IEEE: Wuhan.
45. Olsson, F., *A Literature Survey of Active Machine Learning in the Context of Natural Language Processing.* 2009: Swedish Institute of Computer Science: Kista.
46. Deng, L. and X. Li, Machine learning paradigms for speech recognition: an overview. *IEEE Transactions on Audio, Speech, and Language Processing*, 2013. **21**(5): pp. 1060–1089.
47. Donepudi, P.K., AI and machine learning in banking: a systematic literature review. *Asian Journal of Applied Science and Engineering*, 2017. **6**(3): pp. 157–162.
48. Rao, H. and A. Mukherjee, Artificial neural networks for predicting the macromechanical behaviour of ceramic-matrix composites. *Computational Materials Science*, 1996. **5**(4): pp. 307–322.
49. Reich, Y. and N. Travitzky, Machine learning of material behaviour knowledge from empirical data. *Materials & Design*, 1995. **16**(5): pp. 251–259.
50. Cunningham, P., M. Cord, and S.J. Delany, Supervised learning. In Cord, M., Cunningham, P. (eds) *Machine Learning Techniques for Multimedia. Cognitive Technologies.* 2008, Springer: Berlin. pp. 21–49.
51. Muhammad, L., et al., Supervised machine learning models for prediction of COVID-19 infection using epidemiology dataset. *SN Computer Science*, 2021. **2**(1): pp. 1–13.
52. Das, S., et al., Machine learning in materials modeling-fundamentals and the opportunities in 2D materials. In Yang, E.H., Datta, D., Ding, J., Hader, G. (eds) *Synthesis, Modeling, and Characterization of 2D Materials, and Their Heterostructures.* 2020, Elsevier: Amsterdam. pp. 445–468.
53. Peck, R., C. Olsen, and J.L. Devore, *Introduction to Statistics and Data Analysis.* 2015, Cengage Learning: Boston, MA.
54. Chu, X., et al. Data cleaning: Overview and emerging challenges. In *SIGMOD'16 Proceedings of the 2016 International Conference on Management of Data* (pp. 2201–2206). 2016. Association for Computing Machinery: New York.
55. Kohavi, R. and F. Provost, Glossary of terms. *Machine Learning*, 1998. **30**: pp. 271–274.
56. Chowdhury, A.A., et al., Deepqgho: quantized greedy hyperparameter optimization in deep neural networks for on-the-fly learning. *IEEE Access*, 2022. **10**: pp. 6407–6416.
57. Kirklin, S., B. Meredig, and C. Wolverton, High-throughput computational screening of new Li-ion battery anode materials. *Advanced Energy Materials*, 2013. **3**(2): pp. 252–262.
58. Balachandran, P.V., et al., Importance of feature selection in machine learning and adaptive design for materials. In Lookman, T., Eidenbenz, S., Alexander, F., Barnes, C. (eds) *Materials Discovery and Design*. 2018, Springer: Cham. pp. 59–79.
59. Vezhnevets, A. and O. Barinova. Avoiding boosting overfitting by removing confusing samples. In Kok, J.N., Koronacki, J., Mantaras, R.L., Matwin, S., Mladenič, D., Skowron, A. (eds) *European Conference on Machine Learning.* 2007. Springer: Berlin.

60. Domingos, P., A few useful things to know about machine learning. *Communications of the ACM*, 2012. **55**(10): pp. 78–87.
61. Mehta, P., et al., A high-bias, low-variance introduction to machine learning for physicists. *Physics Reports*, 2019. **810**: pp. 1–124.
62. Prior, F., et al., Open access image repositories: high-quality data to enable machine learning research. *Clinical Radiology*, 2020. **75**(1): pp. 7–12.
63. Zhang, C., L. Cao, and A. Romagnoli, On the feature engineering of building energy data mining. *Sustainable Cities and Society*, 2018. **39**: pp. 508–518.
64. Schütt, K.T., et al., How to represent crystal structures for machine learning: towards fast prediction of electronic properties. *Physical Review B*, 2014. **89**(20): p. 205118.
65. Anowar, F., S. Sadaoui, and B. Selim, Conceptual and empirical comparison of dimensionality reduction algorithms (pca, kpca, lda, mds, svd, lle, isomap, le, ica, t-sne). *Computer Science Review*, 2021. **40**: p. 100378.
66. Himanen, L., et al., DScribe: library of descriptors for machine learning in materials science. *Computer Physics Communications*, 2020. **247**: p. 106949.
67. Feng, P.-M., et al., Naive Bayes classifier with feature selection to identify phage virion proteins. *Computational and Mathematical Methods in Medicine*, 2013. **2013**. doi. 10.1155/2013/530696.
68. Coley, C.W., et al., A graph-convolutional neural network model for the prediction of chemical reactivity. *Chemical Science*, 2019. **10**(2): pp. 370–377.
69. Liu, H., et al., Accurate quantitative structure–property relationship model to predict the solubility of C60 in various solvents based on a novel approach using a least-squares support vector machine. *The Journal of Physical Chemistry B*, 2005. **109**(43): pp. 20565–20571.
70. Schmidt, J., et al., Recent advances and applications of machine learning in solid-state materials science. *npj Computational Materials*, 2019. **5**(1): pp. 1–36.
71. Hoi, S.C., et al. Online feature selection for mining big data. In *Proceedings of the 1st International Workshop on Big Data, Streams and Heterogeneous Source Mining: Algorithms, Systems, Programming Models and Applications* (pp. 93–100). 2012. Association for Computing Machinery: New York.
72. Kohn, W. and L.J. Sham, Self-consistent equations including exchange and correlation effects. *Physical Review*, 1965. **140**(4A): pp. A1133–A1138.
73. Parr, R.G. and Y. Weitao, Aspects of atoms and molecules. In *Density-Functional Theory of Atoms and Molecules*. 1994, Oxford University Press: Cary, NC, USA.
74. Brockherde, F., et al., Bypassing the Kohn-Sham equations with machine learning. *Nature Communications*, 2017. **8**(1): pp. 1–10.
75. Ingle, B.L., et al., Informing the human plasma protein binding of environmental chemicals by machine learning in the pharmaceutical space: applicability domain and limits of predictability. *Journal of Chemical Information and Modeling*, 2016. **56**(11): pp. 2243–2252.
76. Zhang, L., et al., From machine learning to deep learning: progress in machine intelligence for rational drug discovery. *Drug Discovery Today*, 2017. **22**(11): pp. 1680–1685.
77. Yan, S., et al. Regression from patch-kernel. In *2008 IEEE Conference on Computer Vision and Pattern Recognition*, Anchorage, AK, June 23–28, 2008. 2008. IEEE: New York.
78. Honrao, S., et al., Machine learning of ab-initio energy landscapes for crystal structure predictions. *Computational Materials Science*, 2019. **158**: pp. 414–419.
79. Hu, D., et al., Inclusion of machine learning kernel ridge regression potential energy surfaces in on-the-fly nonadiabatic molecular dynamics simulation. *The Journal of Physical Chemistry Letters*, 2018. **9**(11): pp. 2725–2732.

80. Mahmoud, C.B., et al., Learning the electronic density of states in condensed matter. *Physical Review B*, 2020. **102**(23): p. 235130.
81. Zou, J., Y. Han, and S.-S. So, Overview of artificial neural networks. In Livingstone, D.J. (eds) *Artificial Neural Networks*, 2008. pp. 14–22. Springer Link: Berlin
82. Li, H., Z. Zhang, and Z. Liu, Application of artificial neural networks for catalysis: a review. *Catalysts*, 2017. **7**(10): p. 306.
83. Roy, A., D. Dutta, and K. Choudhury, Training artificial neural network using particle swarm optimization algorithm. *International Journal of Advanced Research in Computer Science and Software Engineering*, 2013. **3**(3): pp. 430–434.
84. Xie, T. and J.C. Grossman, Crystal graph convolutional neural networks for an accurate and interpretable prediction of material properties. *Physical Review Letters*, 2018. **120**(14): p. 145301.
85. Hutchinson, M.L., et al., Overcoming data scarcity with transfer learning. arXiv preprint arXiv:1711.05099, 2017.
86. Kim, E., et al., Materials synthesis insights from scientific literature via text extraction and machine learning. *Chemistry of Materials*, 2017. **29**(21): pp. 9436–9444.
87. Nadeau, C. and Y. Bengio, Inference for the generalization error. *Machine Learning*, 2003. **52**(3): pp. 239–281.
88. Mehta, P., et al., A high-bias, low-variance introduction to machine learning for physicists. *Physics Reports*, 2019. **810**: pp. 1–124.
89. Kalogerakis, E., et al., A probabilistic model for component-based shape synthesis. *ACM Transactions on Graphics*, 2012. **31**(4): pp. 1–11.
90. Carrera, G.V., et al., Exploration of quantitative structure-property relationships (QSPR) for the design of new guanidinium ionic liquids. *Tetrahedron*, 2008. **64**(9): pp. 2216–2224.
91. Wang, S.-C., Artificial neural network. In *Interdisciplinary Computing in Java Programming*. 2003, Springer: Boston. pp. 81–100.
92. Altun, F., Ö. Kişi, and K. Aydin, Predicting the compressive strength of steel fiber added lightweight concrete using neural network. *Computational Materials Science*, 2008. **42**(2): pp. 259–265.
93. Dunn, A., et al., Benchmarking materials property prediction methods: the Matbench test set and Automatminer reference algorithm. *npj Computational Materials*, 2020. **6**(1): p. 138.
94. Tawfik, S., et al., Efficient prediction of structural and electronic properties of hybrid 2D materials using complementary DFT and machine learning approaches. *Advanced Theory and Simulations*, 2018. **2**.
95. Majid, A., et al., Lattice constant prediction of cubic and monoclinic perovskites using neural networks and support vector regression. *Computational Materials Science*, 2010. **50**(2): pp. 363–372.
96. Liang, J. and X. Zhu, Phillips-inspired machine learning for band gap and exciton binding energy prediction. *The Journal of Physical Chemistry Letters*, 2019. **10**(18): pp. 5640–5646.
97. Schleder, G.R., C.M. Acosta, and A. Fazzio, Exploring two-dimensional materials thermodynamic stability via machine learning. *ACS Applied Materials & Interfaces*, 2019. **12**(18): pp. 20149–20157.
98. Meredig, B., et al., Combinatorial screening for new materials in unconstrained composition space with machine learning. *Physical Review B*, 2014. **89**(9): p. 094104.
99. Lu, S., et al., Coupling a crystal graph multilayer descriptor to active learning for rapid discovery of 2D ferromagnetic semiconductors/half-metals/metals. *Advanced Materials*, 2020. **32**(29): p. 2002658.
100. Song, Y., et al., Computational discovery of new 2D materials using deep learning generative models. arXiv preprint arXiv:2012.09314, 2020.

101. Zhang, H., et al., Topological insulators in Bi2Se3, Bi2Te3 and Sb2Te3 with a single Dirac cone on the surface. *Nature Physics*, 2009. **5**(6): pp. 438–442.
102. Jin, H., et al., Discovery of novel two-dimensional photovoltaic materials accelerated by machine learning. *The Journal of Physical Chemistry Letters*, 2020. **11**(8): pp. 3075–3081.
103. Lu, S., et al., Rapid discovery of ferroelectric photovoltaic perovskites and material descriptors via machine learning. *Small Methods*, 2019. **3**(11): p. 1900360.
104. Beran, G.J., A new era for ab initio molecular crystal lattice energy prediction. *Angewandte Chemie International Edition*, 2015. **54**(2): pp. 396–398.
105. Sinkov, N.A. and J.J. Harynuk, Cluster resolution: a metric for automated, objective and optimized feature selection in chemometric modeling. *Talanta*, 2011. **83**(4): pp. 1079–1087.
106. Oliynyk, A.O., et al., Classifying crystal structures of binary compounds AB through cluster resolution feature selection and support vector machine analysis. *Chemistry of Materials*, 2016. **28**(18): pp. 6672–6681.
107. Balachandran, P.V., et al., Experimental search for high-temperature ferroelectric perovskites guided by two-step machine learning. *Nature Communications*, 2018. **9**(1): pp. 1–9.
108. Ceder, G., et al., Data-mining-driven quantum mechanics for the prediction of structure. *MRS Bulletin*, 2006. **31**(12): pp. 981–985.
109. Fischer, C.C., et al., Predicting crystal structure by merging data mining with quantum mechanics. *Nature Materials*, 2006. **5**(8): pp. 641–646.
110. Phillips, C.L. and G.A. Voth, Discovering crystals using shape matching and machine learning. *Soft Matter*, 2013. **9**(35): pp. 8552–8568.
111. Liu, R., et al., A predictive machine learning approach for microstructure optimization and materials design. *Scientific Reports,* 2015. **5**(1): pp. 1–12.
112. Hautier, G., et al., Finding nature's missing ternary oxide compounds using machine learning and density functional theory. *Chemistry of Materials*, 2010. **22**(12): pp. 3762–3767.
113. Bergerhoff, G., I. Brown, and F. Allen, *Crystallographic Databases*. Vol. 360. 1987. International Union of Crystallography: Chester. pp. 77–95.
114. Sendek, A.D., et al., Holistic computational structure screening of more than 12000 candidates for solid lithium-ion conductor materials. *Energy & Environmental Science*, 2017. **10**(1): pp. 306–320.
115. Bolstad, W.M. and J.M. Curran, *Introduction to Bayesian Statistics*. 2016: John Wiley & Sons: Hoboken, NJ.
116. Faber, F.A., et al., Machine learning energies of 2 million elpasolite (ABC2D6) crystals. *Physical Review Letters*, 2016. **117**(13): p. 135502.
117. Ward, L., et al., Including crystal structure attributes in machine learning models of formation energies via Voronoi tessellations. *Physical Review B*, 2017. **96**(2): p. 024104.
118. Faber, F., et al., Crystal structure representations for machine learning models of formation energies. *International Journal of Quantum Chemistry*, 2015. **115**(16): pp. 1094–1101.
119. Li, W., R. Jacobs, and D. Morgan, Predicting the thermodynamic stability of perovskite oxides using machine learning models. *Computational Materials Science*, 2018. **150**: pp. 454–463.
120. Zheng, X., P. Zheng, and R.-Z. Zhang, Machine learning material properties from the periodic table using convolutional neural networks. *Chemical Science*, 2018. **9**(44): pp. 8426–8432.
121. Weiss, K., T.M. Khoshgoftaar, and D. Wang, A survey of transfer learning. *Journal of Big Data*, 2016. **3**(1): pp. 1–40.

122. Snyder, J.C., et al., Finding density functionals with machine learning. *Physical Review Letters*, 2012. **108**(25): p. 253002.
123. Thomas, J.K., et al., Battery monitoring system using machine learning. *Journal of Energy Storage*, 2021. **40**: p. 102741.
124. Duraisamy, T. and D. Kaliyaperumal, Machine learning-based optimal cell balancing mechanism for electric vehicle battery management system. *IEEE Access*, 2021. **9**: pp. 132846–132861.
125. Samanta, A., S. Chowdhuri, and S.S. Williamson, Machine learning-based data-driven fault detection/diagnosis of lithium-ion battery: a critical review. *Electronics*, 2021. **10**(11): p. 1309.
126. Nasir, T., et al., Applications of machine learning to friction stir welding process optimization. *Jurnal Kejuruteraan*, 2020. **32**(1): pp. 171–186.
127. Liu, Z., et al., Machine learning with knowledge constraints for process optimization of open-air perovskite solar cell manufacturing. *Joule*, 2022. **6**(4): pp. 834–849.
128. Suzuki, Y., et al. Machine learning approaches for process optimization. In *2018 International Symposium on Semiconductor Manufacturing (ISSM)*, Tokyo, December 10–11 2018. 2018. IEEE: New York.
129. Liu, Q., et al., Machine-learning assisted laser powder bed fusion process optimization for AlSi10Mg: new microstructure description indices and fracture mechanisms. *Acta Materialia*, 2020. **201**: pp. 316–328.
130. Coelho, L.B., et al., Reviewing machine learning of corrosion prediction in a data-oriented perspective. *npj Materials Degradation*, 2022. **6**(1): pp. 1–16.
131. Diao, Y., L. Yan, and K. Gao, Improvement of the machine learning-based corrosion rate prediction model through the optimization of input features. *Materials & Design*, 2021. **198**: p. 109326.
132. Roy, A., et al., Machine-learning-guided descriptor selection for predicting corrosion resistance in multi-principal element alloys. *npj Materials Degradation*, 2022. **6**(1): pp. 1–10.
133. Wen, Y., et al., Corrosion rate prediction of 3C steel under different seawater environment by using support vector regression. *Corrosion Science*, 2009. **51**(2): pp. 349–355.
134. Kamrunnahar, M. and M. Urquidi-Macdonald, Prediction of corrosion behaviour of Alloy 22 using neural network as a data mining tool. *Corrosion Science*, 2011. **53**(3): pp. 961–967.
135. Ossai, C.I., A data-driven machine learning approach for corrosion risk assessment: a comparative study. *Big Data and Cognitive Computing*, 2019. **3**(2): p. 28.
136. Allen, C., et al., Deep learning strategies for addressing issues with small datasets in 2D materials research: microbial corrosion. *Frontiers in Microbiology*, 2022. **13**.
137. Ragi, S., et al., Artificial intelligence-driven image analysis of bacterial cells and biofilms. *IEEE/ACM Transactions on Computational Biology and Bioinformatics*, 2021. **20**(1): pp. 174–184.
138. Meng, J., et al., Overview of lithium-ion battery modeling methods for state-of-charge estimation in electrical vehicles. *Applied Sciences*, 2018. **8**(5): p. 659.
139. Han, Y.F., et al., Prediction of the mechanical properties of forged Ti-10V-2Fe-3Al titanium alloy using FNN. *Computational Materials Science*, 2011. **50**(3): pp. 1009–1015.
140. Fang, S.F., M.P. Wang, and M. Song, An approach for the aging process optimization of Al-Zn-Mg-Cu series alloys. *Materials & Design*, 2009. **30**(7): pp. 2460–2467.
141. Addin, O., et al., A Naïve-Bayes classifier for damage detection in engineering materials. *Materials & Design*, 2007. **28**(8): pp. 2379–2386.
142. Alibagheri, E., B. Mortazavi, and T. Rabczuk, Predicting the electronic and structural properties of two-dimensional materials using machine learning. *CMC-Computers Materials & Continua*, 2021. **67**(1): pp. 1287–1300.

143. Penev, E.S., N. Marzari, and B.I. Yakobson, Theoretical prediction of two-dimensional materials, behavior, and properties. *ACS Nano*, 2021. **15**(4): pp. 5959–5976.
144. Hansen, K., et al., Assessment and validation of machine learning methods for predicting molecular atomization energies. *Journal of Chemical Theory and Computation*, 2013. **9**(8): pp. 3404–3419.
145. Liu, Y., et al., Predicting the onset temperature (Tg) of GexSe1−x glass transition: a feature selection based two-stage support vector regression method. *Science Bulletin*, 2019. **64**(16): pp. 1195–1203.
146. Curtarolo, S., et al., Predicting crystal structures with data mining of quantum calculations. *Physical Review Letters*, 2003. **91**(13): p. 135503.

5 Bacterial Image Segmentation through Deep Learning Approach

Ejan Shakya and Pei-Chi Huang

5.1 INTRODUCTION

For decades, advances in volume scanning electron microscopy (SEM) have contributed to a significant increase in large three-dimensional (3D) images of bacterial cells. Consequently, many deep learning (DL) techniques have been successfully designed as feature extractors that transform the pixel values into a suitable internal representation for learning, making automatic analysis of microscopic images for cellular morphology feasible [1–3]. Observing cell size variability in microbes is an initial step in microbiology research that can provide insights into cellular responses to environmental stimuli through changes in physiology and gene expression [4,5]. Additionally, cell size variation is a fundamental physiological trait that plays a critical role in cellular housekeeping, nutrient transport, environmental adaptation, and cell reproduction [6,7]. Maintaining proper cell size is essential for optimizing regular cell physiology in bacterial cells [8]. Automated cell segmentation techniques in microscopic imaging play a crucial role in measuring cellular characteristics, including changes in size, to assess the effects of environmental changes and growth conditions [9–12]. Quantitative measures such as cell lengths, areas, and densities can provide information about how biofilm growth and material surfaces are intricately related. When analyzing a significant quantity of cells, techniques for automated cell recognition are necessary to differentiate the object of interest or specific region and retrieve quantitative measures accurately. This is important for making informed decisions regarding the accumulation of bacteria on biomaterial surfaces and their resistance to microbial corrosion. Until an equilibrium is reached with the available resources, either reversible or irreversible bacterial adhesion to a surface (or to each other) persists, which in turn affects the growth behavior or survival of material surfaces [13,14]. Also, the methods of biofilm image analysis also can help discover new materials or analyze the biofouling performance of existing materials. Any material—whether natural or engineered—is susceptible to biofilm formation and cell adhesion when exposed to moist environments, such as the surfaces of membranes, pipelines, and ship hulls. Such biofilms can lead to the fouling of material surfaces or microbiologically influenced corrosion issues [15,16]. In some cases, they can serve biotechnology applications. In both scenarios, biofilm image analysis can assist in the assessment of the performance and fate of the materials [17]. To address

DOI: 10.1201/9781003132981-5

these tasks in the biomedical domain, image segmentation of microbial bodies has an important foundational role.

Image segmentation is an active and prolific research problem in the field of computer vision [18]. Multiple scientific domains are utilizing diverse image segmentation techniques, such as semantic segmentation and instance segmentation. For example, in medical fields, automated robotic surgery and computer-assisted diagnosis use image segmentation as a fundamental method for detecting, tracking, and surgical scene understanding [19,20]. Another such domain is biomedical images, where microscopic images are taken into consideration to find the behaviors of the cells and their constituents. Among biomedical images, bacterial cells and biofilms have been a hot topic and a popular research area [21,22]. In general, collecting and organizing enormous digital data remains a critical problem that is time-consuming, expensive, and requires expert involvement. Various traditional image processing approaches have been applied as supervised machine learning methods on visual object counting tasks but require domain experts to provide pseudo-labeling rules from a variety of spatial information [23,24]. To develop advanced, content-based image understanding algorithms, an abundance of annotated examples like ImageNet [25] has been created for training and benchmarking data. These datasets have been used extensively for feature extraction and training contemporary machine learning models.

One of the most favored approaches to automated feature extraction from digital images has been the use of autoencoders and their variants [26]. These methods apply DL models to obtain a set of rich nonlinear representations directly from the input image without assumptions or a priori knowledge. Also, convolutional neural networks (CNNs) are considered the most dominant and effective type of deep neural networks when it comes to processing image data. At each layer, kernels and pooling operations are used to extract more advanced features from the raw pixel values of the input data. The CNNs methods have become the go-to machine learning approach for image-related downstream problems, such as image classification, segmentation, and object detection. It is evident that CNN-based deep learning algorithms require a huge amount of data to train and a greater training time. The data-voracious nature of CNN-based deep learning models demands a high volume of training examples. The reason for this is that the inherent locality of convolution operations limits the ability to model long-range dependencies using low-level features. Transformers, being a neural network architecture that exploits the concepts of global self-attention mechanism in a stack of convolutional encoders and decoders, address this limitation. Recent studies have shown that transformers, which rely solely on attention mechanisms and eliminate the need for recurrence and convolutions, require considerably less training time while producing superior results [27]. On the other hand, the deep CNN-based U-Net architecture [28] has been considered the start-of-the-art implementation for medical image segmentation. Since transformers have been a revelation in machine learning to solve downstream problems in natural language processing, their relevance in image processing and computer vision is novice. Here, we propose a hybrid type of U-Net, called *ViTransUNet*, that uses CNN-vision-transformer-based contraction layers to merge the comprehensive global context captured by transformers with the intricate spatial details provided by CNNs, which requires the creation of layers capable of combining both types of information. The encoded feature representation is upsampled and concatenated with the corresponding encoded layer to learn back image structure lost during pooling in the contraction layer.

The subsequent sections of this chapter are structured as follows. Related work on diverse cell segmentation techniques is presented in Section 5.2. An overview of our approach is provided in Section 5.3, while Section 5.3.3 offers a detailed examination of the *ViTransUNet* network. Section 5.4 elaborates on the experimental setup utilized for the SEM image dataset, and the conclusions of our proposed approach, along with potential avenues for future research, are presented in Section 5.5.

5.2 LITERATURE REVIEW AND RELATED WORK

Bacterial cell segmentation has been addressed by manual and interactive segmentation techniques [29,30]. However, the manual techniques are tedious, labor-intensive, and time-consuming, e.g., overlapping and complex cells in one image. Significant progress has been made in automating the segmentation and quantification of overlapping/touching cells in images through various methodological investigations. This section will cover the following topics: Section 5.2.1 will provide a detailed review of conventional techniques for segmenting overlapping objects; Section 5.2.2 will focus on contour-based methods for detecting corner points; Section 5.2.3 will explore the use of ellipse fitting approaches for object segmentation; finally, Section 5.2.4 will analyze existing methods that use convolutional neural networks (CNNs).

5.2.1 Conventional Approaches for Semantic Segmentation

Traditionally, for image segmentation tasks, histogram-based thresholding techniques [31,32] were employed, wherein clusters were formed to represent homogeneous objects in the image. A set of thresholds were selected so that objects and background pixels can be discriminated against. The selected threshold value was chosen to differentiate each pixel as either a constituent of the background or an object. This conversion of color or grayscale image into a binary image made the image segmentation task easier. With time, many other approaches have come into existence; morphological operations [33,34], watershed segmentation [35], level-set methods [36,37], graph-based approaches [38,39], and their variations [40,41], each of which had their own applicability and limitations in terms of low-level spatial feature extraction. These limitations were addressed and resolved with the breakthrough of CNN and its exceptional representational power.

5.2.2 Contour-Based Methods

The contour-based approach involves the utilization of curvature, skeleton, and polygon approximation to perform curve evolution on segmented regions of the image. This method involves reducing a contour to a collection of discrete vertex coordinates and is regarded as a regression task. This method has primarily been employed in segmenting images with overlapping or touching objects [42–45]. The nature of overlapping objects has been introduced in multiple literatures depending upon the nature and relevancy of overlapping objects. Sliding window-based techniques were utilized by Fernandez et al. and He et al. [46,47] to extract the foreground object and contour from the background. Likewise, Wang et al. [48] suggested a bottleneck detector that identifies a set of splitting points on contours as a means of detecting concave regions. This approach maximizes their distance and minimizes the Euclidean distance transform (EDT). Due to its predisposition to noise [49] and erroneous corner point detection, extensive preprocessing of input images is needed.

5.2.3 Ellipse-Fitting Methods

Ellipses are generally used to address the overlapping or touching regions of multiple cells in a cell cluster. This is a foundational task that profits instance segmentation in cells. There are several related research areas on ellipse-detection algorithms using traditional computer vision methods, e.g., Hough transform with parameter space decomposition and randomized Hough transform (RHT) approach [50,51] to minimize the computation complexity [52,53]. However, these solutions depend on different scenarios and cannot perform well for partially obscured ellipses. Ellipse-fitting methods have garnered significant interest because of their efficacy in dealing with the challenge of segmentation tasks involving elliptical-shaped objects that are in contact with each other [54,55]. An example of utilizing the multiellipse fitting solution can be seen in the segmentation of overlapping elliptical grains [56] and cell nuclei [57]. A minimum threshold for the expected area of each cell is established with this approach, enabling the automatic detection and separation of touching cells. Although this method holds immense potential, its limited ability to generalize for objects with diverse shapes has hindered its widespread adoption in various applications, primarily because it necessitates rules and parameters tailored to the specific task at hand.

Several recent studies have expanded upon the use of ellipse-fitting techniques to enhance the accuracy of segmentation outcomes. One such example is the modified ellipse-fitting approach proposed by Zou et al. [55], which generates candidate ellipses and identifies the most suitable one from the pool of candidates. The particularly useful technique for identifying overlapping elliptical objects in a binary image entails the extraction of concave points through a polygon approximation algorithm. In the research paper, Panagiotaki and Argyros [58] presented an ellipse-fitting algorithm (called DEFA) that eliminates the need for manually set parameters and, instead, utilizes the skeleton of a shape to automatically estimate the parameters and number of ellipse objects. It should be emphasized that this particular method is solely applicable to images featuring elliptical-shaped objects that have undergone binarization and exhibit a significant contrast between their foreground and background. Panagiotaki and Argyros [59] introduced a solution to this limitation with an enhanced version of DEFA, referred to as RFOVE, which utilizes unsupervised learning to optimize the area of shape coverage, and is capable of automatically determining the number of potentially overlapping ellipses even when dealing with previously unknown shapes. Furthermore, Abeyrathna et al. and Panagiotaki et al. [57,60] have been able to effectively utilize these techniques for handling tasks involving the segmentation of overlapping cells and obtaining precise quantitative measures in SEM images. The use of SEM techniques has been demonstrated to be an efficient means of producing high-resolution images of bacterial cells.

5.2.4 CNN-Based Approaches

Because of the remarkable triumph of convolutional neural networks (CNN), they have become the default option for sophisticated segmentation tasks. In recent years, numerous cell segmentation applications and approaches that employ deep learning techniques have gained significant popularity in the field, thanks to their superior feature extraction capabilities and precise segmentation quality [61–63]. Several categories can broadly classify CNN-based methods for object segmentation:

1. Mask R-CNN [64] is an extensively employed neural network architecture designed for detecting multiple objects. It builds upon faster R-CNN [65] by incorporating an additional branch for predicting segmentation masks in addition to the existing branch that locates the bounding box. The method utilizes object detection based on region proposals and generates precise segmentation masks in order to attain instance segmentation results of superior quality. Nonetheless, this approach might not yield optimal results in situations where there is a high overlap among object instances or when objects are located in close proximity. This is primarily due to the utilization of greedy nonmaximum suppression during postprocessing.
2. U-Net [66] has emerged as a popular deep learning architecture for semantic segmentation, which eliminates the need for region proposals or the reuse of pooling indices. Rather than relying on region proposals or the reuse of pooling indices, the U-Net employs an encoder-decoder neural network architecture to generate object segmentation output based on class labels. In the task of the segmentation of overlapping cells, particularly in the medical domain, the U-Net architecture has demonstrated remarkable effectiveness [63,67], largely owing to its innate capability of performing downsampling and upsampling. For instance, research has been conducted to showcase the accurate segmentation of overlapping cervical cells using the U-Net architecture [68,69].
3. UNet3+ [70] is an updated variant of U-Net that leverages full-scale skip connections and deep supervision to enhance its performance. By integrating low-level details with high-level semantics from feature maps of varying scales, the full-scale skip connections enable the model to acquire pixel-level features of the images. Deep supervision, as introduced in UNet3+ [70], facilitates the acquisition of hierarchical representations from fully aggregated feature maps. This approach enhances the efficiency of U-Net models by reducing the network parameters and computational complexity.
4. Vision Transformers (ViT) [71] were proposed to apply transformers, which are generally used in natural language processing, in downstream computer vision tasks. To apply the transformer to an image, the image is initially partitioned into a grid of n patches, and the transformer is subsequently applied to the resulting sequence of patches. Each patch is an individual image that

is linearly transformed into a projected vector of configurable dimensions. To incorporate positional information of the patch, a positional embedding is added to each projected vector, just like sequential information for tokens in text corpora for natural language processing. ViT has emerged as a promising alternative to convolutional neural networks for image recognition tasks, particularly when pretrained on large datasets and fine-tuned to achieve benchmarks such as ImageNet [72] and CIFAR-100 [73].

5. TransUNet [74] is a hybrid implementation of Transformers and CNN, which takes into account the limitations of modeling long-range dependency in CNN models. Since transformers provide a global self-attention mechanism that compliments the drawbacks of a CNN model, TransUnet offers a robust option for medical image segmentation by effectively merging the advantages of two different approaches. The encoding path of TransUnet utilizes a hybrid framework that combines both CNN and transformer techniques. After encoding, the feature representations are upsampled and combined with various high-resolution CNN features that were skipped during the encoding process, allowing for accurate localization. Chen et al. [74] has presented that transformer-based architecture has a better self-attentive feature than the conventional CNN-based self-attention methods.

5.3 METHODOLOGY

This section outlines the key stages of the proposed solution for solving an overlapped cell segmentation challenge in bacterial images, which combines conventional reconstruction techniques with a patch approach. Section 5.3.1 provides an overview of the dataset acquisition process, while Section 5.3.2 explains the image preprocessing methods employed. In Section 5.3.3, we introduce *ViTransUNet*, the neural network that we propose to use, which is based on the transformer model.

5.3.1 Data Collection

Our dataset consists of training and testing samples of curated scanning electronic microscopic images (SEM) of *Geobacillus* genus from the family of *Bacilliceae.* To simulate the microgravity conditions, the bacterial cells were cultivated in a rotating cell culture system at a temperature of 60°C. After 24 hours of growth, the cells were treated with glutaraldehyde to stop their growth and washed three times using alcohol solutions of varying strengths (50%, 70%, and 100%). The resulting diluted cell suspensions were fixed onto a SEM sample mount and left to air dry before being imaged. For the acquisition of SEM images, a Zeiss Supra 40 VP/Gemini Column SEM was utilized [21]. The following microscopy parameters were set: the electron high tension (EHT) voltage was adjusted to 1 kV (also known as accelerating voltage), the SEM type was field emission, and the detector used was SE2 (secondary electron). The comprehensive experimental details can be found in the works of Carlson et al. [75].

The dataset includes 77 grayscale SEM images in terms of their respective masked annotations [21], which are used to validate our proposed bacterial cells segmentation approach. Our dataset includes several types of information for each image, such as magnitude, objective lens focal length (WD), EHT, noise reduction method, chamber status, data, and time. Using the VGG Image annotation software VIA, the surface areas of the bacterial cells were annotated manually to achieve semantic segmentation ground truth masks [21]. To evaluate the cell-detection accuracy, we divided the dataset into 57 training samples and 20 testing samples, using a 3:1 ratio for each.

5.3.2 Image Preprocessing

To improve the accuracy of bacterial image predictions, we perform several preprocessing steps that involve resizing, adjusting contrast, and eliminating unnecessary features, all of which serve to enhance the quality of each individual image. Our objective is to generate a pixelwise label map of size $H \times W \times 1$ that corresponds to an input image $x \in \mathrm{R}^{H \times W \times C}$, which has a spatial resolution of $H \times W$ and consists of C channels. Our following solutions are particularly efficient for the successful quality improvement of bacterial images. Every image is resized into 256×256 pixels. Then, the contrast-limited adaptive histogram equalization (CLAHE) algorithm [76] was utilized to enhance the contrast of foreground and background features during image processing. Furthermore, meta-information (such as methods for reducing noise, scale, magnification, and timestamp of capture including date and time) was removed from the image to ensure that the learning process is optimized for higher accuracy feature detection. Section 5.3.3 provides detailed knowledge about the model and its constituents, and how information is passed through various layers of the model.

5.3.3 ViTransUNet

Although transformers powerfully models global contexts at all stages and achieves superior transferability performance which means the pretrained models are applied using one task for all downstream tasks. However, transformers produce low-resolution features that cannot give sufficient localization information, causing inaccurate segmentation results when using upsampling to recover the full resolution. On the other hand, CNN architecture can learn certain fine spatial details to remedy this shortcoming of transformers effectively by providing complementary information.

To make up for the information loss resulting from the low-resolution features generated by transformers, the proposed neural network—*ViTransUNet*—as shown in Figure 5.1—establishes a hybrid form of CNN-Transformer Encoder where CNN layers are used to not only encode the feature extraction representations but leverage high-resolution spatial information from CNN features. Our approach takes inspiration from Schlemper et al. [77] and Chen et al. [74] who proposed that the u-shaped architectural design can combine the self-attentive features with different high-resolution features brought by CNN, enabling capability of precise localization. *ViTransUNet* extends the architecture of TransUNet in the sense that it includes a vision transformer (ViT) instead of the standard transformer-based encoder. While both are transformer-based models, ViT encoders operate primarily on the 2D grid of image patches generated by the patch encoder, which are processed to generate fixed-length vector representations of the input image. ViT encoders also employ

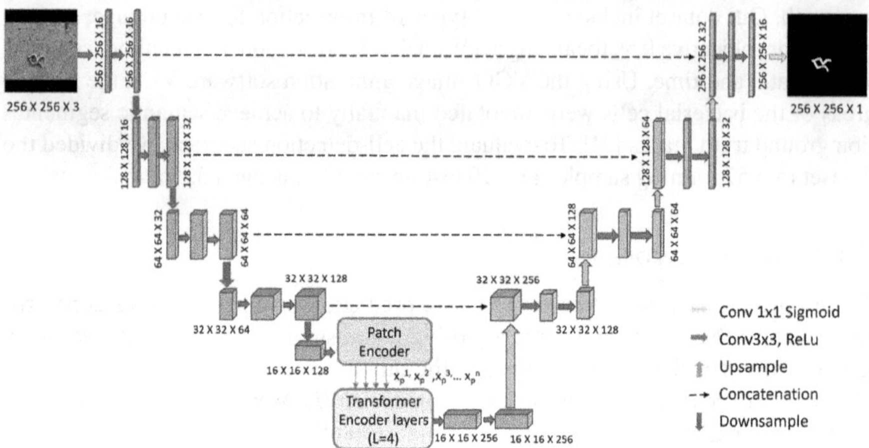

FIGURE 5.1 An example of the *ViTransUNet* architecture. Given an image of 256×256 pixels and its annotated mask as an input, each *grey* box is denoted as $H \times W \times C$, corresponding to a multichannel feature map, where $H \times W$ is the size of the feature map and C is the number of channels. The *arrows* denote the different operations. The *left* contracting path of U-net resizes the map to 16×16 feature map which is fed to the patch encoder to generate n vectorized patches. The encoded image representation output by the transformer is then resized to the initial 16×16 feature map to extract appropriate features and then fed to the *right* expansive path of the U-Net which supports the prediction of the synthesis.

self-attention mechanism that allows the model to learn contextual relationships between different image patches, which in this case encompasses the foreground and background features.

5.3.3.1 CNN-Transformer Encoder

Our encoder is built upon the design as the "left-contracting-part" of the model shown in Figure 5.1 and comprises a series of filtered convolutions and pooling layers, called the "down-block" or the "contraction" block. The downsampling part of the model is the hybrid CNN-Transformer-based encoder which preserves the advantages of transformers and CNN. The downsampler performs multiple feature extraction on the input image in the form of a multichannel feature map.

To extract spatial features or bottom-up features, our feature map is subdivided into patches as mentioned in Section 5.3.1.1 and encoded using a vision transformer as mentioned in Section 5.3.1.2. This step first flattens the patch sequence to latent space and then takes the sequenced raw image as input to the vision transformer and treats the image as a pixel-by-pixel prediction task.

5.3.3.1.1 Patch Encoder

Our solution followed by Dosovitskiy et al. [71] technique, the tokenization is performed by reshaping the input feature maps into flat $2D$ patches, namely, $x_p^1, x_p^2, \ldots\ldots x_p^n \in \mathbb{R}^{p^2}$, where each patch size is $P \times P$ and $n\left(=\dfrac{H \times W}{P^2}\right)$ is a number of generated image patches. For example, if the original input image is of dimension

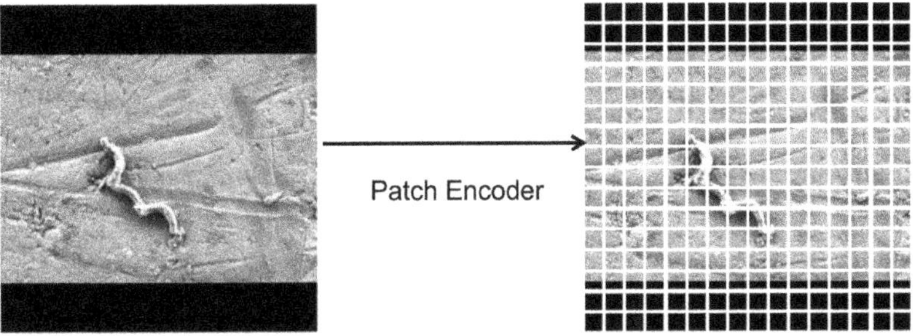

FIGURE 5.2 A sample of patch encoding. A 256×256 image is tokenized into 16×16 patch.

256×256, and P is 16, then 256 patched images of size 256×256 are generated, as shown in Figure 5.2.

Each vectorized patch is denoted as x_p and a dense layer is used to project x_p into a latent D-dimensional space. To encode the patch spatial information, the positional information is added to each patch in the following form of positional embeddings:

$$z_0 = x^0{}_p E + E_{\text{pos}[1]}$$

$$z_1 = x^1{}_p E + E_{\text{pos}[2]}$$

$$\vdots$$

$$z_{n-1} = x^n{}_p E + E_{\text{pos}[n]},$$

where E is the linear projection of patch embedding, and E_{pos} is the positional embedding, and n is the number of the vectorized patches.

5.3.3.1.2 Vision Transformer

In addition, the transformer incorporates a sequence of multihead self-attention (MSA) [27] and multilayer perceptron (MLP) blocks, as shown in Figure 5.3. All transformer models are built upon the self-attention mechanism, which serves as their fundamental building block. In transformers, MSA with multiple attention blocks (called heads) applies a linear transformation to the input matrices and then jointly performs attention multiple times for the learned parameters from different representation subspaces at different positions. On the other hand, MLP with multiple blocks is more than one perception in a deep neural network. Such a network is capable of approximating any continuous function which shows that an unlimited number of neurons in a hidden layer are allowed as the solution for nonlinearly separable functions. Figure 5.3 illustrates only one transformer encoder block, and the total number of transformer encoders includes L layers of MSA and MLP. Therefore, the output of the encoded image representation z_l is generated.

5.3.3.2 Decoder

The "right-expansive-part" of Figure 5.1 denotes the decoder. The resulting features obtained from the CNN-transformer encoder are fed into our decoder block

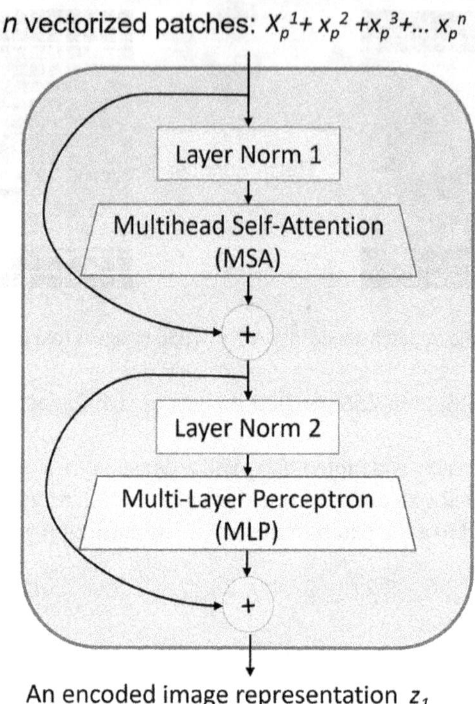

FIGURE 5.3 One transformer layer. The schema of a single transformer layer uses an embedded sequence of vectorized patches, where $1 \leq i \leq n$, obtained from the patch encoder as inputs. The specific positional embeddings are added to the patch internally to encode the spatial information. Every transformer encoder block includes a single layer of multihead self-attention (MSA) and multilayer perceptron (MLP) blocks that are normalized using two-layer normalization operators. For each vectorized patch, an encoded image representation z_1 is generated.

to reconstruct the original image. To generate the hidden features as an input for the ultimate segmentation mask, the decoder block utilizes a cascaded upsampler (CUP) that incorporates several upsampling steps [74]. The primary role of the encoder is to complete the U-shaped-like architecture of the model by restoring the compressed feature map to the original size of the input image. This necessitates an expansion of the feature dimensions, which is achieved through upsampling, also known as transposed convolution or deconvolution. As the upsampling process is performed, the higher resolution feature maps of the encoder are joined with the upsampled features, enhancing the ability of the model to learn representations through convolutions. This sparsely applied operation promotes improved localization. The series of upsampling layers will be concatenated with the corresponding contraction layers in the encoder through the number of skip connections which means some of the layers are skipped in the neural networks and feed one output of these layers into the next input layer. This block is also known as the "expansion" block as the encoded feature representations are expanded to masked spatial representation.

5.4 EXPERIMENTAL DESIGN AND RESULTS

This section presents the design of training and testing experiments for the proposed solution, as applied to the downstream task of image segmentation. Section 5.4.1 first describes the experimental setup, and Section 5.4.2 presents the evaluation metrics for image analysis. The experimental results for training the *ViTransUNet* model and the comparison with the model architectures of conventional U-Net framework are presented in Section 5.4.3.

5.4.1 Experimental Setup

The experimentation, encompassing both training and testing tasks, has been thoroughly examined in this study, used the premium service from Google Colaboratory to run on a GPU-enabled notebook whose components were an Intel Xeon CPU (2.20 GHz), and a GPU of Tesla T4 GPU with 13GB RAM was assigned. The experiments and comparisons were conducted on the SEM dataset for image segmentation in Section 5.3.1.

5.4.2 Evaluation Metrics

Three distinct metrics were employed to assess the instance segmentation task: the pixels accuracy in Section 5.4.2.1; intersection over union in Section 5.4.2.2; and dice coefficient in Section 5.4.2.3.

5.4.2.1 Pixel Accuracy

The accuracy metric is determined by calculating the percentage of pixels in the image that have been correctly classified. It provides a basic measure of the model's performance. However, it does not take into account the spatial relationships between different classes, and so, it may not be sensitive to errors that affect only a small number of pixels. This is computed by dividing the number of accurately classified pixels by the total number of pixels in the image, as represented by the following formula:

$$\text{Pixel Accuracy}(\text{PA}) = \frac{\sum_{j=1}^{k} n_{jj}}{\sum_{j=1}^{k} t_j}, \tag{5.1}$$

where the variable n_{jj} represents the total number of pixels that are both classified and labeled as belonging to class j. To put it another way, n_{jj} denotes the total count of accurate positive predictions made by the classification model for the specific class j. The variable t_j represents the total number of pixels that are labeled as belonging to class j. Given that semantic segmentation involves multiple classes, the mean pixel accuracy (mPA) reflects the average accuracy across all classes, as demonstrated by the following equation:

$$\text{mPA} = \frac{1}{k}\sum_{j=1}^{k} \frac{n_{jj}}{t_j} \tag{5.2}$$

5.4.2.2 Intersection over Union

Intersection over Union (IoU) is a metric that measures the extent of overlap between the predicted masks and the ground truth segmentation mask. This pivot enables us to assign a score to each image, which can be used to predict the accuracy of the predicted segmentation. Along with the overlap between the predicted segmentation and ground truth segmentation, it also penalizes the model for missed detections. It provides more detailed measures of the quality of the segmentation for each class, thus making it more sensitive to errors. In the case of image segmentation, predictions are segmentation masks, and pixel-by-pixel analysis is required, denoted as

$$\mathrm{IoU} = \frac{\mathrm{TP}}{\mathrm{TP} + \mathrm{FP} + \mathrm{FN}}, \tag{5.3}$$

where the true positive (TP) region corresponds to the intersection area between the ground truth and segmentation mask, while the false positive (FP) region corresponds to the predicted area outside of the ground truth. The false negative (FN) region, on the other hand, represents the count of pixels located in the actual target region which the model could not foresee or anticipate.

5.4.2.3 Dice Coefficient

The dice coefficient [78] is an important metric for image segmentation images to evaluate pixel-wise segmentation performance. Simply put, the dice coefficient is calculated by dividing twice the overlap between the two images by the total number of pixels in both images. In other words, the score represents twice the interaction area of overlap between the ground truth label and the predicted segment divided by the total number of pixels that are covered by both the ground truth label and predicted segment. The dice coefficient provides an overall measure of the model's performance, as it considers both true positives and false positives.

Mathematically, it is expressed as

$$\text{Dice Score} = \frac{2\mathrm{TP}}{2\mathrm{TP} + \mathrm{FP} + \mathrm{FN}}, \tag{5.4}$$

where true positive is denoted as TP, false positive as FP, and false negative as FN.

5.4.3 Evaluation Results

To evaluate the effectiveness of the proposed *ViTransUNet* method, we compared its performance with two established instance segmentation methods: the U-Net method [55], which utilizes a U-Net model for region-based fitting of overlapping ellipses, and UNet3+ [60], which based on encoder-decoder structure combines feature maps with different scales through dense skipped connections and deep supervisions. The comparison of the segmentation results for the metrics of dice score, pixel accuracy, and mean IoU is presented in Table 5.1. The samples of the predictions of the three methods based on the trained models are visualized in Figure 5.4. Figure 5.4a is the raw image; (b), (c), and (d) are the output images after using U-Net, UNet3+, and *ViTransUNet*, respectively.

TABLE 5.1
An Evaluation of How Well the Proposed Approach Exhibits in Segmenting Overlapping Objects, in Contrast to Three Alternative Methods

Method	Dice Score (%)	Pixel Accuracy (%)	Mean IoU (%)
U-Net [49]	66.79	91.54	48.76
U-Net 3+ [54]	71.8	90.60	32.6
ViTransUNet	**84.62**	**98.72**	**48.76**

Note: The best results of microaveraged *PixelScore*, *IoU*, and *Dicescore* (mean ± std) are highlighted.

The experiments show convincing results in favor of the proposed *ViTransUNet* model with a dice score of 84.62% and pixel accuracy of over 98%. However, the mean IoU, being more sensitive to errors and more detailed in measurements, displayed a decent performance as compared to the other two model architectures and relatively low accuracy as compared to the other two metrics.

5.5 CONCLUSION AND FUTURE WORK

In summary, our study presents a semantic image segmentation method for bacterial cells in SEM images; the SEM image dataset presents various challenging features, including bacterial cells that intersect or superimpose each other, objects with dissimilar shapes and sizes, and a minimal difference in color between the background and foreground. The method employed in the analysis of biofilm images can aid in the identification of novel materials or the evaluation of the biofouling efficacy of existing materials. Consequently, due to these characteristics, conventional segmentation techniques, e.g., color thresholds, ellipse fitting, or direct instance segmentation methods, exhibit inadequate performance.

This paper suggested a deep semantic segmentation architecture that overcomes these limitations by combining detailed high-resolution spatial information from convolutional neural networks with global context positional information from transformers. The hybrid *ViTransUNet* model achieved similar results if not better than its counterpart U-Net architectures in SEM images. Our approach consists of preprocessing of images, patch encoding and positional embedding of pixels, and transformer-based U-Net semantic segmentation. Based on the experimental results, the hybridization of the transformer-based encoder and deeply convoluted upsampling decoder has been shown to be significant and effective. The training performance of the *ViTransUNet* model as opposed to U-Net architecture is also depicted in Figures 5.5 and 5.6. When compared to other cell overlapping object segmentation methods, such as U-Net and its variants, the proposed approach achieved a dice similarity score of 84.62% for bacterial cell segmentation which demonstrates better than U-Net and UNet3+ architectures with promising performance improvement, given the limited dataset available for the experiment.

In the future, the applicability of this method can be spread out to diverse research in overlapping cell segmentation and quantification. Further, the application of

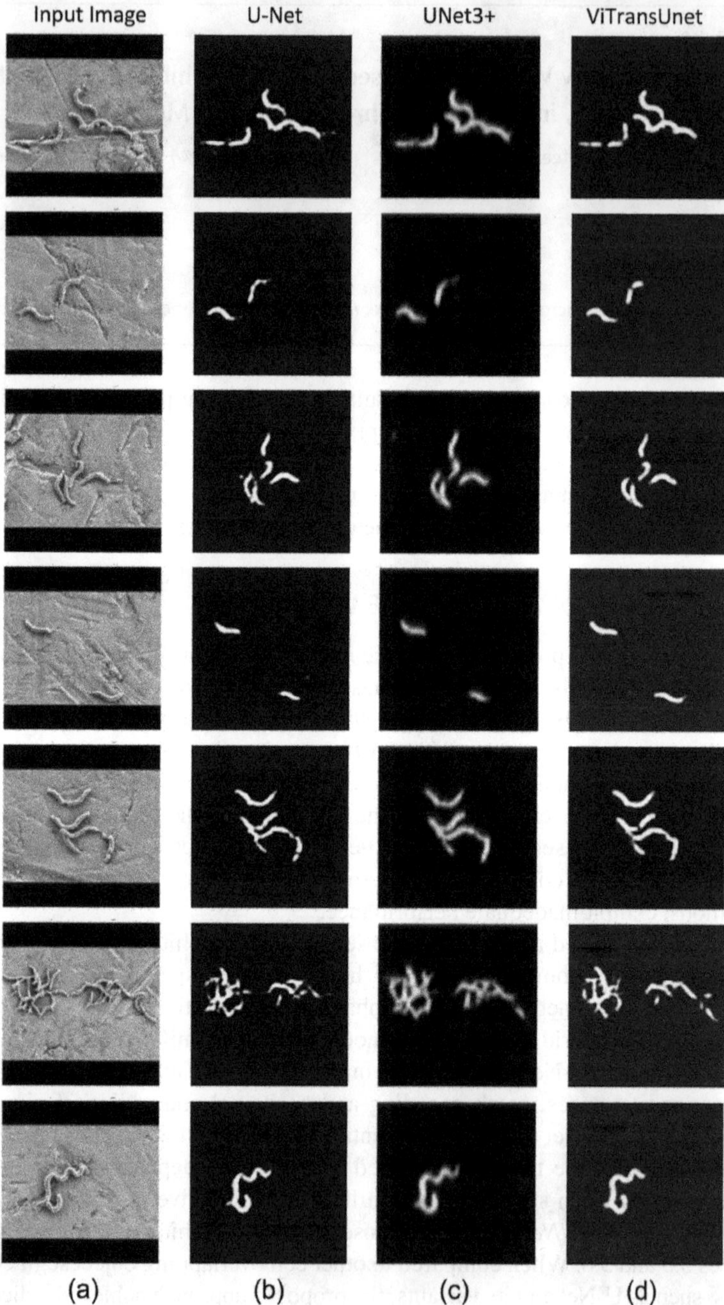

FIGURE 5.4 Qualitative comparisons for U-Net, UNet3+, and *ViTransUNet* predictions on test data: a) The grayscale image used as input for image segmentation. (b) Image segmentation masks obtained from U-Net model prediction. (c) Image segmentation masks obtained from UNet3+ model predictions. (d) Image segmentation masks obtained from trained *ViTransUNet* model predictions.

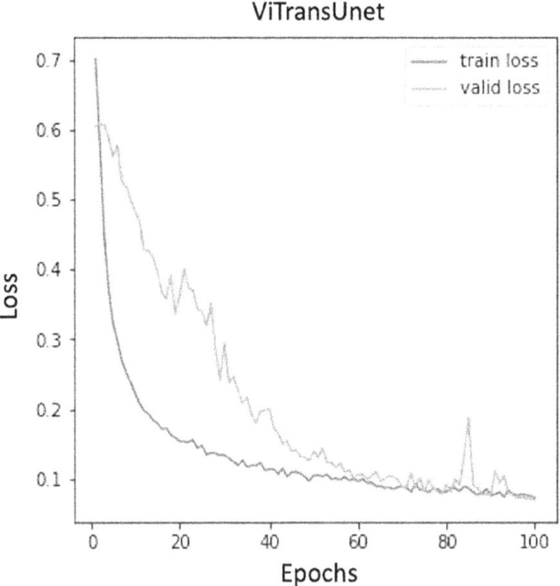

FIGURE 5.5 Training performance. How loss values decrease over epochs is shown for each experiment of the *ViTransUNet* model on our training dataset; calculating loss over 100 epochs.

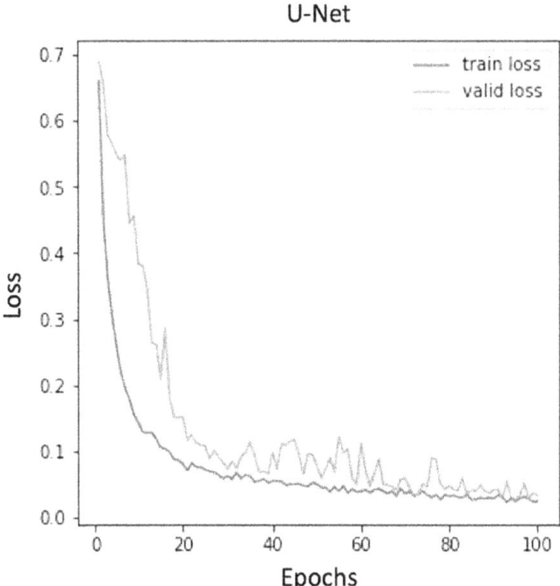

FIGURE 5.6 Training performance. How loss values decrease over epochs is shown for each experiment of U-Net model on our training dataset; calculating loss over 100 epochs.

object segmentation in various fields, such as medicine, engineering, and biology, is expected to yield significant benefits for downstream tasks. Furthermore, we are contemplating the expansion of this approach to meet the increasing demand for three-dimensional cell segmentation tasks in various applications in the fields of medicine and engineering. Accurate segmentation and tracking of cells can enhance our understanding of cell viability, cell signaling, adhesion, and other related factors [79]. The approach of generating patches can provide benefits in identifying distinct types of isolated cell clusters, thereby enhancing the efficiency, accuracy, and overall performance of the proposed method.

REFERENCES

1. M. Helmstaedter, K. L. Briggman, S. C. Turaga, V. Jain, H. S. Seung, and W. Denk, "Connectomic reconstruction of the inner plexiform layer in the mouse retina," *Nature*, vol. 500, no. 7461, pp. 168–174, 2013.
2. M. H. Rahman, M. A. Azam, M. A. Hossen, S. Ragi, and V. Gadhamshetty, "BiofilmScanner: A computational intelligence approach to obtain bacterial cell morphological attributes from biofilm image," *arXiv preprint arXiv:2302.09629*, 2023.
3. S. Ragi, M. H. Rahman, J. Duckworth, J. Kalimuthu, P. Chundi, and V. Gadhamshetty (2021). "Artificial intelligence-driven image analysis of bacterial cells and biofilms," In *IEEE/ACM Transactions on Computational Biology and Bioinformatics*, arXiv preprint arXiv: 2112.01577, pp. 174–184, 2021.
4. J. M. Keegstra, K. Kamino, F. Anquez, M. D. Lazova, T. Emonet, and T. S. Shimizu, "Phenotypic diversity and temporal variability in a bacterial signaling network revealed by single-cell fret," *Elife*, vol. 6, p. e27455, 2017.
5. N. M. V. Sampaio and M. J. Dunlop, "Functional roles of microbial cell-to-cell heterogeneity and emerging technologies for analysis and control," *Current Opinion in Microbiology*, vol. 57, pp. 87–94, 2020.
6. S. Westfall, N. Lomis, I. Kahouli, S. Y. Dia, S. P. Singh, and S. Prakash, "Microbiome, probiotics and neurodegenerative diseases: deciphering the gut brain axis," *Cellular and Molecular Life Sciences*, vol. 74, no. 20, pp. 3769–3787, 2017.
7. D. Serbanescu, N. Ojkic, and S. Banerjee, "Cellular resource allocation strategies for cell size and shape control in bacteria," *The FEBS Journal*, vol. 289, no. 24, pp. 7891–7906, 2022.
8. K. D. Young, "The selective value of bacterial shape," *Microbiology and Molecular Biology Reviews*, vol. 70, no. 3, pp. 660–703, 2006.
9. C. G. Golding, L. L. Lamboo, D. R. Beniac, and T. F. Booth, "The scanning electron microscope in microbiology and diagnosis of infectious disease," *Scientific Reports*, vol. 6, no. 1, pp. 1–8, 2016.
10. A. Paintdakhi, B. Parry, M. Campos, I. Irnov, J. Elf, I. Surovtsev, and C. Jacobs-Wagner, "Oufti: an integrated software package for high-accuracy, highthroughput quantitative microscopy analysis," *Molecular Microbiology*, vol. 99, no. 4, pp. 767–777, 2016.
11. L. K. Harris and J. A. Theriot, "Relative rates of surface and volume synthesis set bacterial cell size," *Cell*, vol. 165, no. 6, pp. 1479–1492, 2016.
12. D. Brahim Belhaouari, A. Fontanini, J.-P. Baudoin, G. Haddad, M. Le Bideau, J. Y. Bou Khalil, D. Raoult, and B. La Scola, "The strengths of scanning electron microscopy in deciphering sars-cov-2 infectious cycle," *Frontiers in Microbiology*, vol. 11, p. 2014, 2020.
13. C. Carrascosa, D. Raheem, F. Ramos, A. Saraiva, and A. Raposo, "Microbial biofilms in the food industry: a comprehensive review," *International Journal of Environmental Research and Public Health*, vol. 18, no. 4, p. 2014, 2021.

14. Z. Li, X. Wang, J. Wang, X. Yuan, X. Jiang, Y. Wang, C. Zhong, D. Xu, T. Gu, and F. Wang, "Bacterial biofilms as platforms engineered for diverse applications," *Biotechnology Advances*, p. 107932, 2022.
15. G. Chilkoor, N. Shrestha, A. Kutana, M. Tripathi, F.C. Robles Hernández, B.I. Yakobson, M. Meyyappan, A.B. Dalton, P.M. Ajayan, M.M. Rahman, and V. Gadhamshetty, "Atomic layers of graphene for microbial corrosion prevention," *ACS Nano*, vol. 15, no. 1, pp. 447–454, 2020.
16. N. Shrestha, A. K. Tripathi, T. Govil, R. K. Sani, M. Urgun-Demirtas, V. Kasthuri, and V. Gadhamshetty, "Electricity from lignocellulosic substrates by thermophilic Geobacillus species," *Scientific Reports*, vol. 10, no. 1, pp. 1–9, 2020.
17. V.K. Upadhyayula and V. Gadhamshetty. "Appreciating the role of carbon nanotube composites in preventing biofouling and promoting biofilms on material surfaces in environmental engineering: a review,". *Biotechnology Advances*, vol. 28, no. 6, pp. 802–816, 2010.
18. H.-D. Cheng, X. H. Jiang, Y. Sun, and J. Wang, "Color image segmentation: advances and prospects," *Pattern Recognition*, vol. 34, no. 12, pp. 2259–2281, 2001.
19. E. Colleoni, P. Edwards, and D. Stoyanov, "Synthetic and real inputs for tool segmentation in robotic surgery," In *Medical Image Computing and Computer Assisted Intervention - MICCAI 2020*, A. L. Martel, P. Abolmaesumi, D. Stoyanov, D. Mateus, M. A. Zuluaga, S. K. Zhou, D. Racoceanu, and L. Joskowicz, Eds. Cham: Springer International Publishing, 2020, pp. 700–710.
20. D. Pakhomov, V. Premachandran, M. Allan, M. Azizian, and N. Navab, "Deep residual learning for instrument segmentation in robotic surgery," In *International Workshop on Machine Learning in Medical Imaging*. Cham: Springer, 2019, pp. 566–573.
21. D. Abeyrathna, S. Rauniyar, R. K. Sani, and P.-C. Huang, "A morphological post-processing approach for overlapped segmentation of bacterial cell images," *Machine Learning and Knowledge Extraction*, vol. 4, no. 4, pp. 1024–1041, 2022.
22. P.-C. Huang, E. Shakya, M. Song, and M. Subramaniam, "Biomdse: A multimodal deep learning-based search engine framework for biofilm documents classifications," In *2022 IEEE International Conference on Bioinformatics and Biomedicine (BIBM)*, Las Vegas, NV, 2022, pp. 3608–3612.
23. V. Lempitsky and A. Zisserman, "Learning to count objects in images," In *Advances in Neural Information Processing Systems*, vol. 23, 2010. Red Hook, NY: Curran Associates Inc., 2010, pp. 1324–1332.
24. Y. Nakano, T. Takeshita, N. Kamio, S. Shiota, Y. Shibata, N. Suzuki, M. Yoneda, Hirofuji, and Y. Yamashita, "Supervised machine learning-based classificationof oral malodor based on the microbiota in saliva samples," *Artificial Intelligence in Medicine*, vol. 60, no. 2, pp. 97–101, 2014.
25. J. Deng, W. Dong, R. Socher, L.-J. Li, K. Li, and L. Fei-Fei, "Imagenet: A largescale hierarchical image database," In *2009 IEEE Conference on Computer Vision and Pattern Recognition*. Miami, FL: IEEE, 2009, pp. 248–255.
26. P. Vincent, H. Larochelle, Y. Bengio, and P.-A. Manzagol, "Extracting and composing robust features with denoising autoencoders," In *Proceedings of the 25th International Conference on Machine Learning (ICML)*. New York, NY: Association for Computing Machinery, 2008, pp. 1096–1103.
27. A. Vaswani, N. Shazeer, N. Parmar, J. Uszkoreit, L. Jones, A. N. Gomez, L. Kaiser, and I. Polosukhin, "Attention is all you need," 2017. https://arxiv.org/abs/1706.03762
28. B. D. S. Gurung, R. Devadig, T. Do, V. Gadhamshetty, and E. Z. Gnimpieba, "U-net based image segmentation techniques for development of non-biocidal fouling-resistant ultra-thin two-dimensional (2D) coatings. In *2022 IEEE International Conference on Bioinformatics and Biomedicine (BIBM)*. Las Vegas, NV: IEEE, 2022, pp. 3602–3604.

29. G. Lin, U. Adiga, K. Olson, J. F. Guzowski, C. A. Barnes, and B. Roysam, "A hybrid 3d watershed algorithm incorporating gradient cues and object models for automatic segmentation of nuclei in confocal image stacks," *Cytometry Part A: the Journal of the International Society for Analytical Cytology*, vol. 56, no. 1, pp. 23–36, 2003.
30. J. Cheng, J. C. Rajapakse *et al.*, "Segmentation of clustered nuclei with shape markers and marking function," *IEEE Transactions on Biomedical Engineering*, vol. 56, no. 3, pp. 741–748, 2008.
31. Z. Al Aghbari and R. Al-Haj, "Hill-manipulation: an effective algorithm for color image segmentation," *Image and Vision Computing*, vol. 24, no. 8, pp. 894–903, 2006.
32. J.-C. Yen, F.-J. Chang, and S. Chang, "A new criterion for automatic multilevel thresholding," *IEEE Transactions on Image Processing*, vol. 4, no. 3, pp. 370–378, 1995.
33. L. Vincent, "Morphological grayscale reconstruction in image analysis: applications and efficient algorithms," *IEEE Transactions on Image Processing*, vol. 2, no. 2, pp. 176–201, 1993.
34. L. A. Cooper, J. Kong, D. A. Gutman, F. Wang, J. Gao, C. Appin, S. Cholleti, Pan, A. Sharma, L. Scarpace *et al.*, "Integrated morphologic analysis for the identification and characterization of disease subtypes," *Journal of the American Medical Informatics Association*, vol. 19, no. 2, pp. 317–323, 2012.
35. L. Vincent and P. Soille, "Watersheds in digital spaces: an efficient algorithm based on immersion simulations," *IEEE Computer Architecture Letters*, vol. 13, no. 06, pp. 583–598, 1991.
36. S. K. Nath, K. Palaniappan, and F. Bunyak, "Cell segmentation using coupled level sets and graph-vertex coloring," In *Medical Image Computing and Computer-Assisted Intervention - MICCAI 2006*, R. Larsen, M. Nielsen, and J. Sporring, Eds. Berlin, Heidelberg: Springer, 2006, pp. 101–108.
37. O. Dzyubachyk, W. Niessen, and E. Meijering, "Advanced level-set based multiple-cell segmentation and tracking in time-lapse fluorescence microscopy images," In *2008 5th IEEE International Symposium on Biomedical Imaging: From Nano to Macro.* Paris: IEEE, 2008, pp. 185–188.
38. H. Chang, J. Han, A. Borowsky, L. Loss, J. W. Gray, P. T. Spellman, and B. Parvin, "Invariant delineation of nuclear architecture in glioblastoma multiforme for clinical and molecular association," *IEEE Transactions on Medical Imaging*, vol. 32, no. 4, pp. 670–682, 2012.
39. Y. Al-Kofahi, W. Lassoued, W. Lee, and B. Roysam, "Improved automatic detection and segmentation of cell nuclei in histopathology images," *IEEE Transactions on Biomedical Engineering*, vol. 57, no. 4, pp. 841–852, 2009.
40. N. Kumar, R. Verma, S. Sharma, S. Bhargava, A. Vahadane, and A. Sethi, "A dataset and a technique for generalized nuclear segmentation for computational pathology," *IEEE Transactions on Medical Imaging*, vol. 36, no. 7, pp. 1550–1560, 2017.
41. D. Jeulin, *Morphological Models of Random Structures.* Cham: Springer, 2021.
42. C. Li, C. Xu, C. Gui, and M. D. Fox, "Distance regularized level set evolution and its application to image segmentation," *IEEE Transactions on Image Processing*, vol. 19, no. 12, pp. 3243–3254, 2010.
43. K. Zhang, H. Song, and L. Zhang, "Active contours driven by local image fitting energy," *Pattern Recognition*, vol. 43, no. 4, pp. 1199–1206, 2010.
44. S. Niu, Q. Chen, L. De Sisternes, Z. Ji, Z. Zhou, and D. L. Rubin, "Robust noise region-based active contour model via local similarity factor for image segmentation," *Pattern Recognition*, vol. 61, pp. 104–119, 2017.
45. J. Zhang, Z. Lu, and M. Li, "Active contour-based method for finger-vein image segmentation," *IEEE Transactions on Instrumentation and Measurement*, vol. 69, no. 11, pp. 8656–8665, 2020.

46. G. Fernandez, M. Kunt, and J.-P. Zryd, "A new plant cell image segmentation algorithm," In *International Conference on Image Analysis and Processing*. Berlin, Heidelberg: Springer, 1995, pp. 229–234.
47. Y. He, Y. Meng, H. Gong, S. Chen, B. Zhang, W. Ding, Q. Luo, and A. Li, "An automated three-dimensional detection and segmentation method for touching cells by integrating concave points clustering and random walker algorithm," *PLoS One*, vol. 9, no. 8, p. e104437, 2014.
48. H. Wang, H. Zhang, and N. Ray, "Clump splitting via bottleneck detection and shape classification," *Pattern Recognition*, vol. 45, no. 7, pp. 2780–2787, 2012.
49. F. Xing and L. Yang, "Chapter 4: machine learning and its application in microscopic image analysis," In *Machine Learning and Medical Imaging*, series. The Elsevier and MICCAI Society Book Series, G. Wu, D. Shen, and M. R. Sabuncu, Eds. Gainesville, FL: Academic Press, University of Florida, 2016, pp. 97–127.
50. J. K. Lee, B. A. Wood, and T. S. Newman, "Very fast ellipse detection using gpu-based rht," In *2008 19th International Conference on Pattern Recognition*. Tampa, FL: IEEE, 2008, pp. 1–4.
51. C. A. Basca, M. Talos, and R. Brad, "Randomized hough transform for ellipse detection with result clustering," In *EUROCON 2005-The International Conference on Computer as a Tool*, vol. 2. Belgrade: IEEE, 2005, pp. 1397–1400.
52. P. Nair and A. Saunders Jr., "Hough transform based ellipse detection algorithm," *Pattern Recognition Letters*, vol. 17, no. 7, pp. 777–784, 1996.
53. I. Abu-Qasmieh, "Novel and efficient approach for automated separation, segmentation, and detection of overlapped elliptical red blood cells," *Pattern Recognition and Image Analysis*, vol. 28, pp. 792–804, 2018.
54. S. Zafari, T. Eerola, J. Sampo, H. Kälviäinen, and H. Haario, "Segmentation of overlapping elliptical objects in silhouette images," *IEEE Transactions on Image Processing*, vol. 24, no. 12, pp. 5942–5952, 2015.
55. T. Zou, T. Pan, M. Taylor, and H. Stern, "Recognition of overlapping elliptical objects in a binary image," *Pattern Analysis and Applications*, vol. 24, no. 3, pp. 1193–1206, 2021.
56. G. Zhang, D. S. Jayas, and N. D. White, "Separation of touching grain kernels in an image by ellipse fitting algorithm," *Biosystems Engineering*, vol. 92, no. 2, pp. 135–142, 2005. https://www.sciencedirect.com/science/article/pii/S1537511005001285
57. C. Panagiotakis and A. A. Argyros, "Cell segmentation via region-based ellipse fitting," In *2018 25th IEEE International Conference on Image Processing (ICIP)*. Athens: IEEE, 2018, pp. 2426–2430.
58. C. Panagiotakis and A. Argyros, "Parameter-free modelling of 2d shapes with ellipses," *Pattern Recognition*, vol. 53, pp. 259–275, 2016.
59. C. Panagiotakis and A. Argyros, "Region-based fitting of overlapping ellipses and its application to cells segmentation," *Image and Vision Computing*, vol. 93, p. 103810, 2020.
60. D. Abeyrathna, T. Life, S. Rauniyar, S. Ragi, R. Sani, and P. Chundi, "Segmentation of bacterial cells in biofilms using an overlapped ellipse fitting technique," In *2021 IEEE International Conference on Bioinformatics and Biomedicine (BIBM)*. Las Vegas, NV: IEEE, 2021, pp. 3548–3554.
61. G. Litjens, T. Kooi, B. E. Bejnordi, A. A. A. Setio, F. Ciompi, M. Ghafoorian, J. A. van der Laak, B. van Ginneken, and C. I. Sánchez, "A survey on deep learning in medical image analysis," *Medical Image Analysis*, vol. 42, pp. 60–88, 2017. https://www.sciencedirect.com/science/article/pii/S1361841517301135
62. L. Yang, Y. Zhang, I. H. Guldner, S. Zhang, and D. Z. Chen, "3d segmentation of glial cells using fully convolutional networks and k-terminal cut," In *International Conference on Medical Image Computing and Computer-Assisted Intervention*. Cham: Springer, 2016, pp. 658–666.

63. H. M. Saleh, N. H. Saad, and N. A. M. Isa, "Overlapping chromosome segmentation using u-net: convolutional networks with test time augmentation," *Procedia Computer Science*, vol. 159, pp. 524–533, 2019.
64. K. He, G. Gkioxari, P. Doll´ar, and R. B. Girshick, "Mask r-cnn," *2017 IEEE International Conference on Computer Vision (ICCV)*, IEEE, Venice, pp. 2980–2988, 2017.
65. S. Ren, K. He, R. Girshick, and J. Sun, "Faster r-cnn: Towards real-time object detection with region proposal networks." In *Advances in Neural Information Processing Systems 28 (NIPS 2015)*. Cambridge, MA: The MIT Press, 2015.
66. O. Ronneberger, P. Fischer, and T. Brox, "U-net: Convolutional networks for biomedical image segmentation," In *Medical Image Computing and ComputerAssisted Intervention - MICCAI 2015*, 2015, p. 234–241.
67. R. L. Hu, J. Karnowski, R. Fadely, and J.-P. Pommier, "Image segmentation to distinguish between overlapping human chromosomes," *arXiv preprint arXiv:1712.07639*, 2017.
68. Kurnianingsih, K. H. S. Allehaibi, L. E. Nugroho, Widyawan, L. Lazuardi, A. S. Prabuwono, and T. Mantoro, "Segmentation and classification of cervical cells using deep learning," *IEEE Access*, vol. 7, pp. 116925–116941, 2019.
69. Z. Lu, G. Carneiro, and A. P. Bradley, "An improved joint optimization of multiple level set functions for the segmentation of overlapping cervical cells," *IEEE Transactions on Image Processing*, vol. 24, no. 4, pp. 1261–1272, 2015.
70. H. Huang, L. Lin, R. Tong, H. Hu, Q. Zhang, Y. Iwamoto, X. Han, Y.-W. Chen, and J. Wu, "Unet 3+: a full-scale connected unet for medical image segmentation," 2020. https://arxiv.org/abs/2004.08790
71. A. Dosovitskiy, L. Beyer, A. Kolesnikov, D. Weissenborn, X. Zhai, T. Unterthiner, M. Dehghani, M. Minderer, G. Heigold, S. Gelly, J. Uszkoreit, and N. Houlsby, "An image is worth 16x16 words: transformers for image recognition at scale," 2020. https://arxiv.org/abs/2010.11929
72. J. Deng, W. Dong, R. Socher, L.-J. Li, K. Li, and L. Fei-Fei, "Imagenet: A largescale hierarchical image database," In *2009 IEEE Conference on Computer Vision and Pattern Recognition*. Miami, FL: IEEE, 2009, pp. 248–255.
73. A. Krizhevsky, G. Hinton *et al.*, "Learning multiple layers of features from tiny images," Technical Report, University of Toronto, Toronto, ON, 2009.
74. J. Chen, Y. Lu, Q. Yu, X. Luo, E. Adeli, Y. Wang, L. Lu, A. L. Yuille, and Y. Zhou, "Transunet: Transformers make strong encoders for medical image segmentation," *arXiv preprint arXiv:2102.04306*, 2021.
75. C. Carlson, N. K. Singh, M. Bibra, R. K. Sani, and K. Venkateswaran, "Pervasiveness of UVC_{254}-resistant Geobacillus strains in extreme environments,". *Applied Microbiology and Biotechnology*, vol. 102, no. 4, 1869–1887. 2018. https://doi.org/10.1007/s00253-017-8712-8
76. C. Lam, D. Yi, M. Guo, and T. Lindsey, "Automated detection of diabetic retinopathy using deep learning," *AMIA Summits on Translational Science Proceedings*, vol. 2018, p. 147, 2018.
77. J. Schlemper, O. Oktay, M. Schaap, M. Heinrich, B. Kainz, B. Glocker, and D. Rueckert, "Attention gated networks: learning to leverage salient regions in medical images," *Medical Image Analysis*, vol. 53, pp. 197–207, 2019.
78. L. R. Dice, "Measures of the amount of ecologic association between species," *Ecology*, vol. 26, no. 3, pp. 297–302, 1945.
79. R. Foresti, S. Rossi, S. Pinelli, R. Alinovi, M. Barozzi, C. Sciancalepore, M. Galetti, C. Caffarra, P. Lagonegro, G. Scavia *et al.*, "Highly-defined bioprinting of long-term vascularized scaffolds with bio-trap: complex geometry functionalization and process parameters with computer aided tissue engineering," *Materialia*, vol. 9, p. 100560, 2020.

6 Self-Supervised Learning-Based Classification of Scanning Electron Microscope Images of Biofilms

Md Ashaduzzaman and Mahadevan Subramaniam

6.1 INTRODUCTION

Machine learning (ML) approaches have made impressive strides in automated analyses of various modalities of information including numeric data, images, text, and audio. More recently, new multi-modal models that can analyze the combinations of these various modalities have begun to emerge and are beginning to exhibit performances comparable to multi-modal information processing by humans. ML models have been developed using a wide variety of approaches including unsupervised, semi-supervised, and supervised learning methods and their variations (see Chapter 1 for an overview). Among these supervised ML approaches, variants have been highly effective in performing image analysis tasks. The success of these approaches typically relies on building ML models by training them on large volumes of labelled data. These models are then used to analyze new test images. However, manually annotating microscopy images is usually very time-consuming and severely limits the amount of available labelled image data. It is a challenging task to develop machine learning models, particularly deep convolutional neural networks (DCNNs), with a limited amount of labelled image data. This can lead to sub-optimal performance of the model on new, unseen images. In this chapter, we describe a case study based on both contrastive and non-contrastive (more details in Section 6.2.1.1) self-supervised learning paradigms for classifying SEM images of biofilms using low volumes of labelled data. A key component of our classification pipeline involves the use of DCNNs to address the heterogeneity and quality of SEM images by using super-resolution. We describe different models that can be used for super-resolution tasks. The best-performing super-resolution model is used to build self-supervised models that can identify cells/cell clusters, byproducts (potentially involving corrosion), and exposed surfaces in small SEM images of biofilms.

DOI: 10.1201/9781003132981-6

6.2 SELF-SUPERVISED LEARNING FOR IMAGE ANALYSES

Self-supervised learning (SSL) involves using unlabelled medical data to pre-train a model and then fine-tuning the pre-trained model for a specific image analysis task using a limited amount of labelled data. This can be an alternative to transfer learning from natural images, as the knowledge learned from the unlabelled medical data is more relevant to the target task. During self-supervised pre-training, surrogate labels are assigned to the unlabelled data and used to train a randomly initialized network.

The main steps in the SSL are as follows.

1. **Prepare Training Set:** Prepare a training set from the unlabelled dataset.
2. **Pretext Task:** Pretext tasks are unsupervised learning tasks that are used to learn the representations. We need to formulate a problem from the training data.
3. **Learned Representation:** Learn the representation of the domain by solving the pretext task. The representations extract the patterns and features from the unlabelled images.
4. **Fine-Tuning:** Using the learned representation, we can fine-tune the model with a small number of labelled images.
5. **Downstream Tasks:** Perform the desired image analysis tasks (classification, detection, or segmentation) using the fine-tuned model.

Preparing the training set usually involves pre-processing the images and is highly dependent on the application. Similarly, fine-tuning is also application specific. We discuss these two steps along with the case study in the next section. Commonly employed pretext tasks are described next.

6.2.1 PRETEXT TASKS

In this section, we will discuss the several types of pretext tasks that are commonly used in building self-supervised models for image analyses.

6.2.1.1 Contrastive and Non-Contrastive Learning

The idea of contrastive learning is to learn the similarities from the positive pair of images and dissimilarities from the negative pair of images. Each image along with additional images generated from that image using data augmentation operations is applied to each image to create positive pair of images. On the other hand, other images and their augmented versions are the negative pair of images. The contrastive predictive coding (CPC) proposed the contrastive loss function *infoNCE* to learn from negative and positive patches (Oord, Li, and Vinyals 2018). SimCLR (Simple Contrastive Learning of Representations) requires a large number of negative samples in a batch to perform well (T. Chen et al. 2020). MoCo (Momentum Contrast) keeps a queue of negative samples and utilizes momentum encoder methodology to learn from the small batch size of negative samples (He et al. 2020). Moreover, non-contrastive methods only learn from positive samples. BYOL (Bootstrap Your Own

Latent) shows that better representation can be learned from only positive samples (Grill et al. 2020). Finally, the SimSiam and Barlow Twins architectures achieve comparative results without using negative samples, large batches, and a momentum encoder (Chen and He 2021).

6.2.1.2 Generative Modelling

SSL can learn useful inherent representations by a powerful pretext task: generative modelling, which can generate plausible samples from a given distribution and reconstruct the original input. One common approach is to use autoencoders or variational autoencoders (VAEs) (Kingma and Welling 2019), which acquire a condensed representation of input data by encoding it into a latent space with fewer dimensions, and subsequently decoding it to recreate the original data. The denoising autoencoder (Vincent et al. 2008) can remove random noise in an image and reconstruct the original image. Generative Adversarial Networks (GANs) (Goodfellow et al. 2020) can be used as a pretext task in SSL by training the discriminator of a GAN to distinguish between real and generated images, while simultaneously training the generator to create images that can fool the discriminator into thinking they are real. The generator in this case can be seen as a self-supervised learner, as it is learning to generate images that match the statistical properties of the real images.

6.2.1.3 Colorization

In this pretext task, we can formulate the problem in such a way that grayscale versions of images are given as inputs, and the model tries to predict the colour of the images (R. Zhang, Isola, and Efros 2016). The loss function tries to minimize the difference between the original colour and the predicted colour. More weight is given to the rare colour bucket in the loss function to prioritize the infrequent colour (object colour) from the frequent colour (background colour). The model learns how to differentiate various objects in an image and predicts the various colours for different parts by solving the pretext task.

6.2.1.4 Jigsaw

Another pretext task can be formulated by creating a puzzle game from an image and learning the representation by solving the puzzle. The jigsaw puzzle paper (Noroozi and Favaro 2016) proposed an SSL technique by creating a puzzle game from the patches of an image. If we take 9 patches from an image, there are 9! possible shuffles available from the patches. To reduce the complexity, the authors choose only 64 possible shuffles with the highest hamming distance for the training dataset. Then they pass each patch to different siamese convolutional layers having shared weights with each other. Finally, all the results from each layer are combined to solve the puzzle. By solving the jigsaw puzzle, the model learns the relative positions of the objects in an image and the contextual information of the objects. We can fine-tune the model to do different downstream tasks.

6.2.1.5 Relative Patch Location

This pretext task is similar to the previous jigsaw puzzle problem. The relative path location paper (Doersch, Gupta, and Efros 2015) formulated the training pairs by randomly taking two neighbouring image patches and predicting relative positions between them. The second patch in a 3×3 image grid can be obtained from 8 neighbouring locations if we start with the first patch from the central location. The authors proposed two siamese ConvNets models for feature extraction from each patch and combine the results to solve 8 classes of classification problems.

6.2.1.6 Inpainting

Image inpainting pretext task is formulated by predicting random missing areas based on the rest of the image (Pathak et al. 2016). The GAN-based architecture can be utilized to reconstruct the missing part from an image. The generator model generates plausible examples for the reconstruction of the missing part, and the discriminator model distinguishes the generated and real images. The networks learn the colour and the structural information of the domain by doing this pretext task.

6.2.1.7 Super-Resolution

SRGAN proposes a pretext task of enhancing the resolution of a low-resolution image (Ledig et al. 2017). The generative network of an SRGAN predicts the high-resolution version of a given down-sample image. The loss function of the generator tries to increase the similarity of the predicted high-resolution image and the original high-resolution image. The discriminator tries to distinguish between the original and fake images. The pretext task learns the semantic features of the images by doing the image super-resolution.

6.2.1.8 Rotation

Rotation (Gidaris, Singh, and Komodakis 2018) is another simple but effective pretext task for self-supervised learning. Training images were created by rotating the images from any dataset to four user-defined degrees (0, 90, 270, and 360). The authors proposed a ConvNet architecture where rotated images are passed to classify the images into 4 classes. The model had to learn semantic information such as the relative positions of the body parts.

6.2.2 Downstream Tasks on Medical Imaging

In this section, we discuss different types of applications done on medical imaging using SSL.

6.2.2.1 Classification

The authors (Jamaludin, Kadir, and Zisserman 2017) proposed a pretext task for the disc generation grading system of four-class from spinal MRI images. They prepared self-supervising training set from the vertebral MRI images of the same patient scanned at different points of time (positive pairs), and images from different patients (negative pairs). A siamese CNN was trained to learn the representation by distinguishing whether the images are from the same patients or different patients. This learning was then transferred to predict a four-class disc generation grading system. The paper (Tajbakhsh et al. 2019) investigated whether the domain-specific pretext task was more effective for weight initialization or weights transferred from unrelated domains. The authors found that pretext tasks like rotation, colorization, and reconstruction were more effective for classification tasks such as nodule detection in chest CT scan images, and diabetic retinopathy classification in fundus images. The paper (Azizi et al. 2021) experimented on dermatology condition classification from digital camera images and multi-label chest X-ray classification using self-supervised learning. They proposed three steps of training to get better classification accuracy: firstly, self-supervised learning on unlabelled ImageNet, then further self-supervised learning domain-specific images, and finally, supervised fine-tuning on labelled medical images. They proposed a novel MICLe method that uses two distinct images directly as positive pair of examples. Self-Path (Koohbanani et al. 2021) proposed a framework for the classification of tissues from pathological images where they utilized a variety of self-supervised pretext tasks. Furthermore, they introduced three novel pathology-specific pretext tasks: magnification prediction, jigmag prediction, and haematoxylin channel prediction.

6.2.2.2 Segmentation

In the paper (Bai et al. 2019), the authors utilized self-supervised learning techniques by defining nine anatomic positions in cardiac images and learning the representation from predicting these anatomical positions. Then transfer learning was used for cardiac MR image segmentation. In the paper (Ouyang et al. 2020), self-supervised learning using the super-pixel method was used for generating pseudo-labels to segment abdominal organs in CT and MRI images. A novel local contrastive learning approach was proposed in the paper (Chaitanya et al. 2020) which is useful for dense predicting tasks such as medical image segmentation.

6.2.2.3 Image Retrieval

In this paper (Gildenblat and Klaiman 2019), the authors proposed a novel self-supervised learning approach by learning the similarity from close patches and dissimilarity from far patches in a whole slide image. This method performed well for retrieval tasks in digital pathology. The authors of the paper (L. Chen et al. 2019) proposed novel context restoration techniques to learn useful semantic features in medical images to improve retrieval. The authors of SMORE (Zhao et al. 2020) proposed a self-supervised, anti-aliasing, and super-resolution technique that doesn't require any external training data. They utilized convolutional neural networks (CNNs) to enhance the resolution and reduce aliasing artefacts in magnetic resonance (MR) images, thereby improving the overall image quality to improve image retrieval.

6.3 USE OF SUPER-RESOLUTION TO ADDRESS THE HETEROGENEITY AND QUALITY OF SEM BIOFILM IMAGES

Image super-resolution (SR) is the process that generates better quality or resolution images from low-resolution images by reconstructing missing details, removing blur and noises, and up-sampling pixels. SR models are trained using a degradation mapping function that down-samples, adds blur and noise, and applies transformations to produce low-resolution input images. The model can estimate the original high-resolution image without prior knowledge of the degradation mapping function by minimizing the dissimilarity between the generated and ground truth high-resolution images using loss functions. The goal is to generate an image that seems more detailed and visually pleasing to the human eye.

In this section, we investigate the use of SR techniques to enhance the performance of semantic segmentation on a biofilm dataset of SEM images. Three distinct generative adversarial learning methods were utilized on the dataset, and their capacity to retain the structural characteristics of the biofilm images was compared. Lastly, the performance of the supervised image segmentation task was evaluated. The results demonstrate that while the degree of preservation of structural features differs across the various SR techniques, their combination in a deep learning pipeline leads to an overall improvement in image segmentation performance. This improvement enables a more precise and quantifiable analysis of SEM images of biofilms.

ESRGAN (Wang et al. 2018) stands for Enhanced Super-Resolution Generative Adversarial Networks which is an improvement over the traditional super-resolution technique SRGAN (Ledig et al. 2017). ESRGAN has shown impressive results in producing more realistic and natural textured super-resolution images by introducing a relativistic discriminator. Modifications were made to the architecture of ESRGAN, including the removal of the batch normalization layer, utilization of a deeper model, and enhancement of the perceptual loss function. See figure 6.1 for the architecture of ESRGAN.

The BSRGAN (Zhang et al. 2021) paper formulated a realistic degradation model for synthesized training data, which is essential for the SR model to perform well in real-world scenarios. The degradation model includes blur, down-sampling, and noise, and to incorporate a broader range of real-world scenarios, random mixture schemes are employed among these factors. This expands the scope of the approach. The paper described how they performed various degradation operations, including Gaussian blur, isotropic and anisotropic blur, down-sampling using one of four interpolation methods, and introducing JPEG compression and camera sensor noise. They emphasized the importance of accurately modelling these degradation operations in order to improve the real-world applications of SR methods. BSRGAN was

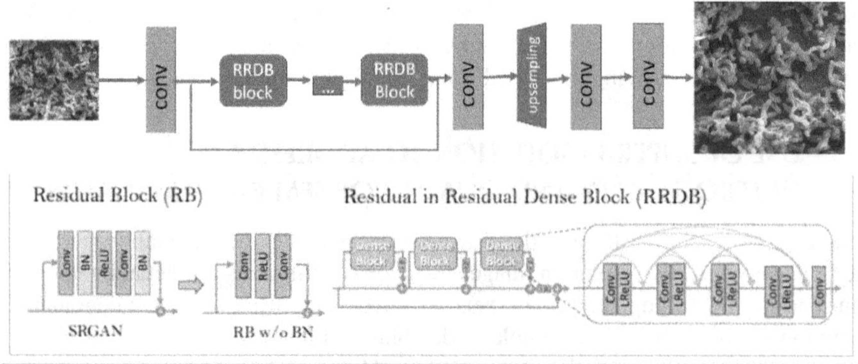

FIGURE 6.1 ESRGAN architecture. (Used with permission from the paper Ashaduzzman, Md et al., 2022 IEEE International Conference on Bioinformatics and Biomedicine (BIBM), 3587–93, 2022.)

trained on synthetic and real-world datasets to improve its general-purpose blind image SR capabilities.

SwinIR (Liang et al. 2021), Shifted Window-based Transformers for Image Restoration, is a deep learning model designed for single image super-resolution, which utilized the idea of Swin Transformer (Liu et al. 2021). The SwinIR architecture consists of two main components: a feature extraction network and a reconstruction network. The feature extraction network is responsible for extracting features from the input image, while the reconstruction network generates the high-resolution output image. The feature extraction network utilized a hierarchical structure of shifted windows to capture information from different parts of the input image. These windows were shifted across the image to ensure full coverage. The high-resolution output image was generated by the reconstruction network using the features extracted from the feature extraction network. The reconstruction network employed a residual block-based architecture, which preserves crucial features and details from the low-resolution input image.

6.3.1 Methodology

This section outlines the process of generating SR images and achieving cell segmentation in biofilm images.

6.3.1.1 Contrast Enhancement

CLAHE (Reza 2004) contrast enhancement process was applied on biofilm SEM images as a pre-processing step to improve object boundary visibility. CLAHE performs histogram equalization locally, enhancing contrast while limiting noise intensification.

6.3.1.2 Applying SR

After applying CLAHE to biofilm SEM images, SR techniques were applied to reconstruct missing details, remove noise and blur, and increase resolution by a scaling factor of $x=4$. The ESGAN, BSRGAN, and SwinIR SR-trained models were employed in the images, and all generated high-resolution images with less blurriness and noise. The original images were processed using the three SR methods, and the resulting outputs are shown in Figure 6.2. However, SwinIR-generated images were not used for segmentation experiments due to alterations in cell morphology and the appearance of random spikes. ESRGAN and BSRGAN results were compared, and BSRGAN was deemed superior in terms of resolution improvement, noise reduction, and blurriness removal.

6.3.1.3 Patch Generation

This section describes the process of applying SR techniques to biofilm SEM images taken at different magnification levels. The images ranged from 435x to 2300x and have different contextual information depending on the level of magnification. Diverse scaling aspects of SR-based models were used for diverse stages of magnification. 2x SR models were used for images magnified at 400x–600x, while 8x SR models were used for images magnified at 2000x–2200x. There were only 6 biofilm

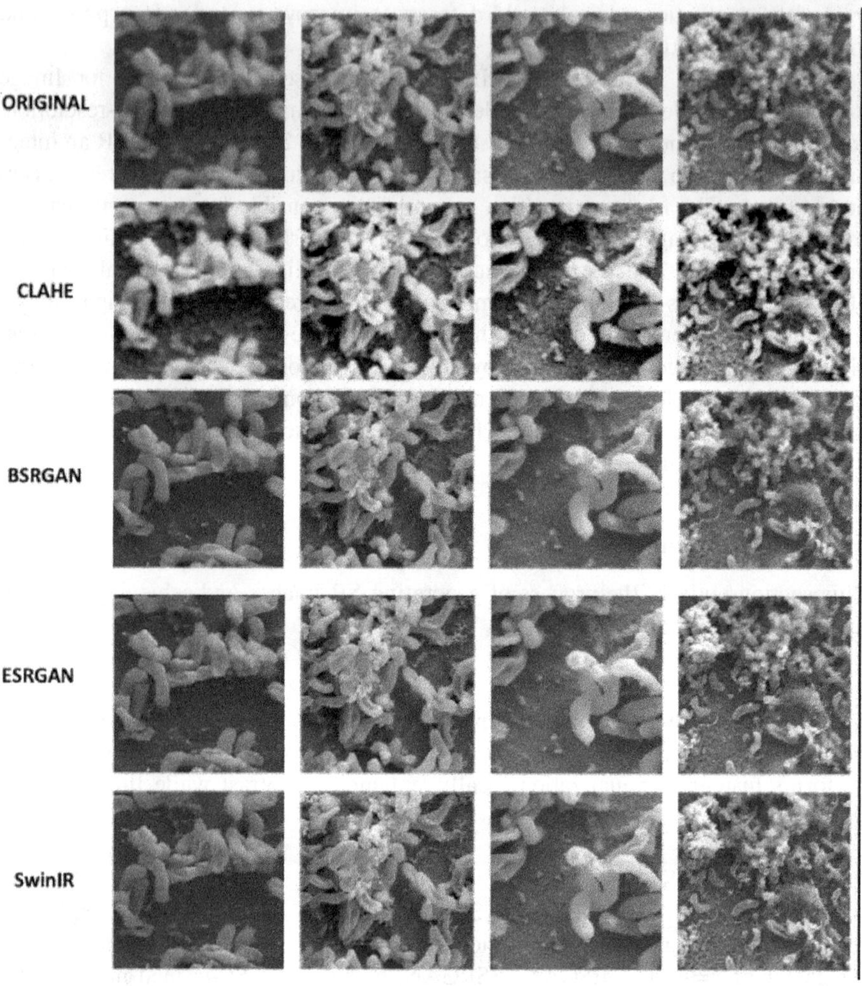

FIGURE 6.2 The illustration depicts the output patches resulting from the utilization of various SR methods on the original images. The *first* row showcases the patches extracted from the original images, while the *second*, *third*, *fourth*, and final rows exhibit the patches generated after the implementation of CLAHE, BSRGAN, ESRGAN, and SwinIR, respectively. (Image used with permission from Ashaduzzman, Md et al., 2022 IEEE International Conference on Bioinformatics and Biomedicine (BIBM), 3587–93, 2022.)

images available in the dataset of size 1024×768, and SR techniques were applied to generate images of 4x resolution, 4096×3072. To address the low volume of input images, patches of 128×128 were created from these images to obtain a higher (4224) number of input images. The chosen patch size was selected as it captures the desired morphological features of the cells, also indicated in Bommanapally et al. (2021).

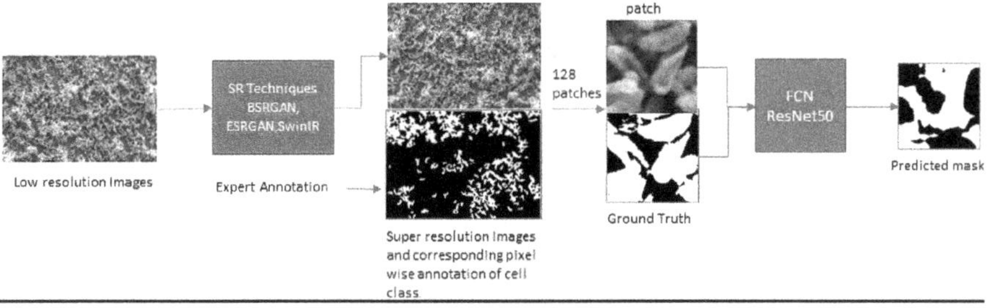

FIGURE 6.3 The procedure used for segmenting cells in biofilm images. (Image used with permission from Ashaduzzman, Md et al., 2022 IEEE International Conference on Bioinformatics and Biomedicine (BIBM), 3587–93, 2022.)

6.3.1.4 Segmentation Downstream Task

The semantic segmentation of cells was done using the popular FCN ResNet50 architecture (Long, Shelhamer, and Darrell 2015), which is pre-trained on ImageNet. During the training stage, the model was trained using the generated patches and their corresponding annotated images, where each pixel in the image was labelled with the corresponding class label. The model learned to recognize patterns and features of the cells in the images and used this information to make predictions about the class of each pixel. The workflow used for segmenting cells in biofilm images is depicted in Figure 6.3.

6.3.2 Experimental Setup and Results

Semantic segmentation results were compared from three types of biofilm input images: (1) original images, (2) SR images by BSRGAN, and (3) SR images by ESRGAN. We discarded the SR images generated by SwinIR as they produced some noises and distortions in the biofilm images. PSNR, commonly used metric to measure the quality of a reconstructed or compressed image or video, was utilized to measure the quality of the generated images. BSRGAN had the highest PSNR value (average of 29), indicating better image quality compared to ESRGAN and SwinIR, which had an average of 25.

We modified the FCN ResNet50 architecture for single-class segmentation. To output a single-channel segmentation mask for the given class, we changed the number of output channels of the last convolutional layer to one, replaced the last pooling layer with a convolutional layer with stride one, removed the fully connected layers at the end of the network, replaced the softmax activation function at the end of the network with a sigmoid activation function, and loaded the pre-trained weights from ImageNet. The backbone pre-trained layers were frozen for fine-tuning purposes. We utilized binary cross-entropy loss as the loss function, stochastic gradient descent for optimization, and a 0.01 learning rate. The model underwent 100 epochs of training. The study utilized cross-validation with 5 folds to improve generalization on the dataset and evaluated the predicted segmentation masks using mIoU scores.

The mIoU score for BSRGAN images had an average value of 0.81 ± 0.01, while ESRGAN and original images had analogous mIoU values of 0.75 ± 0.01 and 0.76 ± 0.02, respectively. Figure 6.4 presents segmentation yields from these experiments.

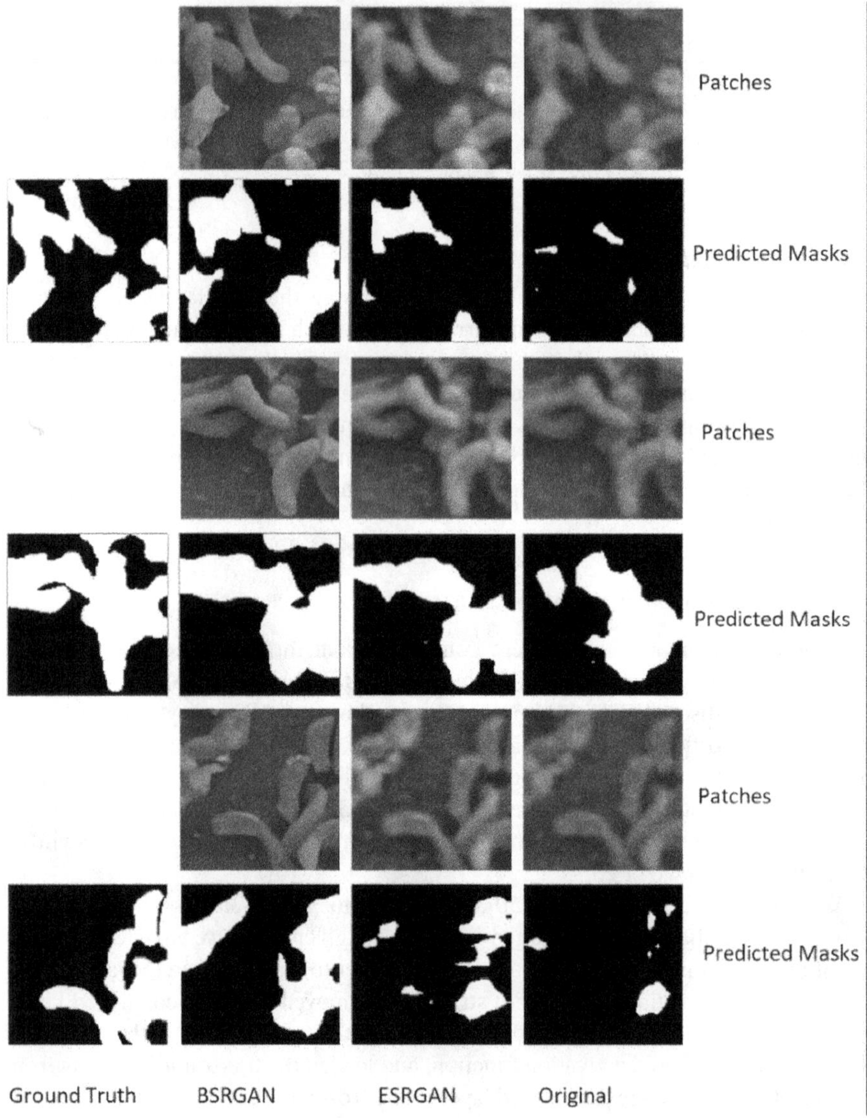

FIGURE 6.4 The output results of the FCN ResNet50 model for different SR techniques were compared to those of the original image. (Image used with permission from Ashaduzzman, Md et al., 2022 IEEE International Conference on Bioinformatics and Biomedicine (BIBM), 3587–93, 2022.)

6.3.3 Summary

The study demonstrates the effectiveness of the SR approaches in enhancing the capabilities of deep-learning models for the quantifiable analysis of SEM biofilm images. Among BSRGAN, ESRGAN, and SwinIR, the BSRGAN approach conserved relevant structures and achieved over 6% improvement in IOU scores compared to those trained on original SEM images. The findings validate that incorporating SR techniques into the workflow of deep learning methods for quantifying SEM biofilm images yields advantageous outcomes.

6.4 CLASSIFICATION OF SEM BIOFILMS USING SSL

Our case study involves the classification of SEM images of biofilms based on the detection of certain objects in these images. This task implements building a self-supervised ML model that solves a multi-label classification problem where the labels—Byproduct, *Cell*, and *Surface* (see Figure 6.8)—are detected in a given SEM image and included in a set of labels assigned to that image. Given the low volume of images, we divide the images into patches and perform the classification on these patches, which can then be combined to assign labels to the original image.

The following AI-based flowchart aims to classify biofilm images using self-supervised approaches. The flowchart comprises gathering SEM biofilm images, pre-processing the images, and manually annotating them by experts. In the image patch generation stage, overlapping miniature image patches are generated. Non-annotated patches are used for representation learning with self-supervised approaches, while annotated patches are used to fine-tune the model for downstream classification. Figure 6.5 illustrates the flowchart with its components.

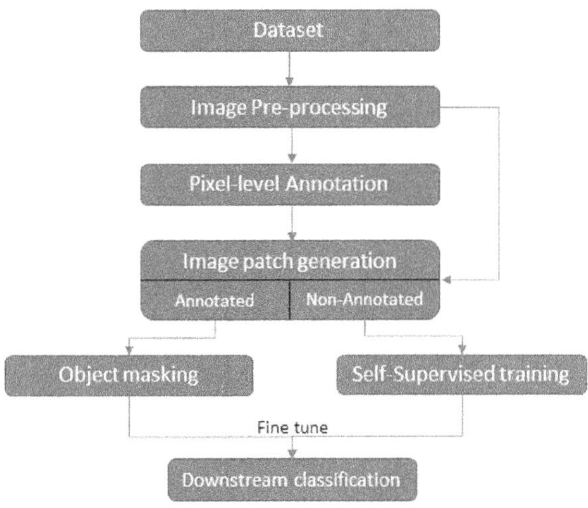

FIGURE 6.5 The flowchart of the approaches to classify biofilm images. (Image used with permission from Abeyrathna, D. et al., Front. Microbiol. 13, 2022.)

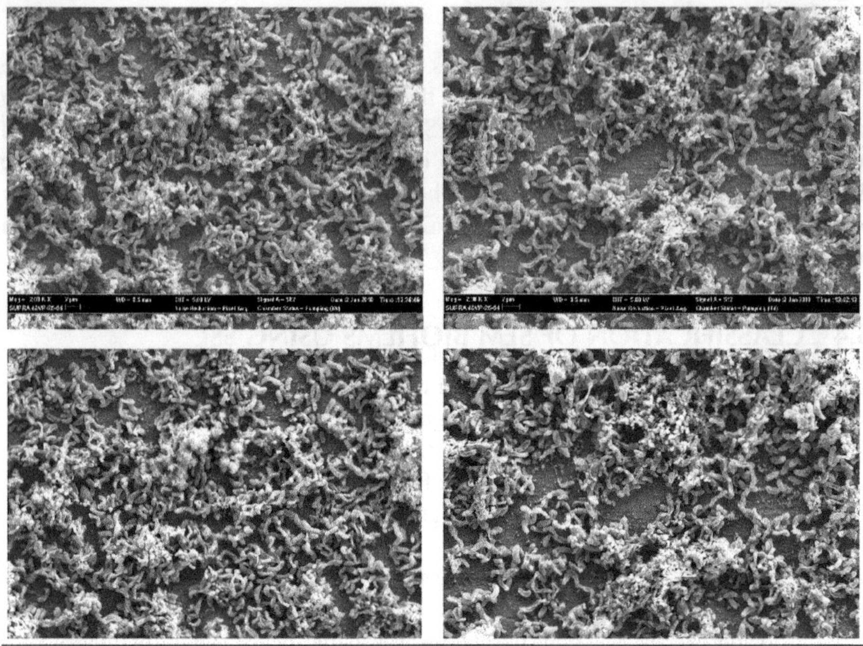

FIGURE 6.6 The *top row* shows two raw images from the biofilm dataset, and the *bottom row* displays the corresponding images after cropping meta information and contrast enhancement. (Image used with permission from Abeyrathna, D. et al., Front. Microbiol. 13, 2022.)

6.4.1 Dataset

For the project, seven SEM images were utilized as biofilm datasets with a resolution of 1024×758 and magnifications ranging from 436X to 2.30KX, covering scale ranges from 2 to 10 micrometres (see two of the raw images in the first row of Figure 6.6).

6.4.2 Image Pre-Processing

The first image pre-processing step was to delete meta-information in a black band by manually clipping it out. Then contrast improvement was pertained to improve image clearness (see the images in the bottom row of Figure 6.6). Some SEM images were captured at different magnification scales, which can cause problems when dividing the images into patches of a similar size to augment data. To address this issue, SR techniques were applied to SEM images that improved details and normalized object sizes. Images with lower scales and higher magnification were not subjected to this process.

We applied three SR approaches, BSRGAN (Zhang et al. 2021), Real-ESRGAN (Wang et al. 2021), and SwinIR (Liang et al. 2021), on the biofilm pre-processed images to get high-quality images. We found out that BSRGAN produced better-quality images at a 4× magnification level (see Section 6.3 for more details).

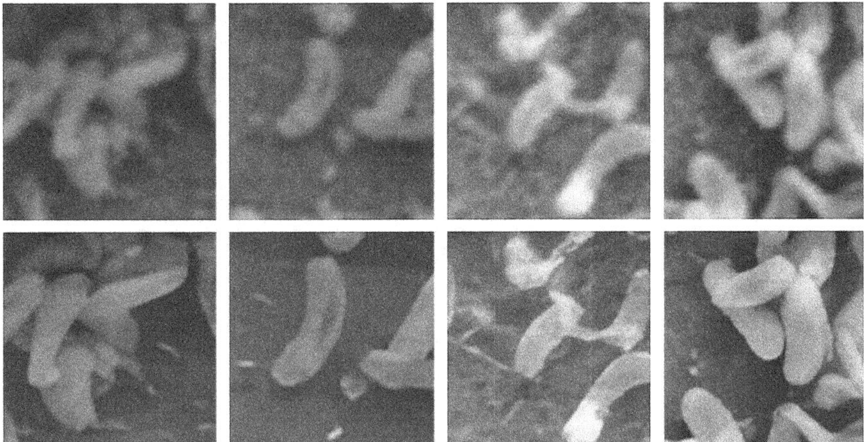

FIGURE 6.7 The *top row* illustrates the output patches before applying the super-resolution method, while the *bottom row* shows that the patches after the super-resolution have been applied. (Image used with permission from Abeyrathna, D. et al., Front. Microbiol. 13, 2022.)

Figure 6.7 shows four random image patches cropped from original images and their high-quality equivalents generated from BSRGAN.

6.4.3 Annotation, Patch Generation, and Object Masking

In this study, experts annotated images in the dataset by assigning class labels (byproduct, cell, and surface) to the images and their components. The image labeller app from MATLAB was used to annotate cells, byproducts, and surfaces using light grey, mid-grey, and dark grey colours, respectively. Figure 6.8 illustrates three random image patches cropped from the original images and their corresponding ground truth annotations.

It required a high volume of data to train any deep neural networks, but biofilm images are difficult to produce in large volumes. To solve the issue, we utilized a method that involves decomposing each image into multiple patches using sliding window mechanism. Then object masking (Li et al. 2004) was applied to generate a better quality set of image patches. Object masking is the process of identifying and isolating specific objects in an image by creating a binary mask that separates the object from its background. This enables machines to recognize and locate individual objects, even when they have similar appearances or are partially hidden by other objects. In Figure 6.9, generated object masks of cells, byproducts, and surfaces are depicted for the corresponding original image patch.

6.4.4 Self-Supervised Training

In this section, we describe an approach to learning the representation of the unlabelled biofilm SEM images using self-supervised training. We selected one contrastive approach and one non-contrastive approach, specifically MoCoV2

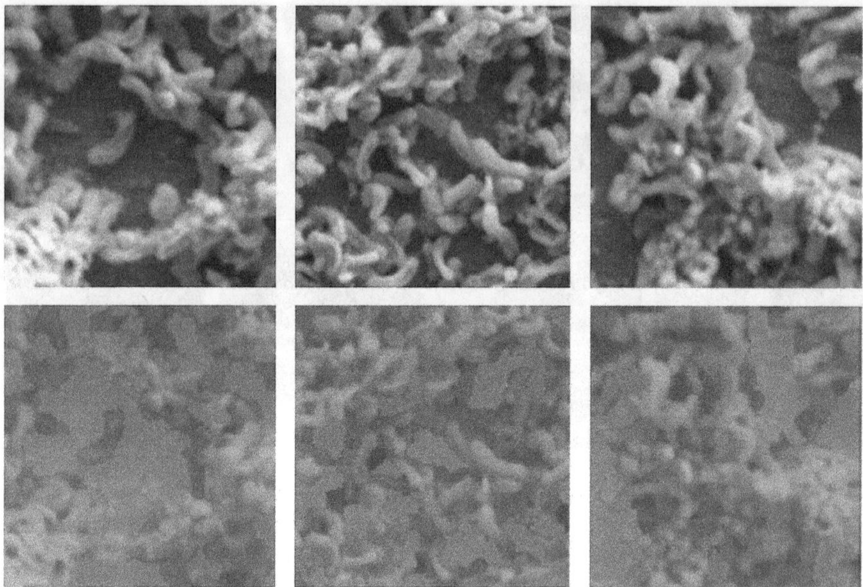

FIGURE 6.8 The *first row* shows sample SEM images, and the *second row* displays the corresponding annotations, where *light grey* represents byproduct, *mid-grey* represents cells, and *dark grey* represents surface. (Image used with permission from Abeyrathna, D. et al., Front. Microbiol. 13, 2022.)

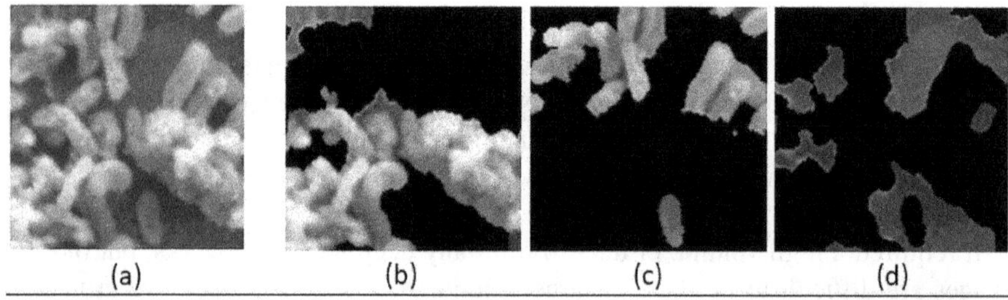

FIGURE 6.9 Generated object mask of byproducts (b), cells (c), and surfaces (d) for a random original biofilm image patch (a). (Image used with permission from Abeyrathna, D. et al., Front. Microbiol. 13, 2022.)

(He et al. 2020) and Barlow Twins (Zbontar et al. 2021) for the self-supervised training. We chose these two approaches as they have several advantages: they can work using mini-batch training samples and have comparable state-of-the-art results.

MoCoV2 stands for "Momentum Contrast V2" and builds on the original Momentum Contrast (MoCo) framework proposed by Facebook AI. MOCOv2 first applies various data augmentation techniques, such as random cropping, colour jittering, or Gaussian blur, to the input image to create two different "views" of each image. MOCOv2 then uses two different encoder networks, a query "q" encoder and a key "k" encoder, to encode each view of the input data into a feature representation (see Figure 6.10a). MoCoV2 then calculates contrastive loss between the positive

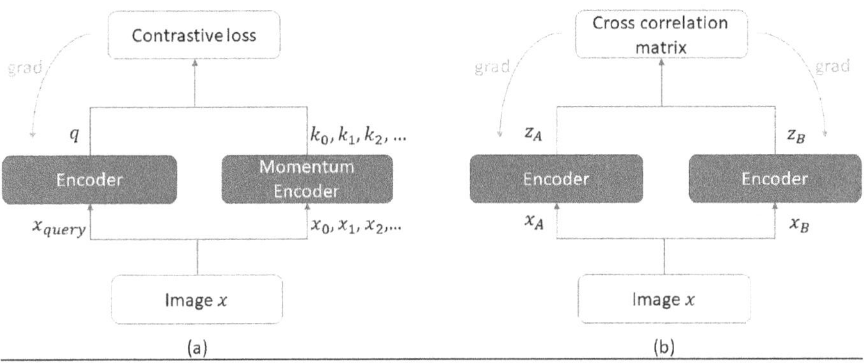

FIGURE 6.10 (a) Self-supervised framework of MoCoV2. (b) Self-supervised framework of Barlow Twins. (Image used with permission from Abeyrathna, D. et al., Front. Microbiol. 13, 2022.)

keys that match with 'q' and negative keys that do not match with 'q'. It uses a queue of keys ($k_{0,1,2,..}$) for this purpose. A temperature parameter T is used to scale the similarity scores. MoCoV2 is designed to work with small mini-batch sizes, and it stores the results in high memory size.

The Barlow Twins self-supervised non-contrastive learning method works by training a neural network to predict the association between two augmented views of the same image data. This method was motivated by the Barlow Twins illusion, as it pursues to learn representations that are invariant to small shifts in input images. It consists of two identical neural networks that share the same set of weights and are trained to encode two slightly different views of the same input data (see Figure 6.10b). The yield of each network is then utilized to calculate a cross-covariance matrix, which is used to measure the similarity between the two views of the data. The aim is then to lessen the distance between the cross-covariance matrix and a target matrix. The framework consists of two components: the invariance term, which guarantees the representation is robust to noise, and the redundancy reduction term, which supports the components of the representation to be independent.

6.4.5 Downstream Task

We propose a method of training a machine learning model for classifying images into "K" classes by converting the problem into "K" binary sub-problems. In this case, K is 3 (byproduct, cell, and surface). We use image patches that are labelled and object-masked to fine-tune separate binary models, one for each class, to predict if the object is present in the image patch or not. The outputs from these binary models are then combined to produce the final classification of an image patch. This method allows for a single patch to be assigned to multiple or even all of the classes.

6.4.6 Experiments

We presented a method for automatically classifying objects in SEM biofilm images using contemporary self-supervised learning approaches. The key focus of the experiments was to examine the feasibility and effectiveness of the method, as well as to differentiate the performance of two different self-supervised learning approaches, MoCoV2 and Barlow Twins. Moreover, the experiments had the objective of assessing the advantages of utilizing self-supervised models regarding expert-annotation workload. Additionally, we aimed to conduct a qualitative assessment of the classification accuracy based on expert input.

We described a pre-processing pipeline in that we used 7 SEM biofilm images for training a machine learning model. We used MATLAB to clip the meta-information from the images. We then applied BSRGAN on images that had a magnification of less than 1KX and processed the images with 4X magnification to normalize the size of objects across the images. We generated image patches of size 64×64 and 128×128 from both annotated and non-annotated images using the sliding windows technique. We used a stride rate of 2 for the sliding window. This process resulted in 24,021 image patches of size 128×128 from all 7 images. They then applied the object-masking process to generate the mask for each object.

We implemented a machine learning model for classifying image patches into 3 classes using two self-supervised learning methods, with ResNet-50 as the base architecture (He et al. 2016), following the recommended configurations (Zbontar et al. 2021; X. Chen et al. 2020) for optimal performance. To obtain a multi-label classification outcome for each patch, we generated three binary classifiers, each for a specific class, and employed the predictions from all three networks. The goal was that an effective classifier should be able to ascribe all 3 classes to a single patch. To ensure the reliability of the outcomes, we carried out the experiments using five cross-validations chosen randomly.

We obtained qualitative feedback from experts by giving them a selection of 10 patches from each image, which varies in terms of difficulty levels for manual classification into the three classes. A user interface had been developed to allow experts to provide their qualitative responses and observations. The experts were shown the original image patch and Class Activation Maps (CAM) (Zhou et al. 2016) generated by the model for three objects. These maps highlight the regions in the image that are most relevant to the predicted class. In addition, the ground truth annotation of the patch, as well as the patch's location in the original image and the model's prediction (True/False), was also given to the experts to indicate the model's certainty in the object's presence.

6.4.7 Evaluation

We conducted evaluations on the self-supervised learning models based on the attribute of the learned representations and the capabilities when fine-tuned for downstream tasks. We first conducted empirical experiments to determine the best configuration for learning representations using different patch sizes, batch sizes, and the number of training epochs on unlabelled data. The learned representations were then used in further experiments.

6.4.7.1 Linear Evaluation for Learned Representation Quality

To evaluate the quality of the learned representations, we performed a linear classification experiment wherein we applied a linear head to the representations.

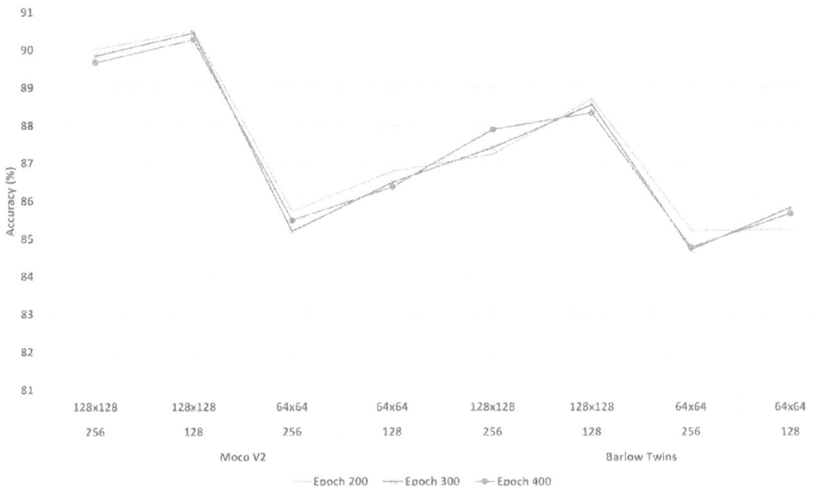

FIGURE 6.11 The comparison of linear evaluation accuracy for the two models is presented, depicting variations in batch sizes and patch sizes. On the y-axis, the *first row* indicates patch sizes, the *second row* indicates batch sizes, and the *third row* indicates the corresponding self-supervised framework. (Image used with permission from Abeyrathna, D. et al., Front. Microbiol. 13, 2022.)

Using standard settings, we trained the linear head with 10% of annotated data while keeping the encoder models fixed. Figure 6.11 illustrates the classification accuracy with various settings, where the x-axis denotes the classification accuracy and the y-axis denotes the different settings. The results showed that both models had the maximum accuracy with a patch size of 128×28, batch size of 128, and 200 training epochs. Random crop and horizontal flip were used as data augmentation during training, and the centre crop was used during testing. The best results were achieved with a patch size of 128×128, batch size of 128, and 200 training epochs. For comparison, we also trained a supervised ResNet-50 model using all the labelled data.

6.4.7.2 Fine-Tuning Evaluation

We evaluated the data efficacy of self-supervised models (MoCoV2 and Barlow Twins) by fine-tuning them on labelled data for classification tasks. We found that using only 10% of labelled data led to significantly better results. We reported accuracy for each binary classification model and overall average. The optimal configuration settings, comprising patch sizes, batch sizes, and data augmentation, were employed to generate all the results.

6.4.8 Results

The dataset underwent a thorough examination to gauge the performance of the two self-supervised learning methods. The robustness of the model's parameters was examined, the models' effectiveness on the dataset was determined, and a qualitative examination of the results was conducted.

6.4.8.1 Robustness of the Model's Parameters

The results indicate that the models utilizing a patch size of 128×28 outperformed those using a patch size of 64x64, regardless of batch size and the number of training epochs. The performance difference was substantial, around 4.5%. The lower information content within a 64×64 patch may be a contributing factor to this difference. As a result, we consistently used a 128×128 patch size in all our subsequent experiments. Additionally, the models exhibited similar performance across varying training epochs of 200, 300, and 400. Therefore, in order to save computational time, we opted for a training epoch of 200 during the representation learning stage.

6.4.8.2 Linear and Fine-Tuning Assessments

Self-supervised models were assessed for the quality of their learned representations using a linear classification head. While the results of Barlow Twins were more consistent, the MoCoV2 model beat it by 2%. Despite not performing as well as a supervised approach, both self-supervised models had accuracy close to the supervised model and required only 10% of the labelled data. In binary classification tasks, the fine-tuned Barlow Twins model acted better (83.18%) than both the supervised baseline (75.01%) and the fine-tuned MoCoV2 model (80.73%), using the same test set as in linear assessment trials. These results suggest that the Barlow Twins model has a superior capability to adapt to downstream tasks after being fine-tuned with inadequate labelled data.

6.4.8.3 Qualitative Results

In the qualitative assessment, experts had highly positive feedback on the classifiers and its potential applicability to various tasks, such as identifying regions in images with specific objects, estimating the correlation and distribution of these objects across patches, etc. They also noted a significant efficiency improvement, estimated to be several orders of magnitude faster, compared to semi-automated approaches using tools like ImageJ. As the number of images grows, relying on manual and semi-automated approaches becomes unfeasible, making scalability crucial. Additionally, the experts were able to identify objects determined by the proposed approach in raw images and agreed with the model's forecasts.

The novel image patch, annotations, and three-class CAM were evaluated by experts. They examined whether the forecasted class existed in the original image patch and was located at the highlighted spot in the CAM (refer to Figure 6.12). Experts identified three reasons for disagreement with machine learning predictions: inappropriate annotations, appropriate annotations with wrong forecasted class, and unclear annotations, which made experts uncertain, particularly in cases of overlapping cells and byproducts.

There was a significant disparity between the evaluations of domain experts and the predictions generated by models. One of the primary reasons for this divergence was attributed to the incapability of the existing class activation map scheme in identifying the presence of a specific class in image patches where entities of the alike class are detached and existed in multiple regions. Additionally, the assessment metric for CAMs is subjective and can lead to discrepancies, especially in image patches

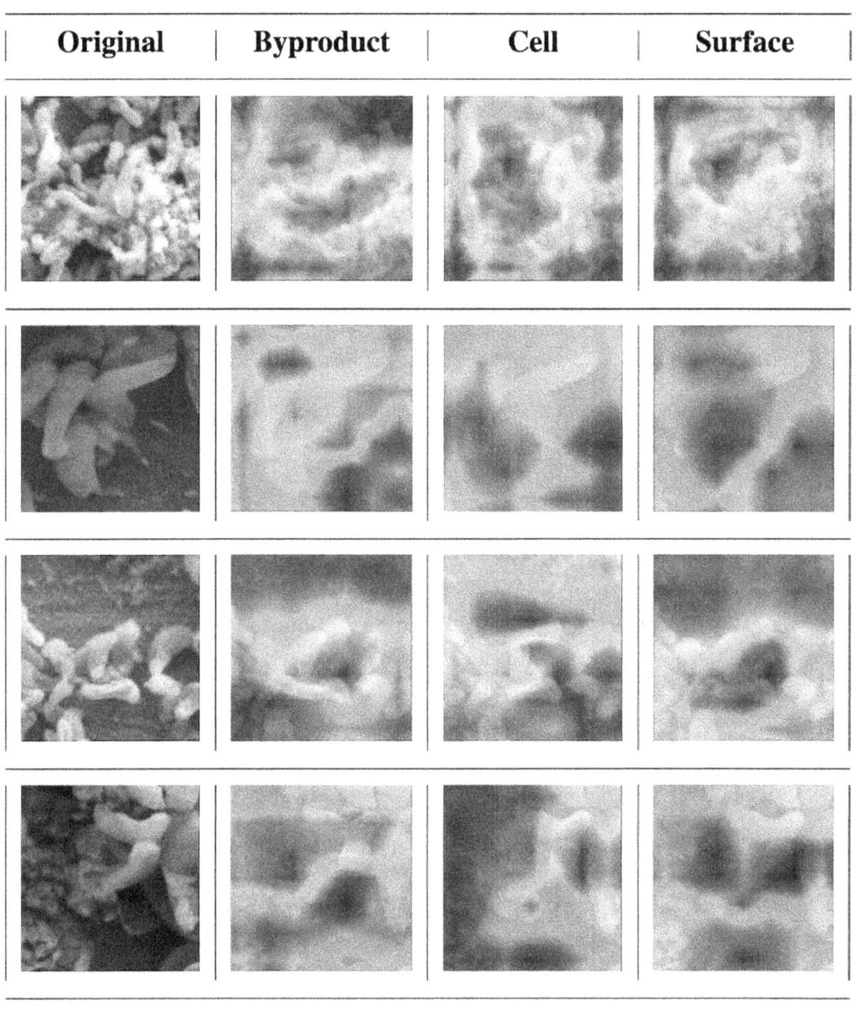

FIGURE 6.12 The comparison of CAM visualization results was obtained from different binary classification models. The visualizations were obtained for the last convolution output of each model. (Image used with permission from Abeyrathna, D. et al., Front. Microbiol. 13, 2022.)

that contain overlying objects from multiple classes. Overall, domain experts agreed with the model predictions of 98% of the time.

6.4.9 Discussion

With only around 10% of annotated data, the suggested workflow based on self-supervised learning attained markedly improved classification accuracy results. The linear assessment experiments showed that both models can retrieve analogous quality representations using the unlabelled dataset, but the Barlow Twins model performed significantly better with fine-tuning and limited labelled data. The performance of both self-supervised models exceeded that of the supervised model. Furthermore, the Barlow Twins model is a preferable choice because it offers higher classification accuracy and requires less computational cost than MoCoV2.

Although multi-label classification is proper for forecasting multiple objects in an image patch, binary classifiers were employed in this case to provide multiple labels to an image patch. This choice was made because of certain characteristics of the dataset, including imbalanced class instance proportions, especially between the Surface class and byproduct class, as well as the low variance between the cell class and the byproduct class, which had comparable visual features. The use of binary classification was deemed more feasible than multi-label approaches like algorithm adaptation. While binary relevance approaches may overlook label correlations, it was believed that this limitation was addressed by the representation learning stage of self-supervised learning models, which could have captured such correlations.

MIC (microbiologically induced corrosion) caused by sulphate-reducing bacteria (SRB) results in billions of dollars in annual costs. MIC is a crucial interfacial process that is dependent on various factors, including microbes, redox potential, dissolved oxygen, salt concentrations, pH, and conductivity. Protective coatings above the metals act as a barrier against corrosive metabolites of both biotic and abiotic forms mainly by passivating the MIC impacts. Two-dimensional (2D) materials including hexagonal-boron nitride (h-BN) and graphene are well-renowned protective coatings due to their excellent barrier property, chemical resistance, and impermeability with thermal stability. Our previous studies showed that both graphene and h-BN coatings regulate SRB biofilms, their attachments, and their electrochemical oxidation when exposed to copper and low-carbon steel. We observed these biofilms on pristine and 2D material-coated copper and low carbon steel, with structural features measured and extracted manually, which is a labour-intensive and expensive task.

Recent advancements in artificial intelligence (AI) and cellular microscopy have created opportunities to collect large amounts of data and analyze/predict cellular structures from biological data. A range of tools had been utilized for extracting and assessing the morphological characteristics of biofilm microstructures, including deep neural networks, BioFilm Analyzer, BiofilmQ, and ImageJ. However, these tools are not suitable for characterizing congested biofilm microstructures and microbial products as they can only handle smooth, homogeneous, and non-overlaying geometric structures.

An investigation was conducted to evaluate the effect of diverse colour spaces, sliding window sizes, and CNN architectures for corrosion detection. They used different colour spaces such as RGB, YCbCr, CbCr, and grayscale to identify the best colour space for corrosion detection. Various architectures were evaluated with the optimal colour space using a sliding window to detect stained areas within an image, and multiple sliding window sizes (128×28, 64×64, and 32×32) were used to classify the areas of an image. Smaller sliding window sizes resulted in more accurate localization of corroded regions but a decrease in the number of attributes a CNN can learn, leading to a decline in the signal-to-noise ratio. Images of 128×128, 64×64, and 32×32 pixels were used to assess the influence of sliding window sizes.

Two distinct sets of microscopy images were utilized to train a CNN in order to create a three-class classification system that can differentiate between the standard, unprotected, and protected states of a surface with copper. Due to the limited size of the dataset, the network's architecture was constrained to only include two convolutional layers, and data augmentation was employed by altering the rotation, shear intensity, and zoom range of the dataset. The approach being proposed tackles a low-volume dataset as well, but it utilizes self-supervised techniques to classify the data while reducing the amount of expert annotation required.

6.4.10 Summary

This study proposes a self-supervised learning-based workflow to classify constituents in biofilm SEM images with limited annotated data. Annotated data are costly and challenging to generate, so the study experimented with image pre-processing and SSL to improve classification accuracy. Super-resolution of SEM images improved the performance of multiple SSL models, and the Barlow Twins SSL model achieved 83.18% classification accuracy with a 90% reduction in required labelled data. This study demonstrates the potential of self-supervised learning to reduce manual annotation requirements and suggests further exploration of self-supervised methods for object segmentation and other tasks.

6.5 CONCLUSION

This chapter is divided into four main sections. The first section provides a detailed overview of modern solutions that address the problem of scarce annotation in medical image segmentation. The second section delves into the practical application of self-supervised learning in medical imaging. The third section delves into the implementation of deep learning-based super-resolution techniques to enhance the effectiveness of diverse downstream tasks on a dataset of biofilm images obtained from a scanning electron microscope (SEM). In the final section, we explore one case study: a method for performing classification tasks on biofilm images that have a low volume of images using SSL.

ACKNOWLEDGEMENTS

The authors thank Vidya Bommanapally, Milind Malshe, Dilanga Abeyrathna, Jawahar Kalimuthu, Ramana Gadhamshetty, and Parvathi Chundi for their earlier contributions without which this work would not be possible. This project was partially supported by NSF EPSCoR RII Track 2 FEC #1920954.

REFERENCES

Abeyrathna, Dilanga, Md Ashaduzzaman, Milind Malshe, Jawaharraj Kalimuthu, Venkataramana Gadhamshetty, Parvathi Chundi, and Mahadevan Subramaniam. 2022. "An AI-Based Approach for Detecting Cells and Microbial Byproducts in Low Volume Scanning Electron Microscope Images of Biofilms." *Frontiers in Microbiology* 13: 996400.

Ashaduzzman, Md, Vidya Bommanapally, Mahadevan Subramaniam, Parvathi Chundi, Jawahar Kalimuthu, Suvarna Talluri, and Ramana Gadhamshetty. 2022. "Using Deep Learning Super-Resolution for Improved Segmentation of SEM Biofilm Images." In *2022 IEEE International Conference on Bioinformatics and Biomedicine (BIBM)*, 3587–93. https://doi.org/10.1109/BIBM55620.2022.9995190.

Azizi, Shekoofeh, Basil Mustafa, Fiona Ryan, Zachary Beaver, Jan Freyberg, Jonathan Deaton, Aaron Loh, Alan Karthikesalingam, Simon Kornblith, and Ting Chen. 2021. "Big Self-Supervised Models Advance Medical Image Classification." In *Proceedings of the IEEE/CVF International Conference on Computer Vision*, 3478–88.

Bai, Wenjia, Chen Chen, Giacomo Tarroni, Jinming Duan, Florian Guitton, Steffen E. Petersen, Yike Guo, Paul M. Matthews, and Daniel Rueckert. 2019. "Self-Supervised Learning for Cardiac Mr Image Segmentation by Anatomical Position Prediction." In *Medical Image Computing and Computer Assisted Intervention–MICCAI 2019: 22nd International Conference*, Shenzhen, October 13–17, 2019, Proceedings, Part II 22, 541–49. Springer.

Bommanapally, Vidya, M Ashaduzzman, M Malshe, Parvathi Chundi, and M Subramaniam. 2021. "Self-Supervised Learning Approach to Detect Corrosion Products in Biofilm Images." In *2021 IEEE International Conference on Bioinformatics and Biomedicine (BIBM)*, Houston, TX, 3555–61. IEEE.

Chaitanya, Krishna, Ertunc Erdil, Neerav Karani, and Ender Konukoglu. 2020. "Contrastive Learning of Global and Local Features for Medical Image Segmentation with Limited Annotations." *Advances in Neural Information Processing Systems* 33: 12546–58.

Chen, Liang, Paul Bentley, Kensaku Mori, Kazunari Misawa, Michitaka Fujiwara, and Daniel Rueckert. 2019. "Self-supervised Learning for Medical Image Analysis Using Image Context Restoration." *Medical Image Analysis* 58: 101539.

Chen, Ting, Simon Kornblith, Mohammad Norouzi, and Geoffrey Hinton. 2020. "A Simple Framework for Contrastive Learning of Visual Representations." In *International Conference on Machine Learning*, 1597–1607. PMLR.

Chen, Xinlei, and Kaiming He. 2021. "Exploring simple Siamese representation learning." In *Proceedings of the IEEE/CVF conference on computer vision and pattern recognition*, 15750–15758.

Chen, Xinlei, Haoqi Fan, Ross Girshick, and Kaiming He. 2020. "Improved Baselines with Momentum Contrastive Learning." *ArXiv Preprint ArXiv:2003.04297.*

Doersch, Carl, Abhinav Gupta, and Alexei A. Efros. 2015. "Unsupervised Visual Representation Learning by Context Prediction." In *Proceedings of the IEEE International Conference on Computer Vision*, 1422–30.

Gidaris, Spyros, Praveer Singh, and Nikos Komodakis. 2018. "Unsupervised Representation Learning by Predicting Image Rotations." *ArXiv Preprint ArXiv:1803.07728.*

Gildenblat, Jacob, and Eldad Klaiman. 2019. "Self-Supervised Similarity Learning for Digital Pathology." *ArXiv Preprint ArXiv:1905.08139.*

Goodfellow, Ian, Jean Pouget-Abadie, Mehdi Mirza, Bing Xu, David Warde-Farley, Sherjil Ozair, Aaron Courville, and Yoshua Bengio. 2020. "Generative Adversarial Networks." *Communications of the ACM* 63 (11): 139–44.

Grill, Jean-Bastien, Florian Strub, Florent Altché, Corentin Tallec, Pierre Richemond, Elena Buchatskaya, Carl Doersch, Bernardo Avila Pires, Zhaohan Guo, and Mohammad Gheshlaghi Azar. 2020. "Bootstrap Your Own Latent-a New Approach to Self-Supervised Learning." *Advances in Neural Information Processing Systems* 33: 21271–84.

He, Kaiming, Haoqi Fan, Yuxin Wu, Saining Xie, and Ross Girshick. 2020. "Momentum Contrast for Unsupervised Visual Representation Learning." In *Proceedings of the IEEE/CVF Conference on Computer Vision and Pattern Recognition*, 9729–38.

He, Kaiming, Xiangyu Zhang, Shaoqing Ren, and Jian Sun. 2016. "Deep Residual Learning for Image Recognition." In *2016 IEEE Conference on Computer Vision and Pattern Recognition (CVPR)*, Las Vegas, NV, 770–78.

Jamaludin, Amir, Timor Kadir, and Andrew Zisserman. 2017. "Self-Supervised Learning for Spinal MRIs." In *Deep Learning in Medical Image Analysis and Multimodal Learning for Clinical Decision Support: Third International Workshop, DLMIA 2017, and 7th International Workshop, ML-CDS 2017*, Held in Conjunction with MICCAI 2017, Québec City, September 14, Proceedings 3, 294–302. Springer International Publishing.

Kingma, Diederik P., and Max Welling. 2019. "An Introduction to Variational Autoencoders." *Foundations and Trends® in Machine Learning* 12 (4): 307–392.

Koohbanani, Navid Alemi, Balagopal Unnikrishnan, Syed Ali Khurram, Pavitra Krishnaswamy, and Nasir Rajpoot. 2021. "Self-Path: Self-Supervision for Classification of Pathology Images with Limited Annotations." *IEEE Transactions on Medical Imaging* 40 (10): 2845–56.

Ledig, Christian, Lucas Theis, Ferenc Huszár, Jose Caballero, Andrew Cunningham, Alejandro Acosta, Andrew Aitken, Alykhan Tejani, Johannes Totz, and Zehan Wang. 2017. "Photo-Realistic Single Image Super-Resolution Using a Generative Adversarial Network." In *Proceedings of the IEEE Conference on Computer Vision and Pattern Recognition*, Montrel, BC, 4681–90.

Liang, Jingyun, Jiezhang Cao, Guolei Sun, Kai Zhang, Luc Van Gool, and Radu Timofte. 2021. "Swinir: Image Restoration Using Swin Transformer." In *2021 IEEE/CVF International Conference on Computer Vision Workshops (ICCVW)*, 1833–44.

Li, Yin, Jian Sun, Chi-Keung Tang, and Heung-Yeung Shum. 2004. "Lazy snapping." *ACM Transactions on Graphics (ToG)* 23 (3): 303–8.

Liu, Ze, Yutong Lin, Yue Cao, Han Hu, Yixuan Wei, Zheng Zhang, Stephen Lin, and Baining Guo. 2021. "Swin Transformer: Hierarchical Vision Transformer Using Shifted Windows." In *Proceedings of the IEEE/CVF International Conference on Computer Vision (ICCV)*, 10012–22.

Long, Jonathan, Evan Shelhamer, and Trevor Darrell. 2015. "Fully Convolutional Networks for Semantic Segmentation." In *Proceedings of the IEEE Conference on Computer Vision and Pattern Recognition*, 3431–40.

Noroozi, Mehdi, and Paolo Favaro. 2016. "Unsupervised Learning of Visual Representations by Solving Jigsaw Puzzles." In *European Conference on Computer Vision*, Amsterdam, The Netherlands, 69–84. Springer.

Oord, Aaron van den, Yazhe Li, and Oriol Vinyals. 2018. "Representation Learning with Contrastive Predictive Coding." *ArXiv Preprint ArXiv:1807.03748.*

Ouyang, Cheng, Carlo Biffi, Chen Chen, Turkay Kart, Huaqi Qiu, and Daniel Rueckert. 2020. "Self-Supervision with Superpixels: Training Few-Shot Medical Image Segmentation without Annotation." In *European Conference on Computer Vision*, Glasgow, UK, 762–80. Springer.

Pathak, Deepak, Philipp Krahenbuhl, Jeff Donahue, Trevor Darrell, and Alexei A. Efros. 2016. "Context Encoders: Feature Learning by Inpainting." In *Proceedings of the IEEE Conference on Computer Vision and Pattern Recognition*, Las Vegas, Nevada, USA, 2536–44. IEEE.

Reza, Ali M. 2004. "Realization of the Contrast Limited Adaptive Histogram Equalization (CLAHE) for Real-Time Image Enhancement." *Journal of VLSI Signal Processing Systems for Signal, Image and Video Technology* 38 (1): 35–44.

Tajbakhsh, Nima, Yufei Hu, Junli Cao, Xingjian Yan, Yi Xiao, Yong Lu, Jianming Liang, Demetri Terzopoulos, and Xiaowei Ding. 2019. "Surrogate Supervision for Medical Image Analysis: Effective Deep Learning from Limited Quantities of Labeled Data." In *2019 IEEE 16th International Symposium on Biomedical Imaging (ISBI 2019)*, Venice, Italy, 1251–55. IEEE.

Vincent, Pascal, Hugo Larochelle, Yoshua Bengio, and Pierre-Antoine Manzagol. 2008. "Extracting and Composing Robust Features with Denoising Autoencoders." In *Proceedings of the 25th International Conference on Machine Learning*, New York, NY, United States, 1096–1103. Association for Computing Machinery.

Wang, Xintao, Ke Yu, Shixiang Wu, Jinjin Gu, Yihao Liu, Chao Dong, Yu Qiao, and Chen Change Loy. 2018. "Esrgan: Enhanced Super-Resolution Generative Adversarial Networks." In *Proceedings of the European Conference on Computer Vision (ECCV) Workshops*, Munich, Germany, 63–79. Springer.

Wang, Xintao, Liangbin Xie, Chao Dong, and Ying Shan. 2021. "Real-Esrgan: Training Real-World Blind Super-Resolution with Pure Synthetic Data." In *Proceedings of the IEEE/CVF International Conference on Computer Vision (ICCV) Workshops*, 1905–14. IEEE, Virtual.

Zbontar, Jure, Li Jing, Ishan Misra, Yann LeCun, and Stéphane Deny. 2021. "Barlow Twins: Self-Supervised Learning via Redundancy Reduction." In *Proceedings of the 38th International Conference on Machine Learning*, 12310–20. PMLR, Virtual.

Zhang, Kai, Jingyun Liang, Luc Van Gool, and Radu Timofte. 2021. "Designing a Practical Degradation Model for Deep Blind Image Super-Resolution." In *Proceedings of the IEEE/CVF International Conference on Computer Vision (ICCV)*, 4791–4800. IEEE, Virtual.

Zhang, Richard, Phillip Isola, and Alexei A. Efros. 2016. "Colorful Image Colorization." In *European Conference on Computer Vision*, Amsterdam, The Netherlands, 649–66. Springer.

Zhao, Can, Blake E Dewey, Dzung L Pham, Peter A Calabresi, Daniel S Reich, and Jerry L Prince. 2020. "SMORE: A Self-Supervised Anti-Aliasing and Super-Resolution Algorithm for MRI Using Deep Learning." *IEEE Transactions on Medical Imaging* 40 (3): 805–17.

Zhou, Bolei, Aditya Khosla, Agata Lapedriza, Aude Oliva, and Antonio Torralba. 2016. "Learning Deep Features for Discriminative Localization." In *Proceedings of the IEEE Conference on Computer Vision and Pattern Recognition (CVPR)*, Las Vegas, Nevada, USA, 2921–29. IEEE.

7 Quorum Sensing Mechanisms, Biofilm Growth, and Microbial Corrosion Effects of Bacterial Species

Vaibhav Handa, Saurabh Dhiman, Kalimuthu Jawaharraj, Vincent Peta, Alain Bomgni, Etienne Z. Gnimpieba, and Venkataramana Gadhamshetty

DEFINITIONS

- **Autoinduction:** Cell-to-cell communication that enables population density-based control of gene transcription. This is done via the production, release, and sensing of low-molecular-weight compounds.
- **Biofilm:** Surface-attached microbial communities that are embedded within a self-produced extracellular matrix consisting of polysaccharides and DNA.
- **Quorum sensing:** A mechanism where bacteria use signaling molecules for regulating gene expression, typically based on population density.
- **Sessile cells:** Cells that are encapsulated within the extracellular polymeric substance component of biofilms.
- **Signaling:** The ability to detect and respond to cell population density by gene regulation.
- **SRB:** Sulfate-reducing bacteria.
- **Stress:** Adverse and fluctuating conditions in the immediate surroundings of bacteria.
- **Stress response:** Mechanisms used by bacteria to survive stressful environmental conditions.
- **Virulence:** Ability to invade and multiply within the host; it is defined in terms of the degree of pathogenicity.

DOI: 10.1201/9781003132981-7

ACRONYMS

AI-2	Autoinducer-2
DSF	Diffusible signal factor
GCL	Gamma-caprolactone
HTH	Helix-turn-helix
Mpy	Mils per year
MBR	Membrane bioreactor
QQ	Quorum quenchers
XAC	Citrus canker

7.1 INTRODUCTION

Quorum sensing (QS) is a fascinating mechanism used by bacterial cells to communicate with each other and jointly regulate their activities in sociality. Bacterial cells synthesize diffusible molecules—known as autoinducers (AIs)—for creating intercellular signaling mechanisms and controlling their social network. Such QS-related AIs are implicated in cross talk among diverse bacterial cells and hosts (e.g., plants and human intestines). QS mechanisms can regulate diverse stages of biofilm formation. They can modify the surface topography of exposed substrates, binding of cells with the substrates, surface-induced responses of the adhering cells, and secretion of extracellular polymeric substance (EPS). Examples of impacted substrates include metals, polymers, soil particles, medical implants, and biological tissues. Thus, QS mechanisms can influence biofilm growth in diverse domains including agricultural, industrial, and commercial.

7.2 QUORUM SENSING

QS is a cell-to-cell communication process that involves the secretion and sensing of extracellular signaling molecules called autoinducers (AIs). These communications can occur via AI in both inter- and intra-bacterial species [1,2,3]. They are more dominant in populations based on identical organisms, especially when the population density exceeds a threshold level [1,4–6]. QS controls diverse bacterial functions including antibiotic production, biofilm formation, bioluminescence, competence, sporulation, swarming motility, and virulence factor secretion. It can even alter the behavior of the entire bacterial population [7,8].

Bacterial cells can sense levels of the signaling molecules to determine the number of other cells present within the same environment [1,4,5]. When a threshold level of cells is reached, the population density is said to achieve a quorum [2,3,6]. Quorum-dependent genes are expressed by autoinduction. Bacterial species produce a range of AI molecules that regulate genes and control characteristics that are exhibited above the critical population or threshold population density [2,3,5,6,9,10]. Detection of a minimum threshold stimulatory concentration of AI leads to an alteration in gene expression. Both gram-positive and gram-negative bacteria use QS communications to regulate their physiological activities [11–14]. However, gram-negative bacteria use acyl homoserine lactones (AHL) as an AI while gram-positive bacteria use peptides for communication [13].

7.3 KEY QUORUM SENSING MOLECULES AND THEIR SIGNALING MECHANISMS

Examples of signaling molecules in QS systems include homoserine lactone (HSL), AHL, and autoinducing peptides. Different molecules have been observed in different species, and they all display different genotypical and phenotypical effects (Table 7.1). The chemical structures of these molecules listed in Table 7.1 are shown in Figure 7.1.

TABLE 7.1
Quorum Sensing Molecules and Their Classes of Action

Name	Chemical Structure	Class of Action	Name of Receptor	Ref
LuxS	Figure 7.1a [15]	Transportation of the QS signal AI-2 by enhancing its secretion. Consequently, it represses biofilm formation and motility [16]	Histidine **protein** kinase	[15]
LuxR	Figure 7.1b [17]	**LuxR-type** is a DNA-binding helix-turn-helix (HTH) domain consisting of about 65 amino acids. It participates in the transcriptional regulators in the LuxR family of response regulators [18]	N-terminal receptor site of the proteins	[17]
LuxO	Figure 7.1c [19]	LuxO acts indirectly on virulence gene expression by repressing hapR gene. This leads to the expression of virulence factors. In strain El Tor N16961, the hapR gene is inactive due to a natural frameshift mutation [17]	LuxR-type DNA-binding HTH domain	[19]
LuxQ	Figure 7.1d [16]	In SRB, the binding of AI-2 to the periplasmic receptor LuxP modulates the activity of the inner membrane sensor kinase LuxQ, transducing the AI-2 information into the cytoplasm [15]	apoLuxP	[16]
Diffusible Signal Factor (DSF)	Figure 7.1e [20]	DSF-mediated QS regulation of *X. citri* subsp. *citri* (Xac), the causal agent of citrus canker. DSF-mediated QS specifically modulates bacterial adaptation, nutrition uptake and metabolism, stress tolerance, virulence, and signal transduction to favor host infection [21]	RpfC receptor of histidine Kinase [21]	[22]

(*Continued*)

TABLE 7.1 (*Continued*)
Quorum Sensing Molecules and Their Classes of Action

Name	Chemical Structure	Class of Action	Name of Receptor	Ref
Furanone C-30	Figure 7.1f [23]	Diminished swarming motility was observed in the presence of furanone C-30. The wild-type strain exhibited swarming across the soft agar, but its motility was markedly inhibited in the presence of furanone C-30. These results suggest that *P. aeruginosa* surface colonization is controlled by furanone C-30 which effectively inhibits the C4-HSL-mediated QS system (QS system in the bacteria) reducing the virulence property of the bacteria [24]	pqsR receptor [24]	[24]

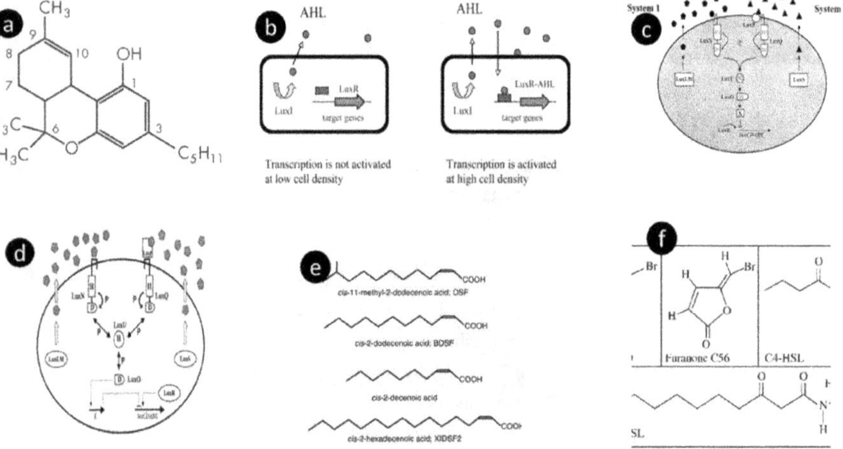

FIGURE 7.1 Chemical structures of quorum sensing molecules. (a) LuxS; (b) LuxR; (c) LuxO; (d) LuxQ; (e) Diffusible Signal Factor; (f) Furanone C-30.

7.4 QUORUM SENSING IN RELATION TO STRESS RESPONSE

Stress response mechanisms allow microbial species to survive adverse and fluctuating conditions in their immediate surroundings. Bacterial cells respond to stressors by leveraging multiple stress response systems that interact via complex global regulatory networks [7,8,10,25–28]. The significance of QS in regulating the stress response with respect to stressors [e.g., heat, heavy metals in the case of microbiologically influenced corrosion (MIC)] has been well reported in the literature [26–30]. For example, QS mechanisms improve the viability of *Vibrio cholerae* under stressful conditions by regulating the expression of the RNA polymerase sigma S gene. Such regulation takes place via the HapR gene, highlighting the roles of QS-enhanced stress responses in *V. cholerae* when exposed to oxidative and nutritional stresses [27]. Given the wide range of environmental stressors in nature, it is likely that QS-enhanced stress tolerance allows the microbial cells to counteract QS inhibition [31] and invasion by other virulent species having a broader impact on bacterial ecology.

7.5 BACKGROUND ON BIOFILMS WITH FOCUS ON ITS ECOLOGY IN NATURAL ECOSYSTEMS

Here, we provide a generic overview of biofilms using their microbial life in stream ecosystems as a practical example. Biofilms exist as matrix-enclosed and surface-attached microbial communities that are highly active at streambed interfaces. Such biofilm modes allow bacterial populations to sustain challenges posed by a fast flow of water and the need for continuous export of nutrients and organic matter. Biofilms in streams are considered a 'microbial skin' that allows the sessile cells to process and export nutrients along with the organic matter from the structure. Fluid dynamics influence the dispersal of microbes and their biodiversity dynamics at the scale of stream networks [30, 32]. Interactions among parameters related to biofilm growth, stream water flow, and substrate chemistry are responsible for environmental complexity in the streambed. Species like Proteobacteria and Bacteroidetes dominate the communities of stream biofilms [15–19, 33], including those based on Flavobacteria and Sphingobacteriia. The biodiversity present in stream biofilms is supported by the continuous input of microbes [34].

Biofilms exert both negative and positive roles. For example, beneficial biofilms that live inside the gut ensure the normal functioning of human beings as well as animals. Beneficial biofilms have been implicated in their roles in mitigating the negative effects of obesity, autism or cancer, and infectious diseases. Biofilms are not only essential for the normal functioning of ecosystems (e.g., providing oxygen and food for many organisms using solar energy as plants do) but also for protecting health by degrading pollutants in water and soil, limiting erosion, and ensuring soil fertility, among other things [35–39]. Negative biofilms are implicated in terms of their roles in improving the resiliency of microorganisms involved in the pathogenesis and MIC.

7.6 QUORUM SENSING, BIOFILM GROWTH, AND MICROBIOLOGICALLY INFLUENCED CORROSION

QS mechanisms can influence different stages of biofilm formation (Figure 7.1). Given that sessile cells within biofilms are known to aggravate MIC, we considered MIC in this study (Figure 7.2) [10,14,25,31].

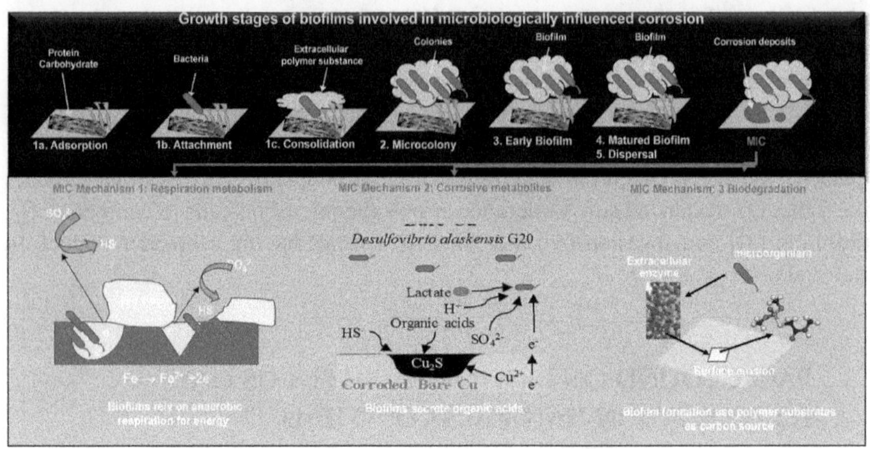

FIGURE 7.2 Top panel: Different stages of biofilm growth that may influence microbial corrosion. Bottom panel: Three major types of microbial corrosion to attack metals and plastics.

7.6.1 QS, Biofilm Growth, and MIC

We explain the role of QS on biofilm formation by sulfate-reducing bacteria (SRB) that are widely implicated in MIC, a problem that contributes to $5 billion in corrosion costs. The MIC costs include resources for addressing corrosion issues that call for maintenance, repairs, and lost time for delays, failures, outages, litigation, and taxes. MIC can occur during any of the five stages of biofilm growth (Figure 7.1). These stages include (1a) the conditioning phase where self-secreted molecules (e.g., proteins and carbohydrates) are adsorbed by the underlying surfaces; (1b) the attachment phase where planktonic cells are immobilized on the polymer matrix; (1c) the consolidation of sessile cells within the EPS; (2) formation of microcolonies within the EPS of biofilms; (3) growth of early biofilm; (4) growth of matured biofilm; and (5) dispersal [18,39–42]. SRB biofilms can influence MIC in all five stages (Figure 7.1). SRB biofilms use three different types of mechanisms to influence MIC. In the Type I mechanism, they use metal as an electron donor under nutrient-limiting conditions. In the Type II mechanism, biofilms secrete metabolites that generate terminal electron acceptors (e.g., protons) that support cathodic reduction reactions involved in corrosion [7,8,10,26,27]. Type III mechanisms are used to degrade nonmetals by using them as carbon sources. Readers are encouraged to review [18,37,42–47] to get an in-depth understanding of these mechanisms.

Desulfovibrio (D) vulgaris and *Desulfobacterium* (*Db*) *corrodens* spp. are commonly studied model organisms in MIC studies. *D. vulgaris* is a thoroughly studied SRB with its entire genome sequenced. *Db. corrodens* is an SRB whose genome has been well annotated but with zero evidence in the presence of QS-based gene homologs [47]. In a recent study [45], both these species were grown in either saline or freshwater media. Here, saline conditions represent an example of stressful environments discussed earlier. They used lactate and sodium sulfate (Na_2SO_4) as sources of electron donors and acceptors, respectively. Increased potentials of sulfate reduction,

TABLE 7.2
Quorum Quenchers and Its Effect on Biofilm Formation

Quorum Quencher	Dosage (μM)	Observed effect on biofilm formation	References
Bromo furanone	80	Decreases specific sulfate reduction rate of *D. vulgaris* and subsequently its biofilm formation	[35,45,49]
Butanamide, 3-oxo-N-phenyl	40	Decreases specific growth rates of *D. vulgaris* and *Db. corrodens*, discouraging biofilm formation	[21]
GABA (Gamma-aminobutyric acid)	40	Compromises the specific growth rate of *Db. corrodens* and biofilm formation by *D. vulgaris* and *Db. corrodens*	[9, 50]

AHL production, and biofilm formation by *D. vulgaris* and *Db. corrodens* were observed under saline conditions [45]. As mentioned earlier, AHL is a typical signaling molecule encountered in many QS systems. To analyze the effects of salinity at the genetic expression level, quantification of transcript levels of genes responsible for sulfate reduction, carbon utilization, biofilm formation-based hydrogenases, as well as histidine kinases involved in cell–cell communication was analyzed. Transcript levels of all relevant genes were found to be upregulated under saline conditions. Hence, saline conditions have a pronounced effect on sulfate reduction, biofilm formation, and AHL production at the genetic level by both planktonic cells and biofilms of SRB [7]. As shown in Table 7.3, QS mechanisms can be involved in different stages of biofilm growth. Thus, QS mechanisms can be expected to influence the growth of biofilms that are involved in metallic corrosion. Such mechanisms can be considered for developing effective MIC prevention mechanisms, for example by developing protective coatings that release QS-inhibiting molecules. We can also incorporate quorum quenching supplements for inhibiting QS communications (Table 7.2).

7.6.2 Bioinformatics Analysis

We selected several QS-associated proteins to check for any discernible consensus sequences, analyze phylogenetically, and determine which biosynthetic pathways these proteins belong to. Using a MAFFTT workflow [49], the proteins were aligned and a maximum likelihood of a phylogenetic tree was also created using the output of the previous workflow with the addition of RAxML ver. 8 [35] with default parameters, LG+G4 [36] model and 100 bootstrap iterations. The multiple sequence alignment (MSA) (Figure 7.3) for the selected QS proteins did not yield a discernible consensus sequence. The phylogenetic tree showed that the LuxO and LasR proteins from *Vibrio harveyi* and *Pseudomonas aeruginosa* [44, 51], respectively, have a common ancestor. Likewise, the protein AHL synthesis from both *P. aeruginosa* and *Aeromonas hydrophila*, also shares a common ancestor, with different gene origins [43]. AHL synthesis from *Burkholderia vietnamiensis* shares a common ancestor

TABLE 7.3
Role of Quorum Sensing (QS) Molecules at Different Stages of Biofilm Formation

#	Biofilm growth stage	Role of QS	Examples of QS	Ref
1	Attachment	Initial attachment of the mutant bacteria seems to be affected due to the presence of the QS system, rendering them more adherent to the underlying surfaces	cepIR and cciIR QS systems	[40]
2	Microcolony	The interaction of the ahyI protein and C4 HSL with the ahyrI locus receptor improved the development of microcolonies	AhyI protein and C4 HSL QS systems	[41]
3	Early biofilm growth	The ratio of LasI and RhiI is critical while governing biofilm formation. The production of RhiI is seen to be less during the log phase of biofilm formation, increasing the production of LasI, and further contributing to the initial stages of biofilm formation	LasI quorum sensing genes	[42]
4	Mature biofilm	LuxO is involved in the downstream phosphorylation cascade reactions, upregulation, or repression of QS-associated genes. Involved in many phenotypic traits, including mature biofilm formation	LuxO protein	[18]
5	Dispersion	Bacterial species use QS to coordinate the disassembly of the biofilm community. Biofilm dispersion allows cells to escape the current environment where nutrients are depleted and waste products are accumulated. This allows the cells to colonize new niches	LasI/LasR quorum sensing system	[43]

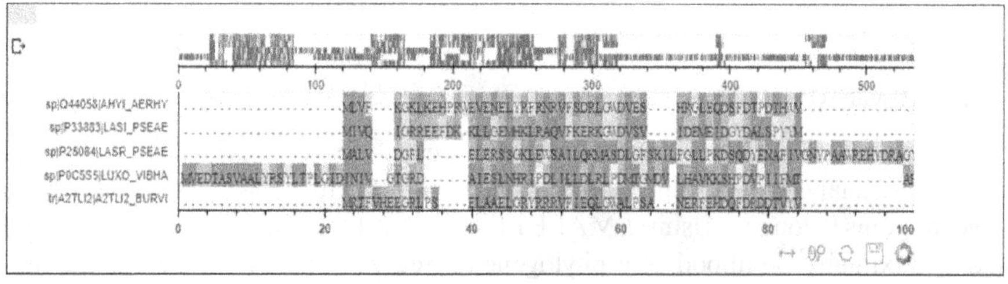

FIGURE. 7.3 MSA of selected quorum sensing proteins from different bacterial taxa.

with all group nodes (Figure 7.4). Proteins were determined to belong to the QS pathway. These results tell us that even though these proteins are all associated with QS, there is no grouping of apparent similar proteins throughout. Our analysis also showed that there were no sequence motifs that would help differentiate a protein from a QS protein. In the future, a larger set of QS proteins could be used to find a

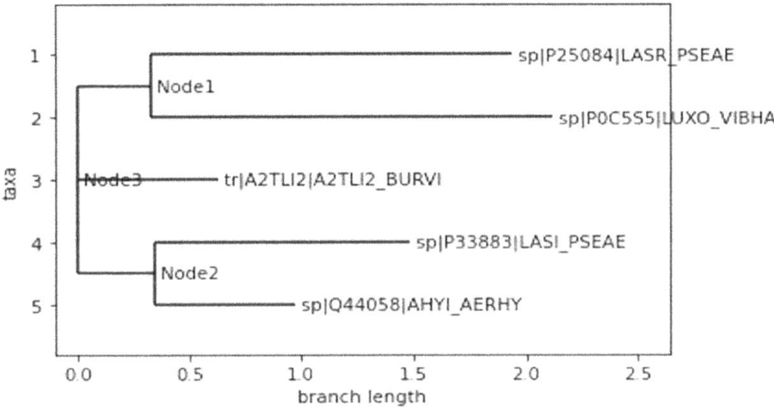

FIGURE 7.4 Maximum likelihood of phylogenetic tree produced from MSA of selected quorum sensing proteins.

consensus sequence as well as any sequence or structural motifs that could be used to determine new QS proteins not found before.

7.7 ADHESION-INDUCED EMERGENT PROPERTIES IN BIOFILM

The properties of EPS depend on microbial origin and growth conditions, for example, the availability of nutrients and hydrodynamics. In addition, the release of EPS has been reported to be controlled by QS mechanisms. Several fractions of EPS can be distinguished such as capsular EPS wrapping the single cell and EPS of the biofilm. It is also important to discuss the differences between EPS properties of planktonic and sessile cells. Due to the complexity of EPS, its analysis depends on the methods used to extract EPS. For instance, certain harsh methods destroy cell walls and introduce cell material into the medium. Some bacterial EPSs exhibit a higher ability to bind metal ions and thus are known to promote corrosion as in the case of EPS extracted from SRB [37]. An important feature of biofilms is the extracellular matrix—a complex mixture of biomolecules termed EPS—which contributes to reduced antimicrobial properties. Nanoparticles (NPs) play a very important role in the form of 'carriers' of EPS matrix disruptors leading to several approaches that have recently been proposed. Little relevance is also given to the application of NPs as an antibiofilm technology with more emphasis on the function of the EPS matrix in the physicochemical regulation of the nanoparticle–biofilm interaction. We highlight the use of NPs as a platform for the new generation of antibiofilm approaches [38–40].

7.8 METHODS TO INHIBIT QUORUM SENSING

As discussed earlier, quorum quenchers can inhibit the QS mechanisms, shunt cell-to-cell communications, and discourage bacterial cells from sharing information about cell density and associated gene expression [15,17]. Below, we present an overview of other known methods for inhibiting QS mechanisms (Table 7.4).

TABLE 7.4
Methods for Inhibiting Quorum Sensing

#	Method	Effect	Comments	Examples
1	QS inhibition	Cut off the QS communication and inhibit biofilm formation [38]	Strategies that include the discovery of QS-inhibiting agents and the current applications of QS-inhibiting agents in several fields to provide insight into the development of effective drugs to control pathogenic bacteria [52]	Levamisole [38]
2	Chemical inhibition	Disrupt the cellular communication and inhibit biofilm formation [37]	Development of certain quorum quenching inhibitors chemically that would inhibit or control the pathogenic activity of bacteria [24]	Savarin [37]
3	Sequestration by antibodies	Specific antibodies target quorum sensing pathway within the bacteria and terminate the effect of cellular communication within the bacteria, e.g., RS2-1G9 generated against a 3-oxo-dodecanoyl homoserine lactone analog to hapten was able to protect murine bone marrow–derived macrophages from the cytotoxic effect [36]	Development of certain antibiotics would inhibit the effect of certain quorum sensing agents which would generate certain cytotoxic effects in bacteria [23]	RS2-1G9 [36]
4	Quorum quenching enzymes	Quorum quenchers (QQ) are often used in the form of enzymes which nullify the effect of quorum sensing molecules within the bacteria. Some enzymes are involved in reduction while others terminate the effect completely. Lactonases and acylases hydrolyze N-acyl homoserine lactone (AHL)-signaling molecules have been investigated most intensively and nullify the effect of quorum sensing molecules [35]	These approaches have been assessed which aim at alleviating virulence, or biofilm formation, by reducing the signal concentration in the bacterial environment [21]	Lactonases and acylases [35]

(*Continued*)

TABLE 7.4 (*Continued*)
Methods for Inhibiting Quorum Sensing

#	Method	Effect	Comments	Examples
5.	Biostimulation	Biostimulation is a phenomenon in which rate-limiting nutrients or electron acceptors are added to the environment to stimulate indigenous bacteria capable of bioremediation. Instead of immobilizing QQ bacteria in any kind of media, biostimulation was used to augment the population of QQ bacteria in the MBR (membrane bioreactor) [53]	Gamma-caprolactone (GCL), which is structurally like AHL, was used to specifically stimulate QQ (AHL-degrading) bacteria. When the GCL consortia were injected into MBR and GCL was continuously dosed, the secretion of EPS decreased, and biofouling was effectively controlled [53]	AHL-lactonase [53]

7.9 CONCLUSION

Quorum sensing (QS) allows bacterial cells to communicate with each other, allowing them to jointly alter phenotypical changes, including biofilm growth, virulence, and MIC. This chapter highlighted the roles of QS at different stages of biofilm growth, including effects on adhesion-induced properties, formation of exopolysaccharides (EPS), and maturation and formation of biofilm. Although we primarily focused on QS effects on SRB, these mechanisms are equally important in other gram-negative bacteria that are implicated in biotechnology applications. Furthermore, we discussed different types of QS inhibition methods that can be used to control biofilm growth in engineering applications. However, such methods may not be viable for field- scale environmental biotechnology applications, especially those that entail the presence of mixed microbial populations and complex environmental conditions. From the microbial corrosion prevention standpoint, it is important to develop protective coatings that can intercept the QS signaling mechanisms in bacterial cells that adhere to the corroding metal surfaces. Moreover, this classification will constitute a baseline dataset to develop a machine learning model for biofilm developmental-stage gene marker prediction.

ACKNOWLEDGMENTS

The authors acknowledge the funding support from the National Science Foundation (award #1736255, #1849206, #1454102), and the Institutional Development Award (IDeA) from the National Institute of General Medical Sciences of the National Institutes of Health P20GM103443.

REFERENCES

[1] Dunny GM, Leonard BA (1997) Cell-cell communication in gram-positive bacteria. *Annu Rev Microbiol*, 51:527–564.

[2] Williamson LL, Borlee BR, Schloss PD, Guan C, Allen HK, Handelsman J (2005) Intracellular screen to identify metagenomic clones that induce or inhibit a quorum-sensing biosensor. *Appl Environ Microbiol* 71:6335–6344. 10.1128/AEM.71.10.6335-6344.2005.

[3] Kawaguchi T, Chen YP, Norman RS, Decho AW (2008) Rapid screening of quorum-sensing signal N-acyl homoserine lactones by an in vitro cell-free assay. *Appl Environ Microbiol* 74:3667–3671. 10.1128/AEM.02869-07.

[4] Federle MJ, Bassler BL (2003) Interspecies communication in bacteria. *J Clin Invest*, 112:1291–1299.

[5] Schauder S, Bassler BL (2001) The languages of bacteria. *Genes Dev* 15:1468–1480

[6] Scarascia G, Wang T, Hong P-Y (2016) Quorum sensing and the use of quorum quenchers as natural biocides to inhibit sulfate-reducing bacteria. *Antibiotics* 5:39. 10.3390/antibiotics5040039.

[7] Heggendorn FL, Fraga AGM, Ferreira DDC, Gonçalves LS, Lione VDOF, Lutterbach MTS (2018) Sulfate-reducing bacteria: biofilm formation and corrosive activity in endodontic files. *Int J Dent* 2018:8303450.

[8] Tsuneda S, Aikawa H, Hayashi H, Yuasa A, Hirata A (2003) Extracellular polymeric substances responsible for bacterial adhesion onto solid surface. *FEMS Microbiol Lett* 223:287–292.

[9] Sivakumar K, Scarascia G, Zaouri N, Wang T, Kaksonen AH, Hong P (2019) Salinity-mediated increment in sulfate reduction, biofilm formation, and quorum sensing: a potential connection between quorum sensing and sulfate reduction? *Front Microbiol* 10:188. 10.3389/fmicb.2019.00188.

[10] Clark ME, He Z, Redding AM, Joachimiak MP, Keasling JD, Zhou JZ, Arkin AP, Mukhopadhyay A, Fields MW (2012) Transcriptomic and proteomic analyses of Desulfovibrio vulgaris biofilms: carbon and energy flow contribute to the distinct biofilm growth state. *BMC Genom* 13:1–17.

[11] Magana M, Sereti C, Ioannidis A, Mitchell CA, Ball AR, Magiorkinis E, Tegos GP (2018) Options and limitations in clinical investigation of bacterial bioflms. *Clin Microbiol Rev* 31:e00084

[12] Zhuang WQ, Tay JH, Maszenan A, Tay S (2002) Bacillus naphthovorans sp. nov. from oil-contaminated tropical marine sediments and its role in naphthalene biodegradation. *Appl Microbiol Biotechnol* 58:547–554.

[13] Padder SM, Prasad R, Shah AH (2018) Quorum sensing: a less known mode of communication among fungi. *Microbiol Res* 210:51–58.

[14] Pagès JM, Amaral L, Fanning S (2011) An original deal for new molecule: reversal of efflux pump activity, a rational strategy to combat gram-negative resistant bacteria. *Curr Med Chem* 18:2969–2980.

[15] Hao Y, Winans SC, Glick BR, Charles TC (2010) Identification and characterization of new LuxR/LuxI-type quorum sensing systems from metagenomic libraries. *Environ Microbiol* 12:105–117.

[16] Norsworthy AN, Visick KL (2013) Gimme shelter: how Vibrio fischeri successfully navigates an animal's multiple environment. *Front Microbiol* 4:356.

[17] Li YH, Tian X (2012) Quorum sensing and bacterial social interactions in biofilms. *Sensors* 12:2519–2538.

[18] Tomlin KL, Malott RJ, Ramage G, Storey DG, Sokol PA, Ceri H (2005) Quorum-sensing mutations affect attachment and stability of Burkholderia cenocepacia biofilms. *Appl Environ Microbiol* 71:5208–5218.

[19] Miller MB, Skorupski K, Lenz DH, Taylor RK, Bassler BL (2002). Parallel quorum sensing systems converge to regulate virulence in Vibrio cholera. *Cell* 110:303–314.
[20] Heimann K, Karthikeyan OP, Muthu SS (2017) *Biodegradation and Bioconversion of Hydrocarbons*. Springer. Environmental Footprints and Eco-design of Products and Processes. DOI: 10.1007/978-981-10-0201-4
[21] Jia R, Yang D, Rahman HBA, Gu T (2018) Investigation of the Impact of an Enhanced Oil Recovery Polymer on Microbial Growth and MIC. In *CORROSION 2018*. NACE International.
[22] Fu W (2013) Investigation of Type II of Microbiologically Influenced Corrosion (MIC) Mechanism and Mitigation of MIC Using Novel Green Biocide Cocktails. Doctoral Dissertation, Ohio University.
[23] Li L, Li J, Zhang Y, Wang N (2019) Diffusible signal factor (DSF)-mediated quorum sensing modulates expression of diverse traits in Xanthomonas citri and responses of citrus plants to promote disease. *BMC Genom* 20:55. https://pubmed.ncbi.nlm.nih.gov/30654743/.
[24] Dow JM (2017) Diffusible signal factor-dependent quorum sensing in pathogenic bacteria and its exploitation for disease control. *J Appl Microbiol* 122:2–11. https://sfamjournals.onlinelibrary.wiley.com/doi/full/10.1111/jam.13307
[25] Viveiros M, Martins M, Rodrigues L, Machado D, Couto I, Ainsa J, Amaral L (2012) Inhibitors of mycobacterial efflux pumps as potential boosters for anti-tubercular drugs. *Expert Rev Anti-Infect Therap* 10:983–998.
[26] George SE, Hrubesch J, Breuing I, Vetter N, Korn N, Hennemann K, Bleul L, Willmann M, Ebner P, Götz F, Wolz C (2019) Oxidative stress drives the selection of quorum sensing mutants in the Staphylococcus aureus population. *Proc Nat Acad Sci* 116:19145–19154.
[27] Joelsson A, Kan B, Zhu J (2007) Quorum sensing enhances the stress response in Vibrio cholera. *Appl Environ Micobiol* 73:3742–3746.
[28] Kumar A, Rahal A, Sohal JS, Gupta VK (2021) Bacterial stress response: understanding the molecular mechanics to identify possible therapeutic targets. *Expert Rev Anti-infect Therapy* 19:121–127.
[29] Miller MB, Bassler BL (2001) Quorum sensing in bacteria. *Annu Rev Microbiol* 55:165–199.
[30] Battin TJ, Besemer K, Bengtsson MM, Romani AM, Packmann AI (2016) The ecology and biogeochemistry of stream biofilms. *Nat Rev Microbiol* 14:251–263.
[31] Szabó MÁ, Varga GZ, Hohmann J, Schelz Z, Szegedi E, Amaral L, Molnár J (2010) Inhibition of quorum-sensing signals by essential oils. *Phytotherap Res* 24:782–786.
[32] Dunsmore BC, Jacobsen A, Hall-Stoodley L, Bass CJ, Lappin-Scott HM, Stoodley P (2002) The influence of fluid shear on the structure and material properties of sulphate-reducing bacterial biofilms. *J Ind Microbiol Biotechnol* 29:347–353. https://academic.oup.com/jimb/article/29/6/347/5989314?login=true
[33] Passos da Silva D, Schofield MC, Parsek MR, Tseng BS (2017) An update on the socio-microbiology of quorum sensing in gram-negative biofilm development. *Pathogens* 6:51.
[34] Donlan RM (2002) Biofilms: microbial life on surfaces. *Emerg Infect Dis* 8:881.
[35] Stamatakis A (2014) RAxML version 8: a tool for phylogenetic analysis and post-analysis of large phylogenies. *Bioinformatics* 30(9):1312–1313.
[36] Le SQ, Gascuel O (2008) An improved general amino acid replacement matrix. *Mol Biol Evol* 25(7):1307–1320.
[37] Solano C, Echeverz M, Lasa I (2014) Biofilm dispersion and quorum sensing. *Curr Opin Microbiol* 18:96–104.
[38] Fetzner S (2015) Quorum quenching enzymes. *J Biotechnol* 201:2–14.

[39] Kaufmann GF, Park J, Mee JM, Ulevitch RJ, Janda KD (2008) The quorum quenching antibody RS2-1G9 protects macrophages from the cytotoxic effects of the Pseudomonas aeruginosa quorum sensing signalling molecule N-3-oxo-dodecanoyl-homoserine lactone. *Mol Immunol* 45:2710–2714.

[40] Sully EK, Malachowa N, Elmore BO, Alexander SM, Femling JK, Gray BM, DeLeo FR, Otto M, Cheung AL, Edwards BS, Sklar LA (2014) Selective chemical inhibition of agr quorum sensing in Staphylococcus aureus promotes host defense with minimal impact on resistance. *Plos Pathog* 10:e1004174.

[41] Zhou L, Zhang Y, Ge Y, Zhu X, Pan J (2020) Regulatory mechanisms and promising applications of quorum sensing-inhibiting agents in control of bacterial biofilm formation. *Front Microbiol* 11:589640.

[42] Unosson E (2015) Antibacterial Strategies for Titanium Biomaterials. PhD Dissertation, Acta Universitatis Upsaliensis.

[43] Lynch MJ, Swift S, Kirke DF, Keevil CW, Dodd CE, Williams P (2002) The regulation of biofilm development by quorum sensing in Aeromonas hydrophila. *Environ Microbiol* 4:18–28.

[44] Mangwani N, Kumari S, Das S (2015) Involvement of quorum sensing genes in biofilm development and degradation of polycyclic aromatic hydrocarbons by a marine bacterium Pseudomonas aeruginosa N6P6. *Appl Micorbiol Biotechnol* 99:10283–10297.

[45] Tripathi AK, Thakur P, Saxena P, Rauniyar S, Gopalakrishnan V, Singh RN, Gadhamshetty V, Gnimpieba EZ, Jasthi BK, Sani RK (2021) Gene sets and mechanism of sulafte reducing bacteria biofilm formation and quorum sensing with imapct on corrosion. *Front Micorbiol* 12:3120.

[46] Chilkoor G, Shrestha N, Karanam SP, Upadhyayula VK, Gadhamshetty V (2019) Graphene coatings for microbial corrosion applications. *Encycl Water: Science, Technology, and Society*, pp.1–25. doi.org/10.1002/9781119300762.wsts0125

[47] Puigdomenech I, Taxén C (2000) *Thermodynamic Data for Copper. Implications for the Corrosion of Copper under Repository Conditions.* Swedish Nuclear Fuel and Waste Management Co: Stockholm.

[49] Katoh K, Misawa K, Kuma KI, Miyata T (2002) MAFFT: a novel method for rapid multiple sequence alignment based on fast Fourier transform. *Nucl Acids Res* 30:3059–3066.

[50] Oh HS, Lee CH (2018) Origin and evolution of quorum quenching technology for biofouling control in MBRs for wastewater treatment. *J Membr Sci* 554:331–345.https://www.sciencedirect.com/science/article/pii/S0376738817335135.

[51] Markus V, Golberg K, Teralı K, Ozer N, Kramarsky-Winter E, Marks RS, Kushmaro A (2021) Assessing the molecular targets and mode of action of furanone C-30 on pseudomonas aeruginosa quorum sensing. *Molecules* 26:1620. https://www.academia.edu/es/61086761/

[52] Cai Z, Yuan ZH, Zhang H, Pan Y, Wu Y, Tian XQ, Wang FF, Wang L, Qian W (2017) Fatty acid DSF binds and allosterically activates histidine kinase RpfC of phytopathogenic bacterium Xanthomonas campestris pv. campestris to regulate quorum-sensing and virulence. *J Plos Pathog* 13:e1006304. https://www.ncbi.nlm.nih.gov/pmc/articles/PMC5391125/

[53] Wu H, Song Z, Hentzer M, Andersen JB, Molin S, Givskov M, Høiby N (2004) Synthetic furanones inhibit quorum-sensing and enhance bacterial clearance in Pseudomonas aeruginosa lung infection in mice. *J Antimicrob Chemother* 53:1054–1061. https://www.researchgate.net/publication/8587214

8 Data-Driven 2D Material Discovery Using Biofilm Data and Information Discovery System (Biofilm-DIDS)

Tuyen Do, Alain Bomgni, Shiva Aryal, Venkataramana Gadhamshetty, Diing D. M. Agany, Tim Hartman, Bichar D. Shrestha Gurung, Carol M. Lushbough, and Etienne Z. Gnimpieba

8.1 INTRODUCTION

The study of microbe–material systems (biointerfaces) is of great interest for various applications such as infrastructure (e.g., corrosion study), biomedical science (cell implant study), and environmental health.[1] The complex biointerface system involves both the material system (non-living) and biosystem (systems biology).[2]

8.1.1 Microbial Community, Biofilm, and Material–Biofilm Interaction

Biofilms grow on practically every surface exposed to aqueous environments including but not limited to metals, polymers, living tissues, and medical implants.[3] They are widely researched in agricultural, industrial, and life science domains. Biofilms can be incredibly beneficial or exceedingly harmful. For example, detached cells from pathogenic biofilms are known to transmit pathogens in food production facilities, water pipelines, and medical devices.[4–6] The United States alone spends about $90 billion/year to deal with the associated infection challenges.[4,7,8]

Sulfate-reducing bacteria (SRB), a special class of microorganisms, are adept in colonizing and growing on metal surfaces. Furthermore, they play a pivotal role in accelerating the corrosion of these surfaces and use the oxidizing power to meet their metabolic needs. This special class of corrosion, known as microbiologically influenced corrosion, is responsible for the expenditure of about $4 billion/year in the United States. Many other biofilms have been reported to thrive in the most well-known harsh conditions including hot environments in deep biospheres (e.g., abandoned gold mines) as well as the hot springs of Yellowstone National Park. For these

DOI: 10.1201/9781003132981-8

vexing problems to be solved, there is a need to develop focused transdisciplinary collaborations that cross typical disciplinary and organizational boundaries.

8.1.2 Complex System Design: SDLC and Agile Methodology Meets Big Data

Complex system design requires rigorous methodology and assessment tools to guide engineers and scientists toward a viable solution. The complexity of microbe–material systems cannot be handled with one domain's methodology. The integration of labs' experimental design roadmaps, data science, knowledge discovery processes, and system design methodologies will allow for the development of a transdisciplinary convergence solution. In that context, the Agile methodology offers an incremental approach based on use cases and user stories to connect a scientific hypothesis to a computing solution using the system design life cycle.[9–11] By achieving this integration and adding the data mining process to the loop, we will provide the scientist with a roadmap based on previous knowledge to inform new knowledge discovery.[12–14]

8.1.3 Big Data Mining and Knowledge Discovery

Current advancements in data acquisition technologies both in material and biological science have led to the accumulation of a large number of dataset scatters across various sources.[12,15] This big data accumulation is facing diverse issues before it can be leveraged by researchers. Among these issues is a lack of standard and proper annotation. Big data mining is the process of identifying and facilitating the retrieval of data that is so large that traditional methods of analysis are unable to handle it.[13,16,17] In contrast, knowledge discovery is a process of gaining new information from analyzing this extensive data. One of the most relevant methods in data mining that can bring the dataset closer to the scientific problem is text mining (TM) and natural language processing (NLP). And recently, generalization of these large language models (LLM) is leading a new generation of data modeling as revealed by OpenAI in early 2023. The NLP method tries to learn from human language to bridge the gap between the user question and the dataset entry in data sources. However, most material and biological databases do not have that technology implemented at the time of this study. Here, we present how to use the Biofilm Data and Information Discovery System (Biofilm-DIDS) to answer biointerface questions using NLP and the first generation of LLMs such as BioBERT (Abstract Figure 8.1).[18–21] This chapter presents the biointerface system design (Section 8.2), the data mining and knowledge discovery of biointerface (Section 8.3), an overview of Biofilm-DIDS (Section 8.4), and the uses of Biofilm-DIDS for biointerface question resolution.

8.2 Interface between the Living and the Non-Living: A System Thinking Approach

8.2.1 System Understanding of Biointerface

A biointerface could be defined in different contexts based on which system is being studied.[1] However, the end goal of a system-level study is to understand the target

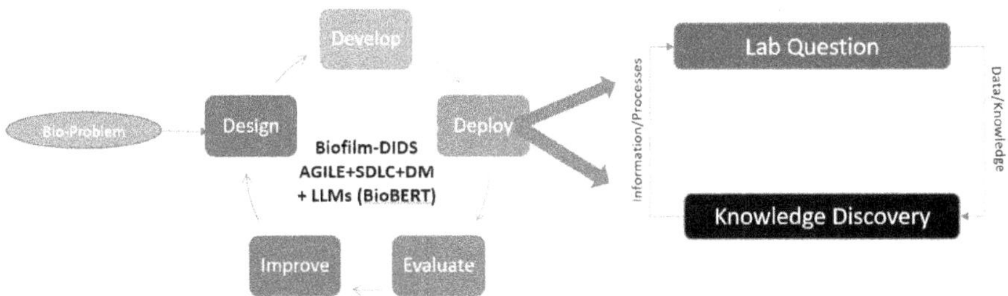

FIGURE 8.1 Leveraging Agile, SDLC, big data mining, and knowledge discovery to assist scientists in addressing complex biointerface lab questions.

biointerface as a whole. In this context, we define a biointerface as actions within a community of microbes making contact and interacting on a molecular level with a material such as biological tissue, cell membrane, living organism, or other materials.[22] Instead of looking at the individual effects of the microbe on the material and vice versa, we look at the complex contribution of all interactions between microbial communities and the materials at-large to understand their effects and gain a holistic picture of the system including subsystem interactions within the main system.[2,9]

8.2.2 Big Data in Biointerfaces

Materials in their natural forms show fascinating properties as they are either formed by or interact with living cells, which sense and process environmental cues and conditions through signaling and genetic programs to control the biosynthesis, remodeling, functionalization, or degradation of the natural material.[23] In an era of big data, material production could benefit from modeling material properties from system-level data. Big data can be obtained from the biointerface by mining existing big data and leveraging knowledge discovery to engineer a living system that mimics the natural process explained above.[24–27] In Figure 8.2, data mining methodologies and techniques of knowledge discovery are described as well as machine learning processes in which biointerface data—both at material and biological system levels—are retrieved from a variety of databases in different modalities and many omics layers (genomics, transcriptomics, etc.). This is then integrated and selected for preprocessing and is subject to feature engineering. Next, machine learning tasks

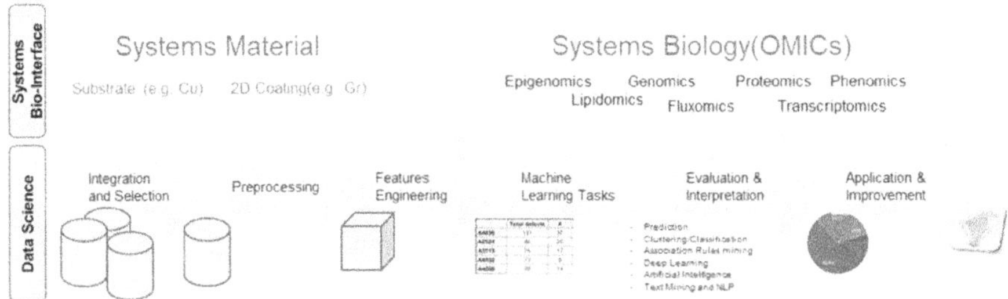

FIGURE 8.2 System biointerface overview.

are performed on the features by training models to be used in the subsequential prediction of new materials and properties for the production of biosensors, biocompatible devices, drug delivery systems, building materials, corrosion-resistant surfaces, and bioremediation.[28] The data mining and knowledge discovery process depicted in Figure 8.2 to model material properties could be categorized into three major categories depending on the tasks:

1. *Supervised learning*, which is further divided into tasks of classification or regression, both of these labeled examples are used to train the models or algorithms. Examples include K-nearest neighbor, multiple linear regression, logistic regression, support vector machine, random forest, artificial neural network, decision tree, and Bayesian network.
2. *Unsupervised learning*, in which the algorithm learns directly from data by discovering the patterns from datasets and grouping them based on specific rules or associations. The example of unsupervised learning includes tasks such as clustering to which principal component analysis, independent component analysis, and K-means algorithms are applied to build models.
3. *Reinforcement learning*, in which the agents learn from their environment through rewards.[29–33]

8.3 BIOFILM-DIDS OVERVIEW

The Biofilm-DIDS (https://biofilmdids.bicbioeng.org/) architecture is comprised of modules that mine, map, annotate, and index biofilm and material metadata to enable data discovery through a free text searcher. These modules and submodules include:

1. REX, a resource extraction module that gathers and mines data with metadata for request data sources,
2. REMAP, a resource mapper module that connects data-mined publication data and other metadata definitions,
3. RONER, a resource annotation module that leverages domain ontologies,
4. BioBERT, a pre-trained biomedical language representation model for biomedical text mining,
5. Generative pre-trained transformer models by OpenAI,
6. REIS, a resource-indexed system, and
7. RAPI, a resource application program interface (API) providing programmatic access to the Biofilm-DIDS database (Figure. 8.3).

The biofilm-data fusion module retrieves, curates, annotates, and indexes metadata from public data sources. The indexes integrate with the experimental datasets. Working with biointerface scientists, we identified numerous data sources (Table 8.2) to develop the datasets of 2D materials, transcriptomics, proteomics, metabolomics, methylomes, and phenotypic information for our use case collection. Table 8.1 presents a snapshot of our reference use cases on sulfate-reducing DA-G20 biofilm. The fusion module locally stores metadata describing the requisite datasets and uses them to build searchable indices that can be accessed by the other three modules via the

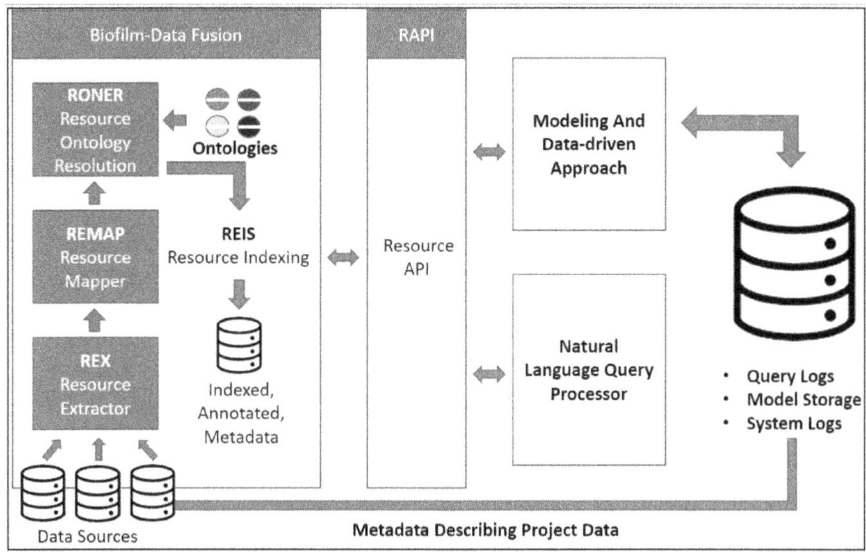

FIGURE 8.3 Core modules of Biofilm-DIDS.

application program interface resource interfaces (Figure 8.3). NLP allows users to enter their queries using free text. For example, they could enter their query in the form of "As a…I would like to…so that…" query structure. For example, [as a] biofilm researcher developing a new class of 2D materials, [I would like to] identify known genes and predict unknown gene sets in DA-G20 that represent copper stress resistance induced by the defective 2D coatings on Gr/Cu-aggravating biocorrosion, [so that] I can design an experiment to evaluate material properties that trigger genes responsible for stress response and biocorrosion, all with an accurate and reliable gene list (reproducible research). The queries are parsed using the NLP module and annotated using the integrated ontologies in order to provide the most relevant results. The modeling and data-driven approach module use information extracted from the biofilm–data fusion module to retrieve the requisite datasets it will use as input into its processes. The query itself, the query result, and the predictive models are stored in a system log. The performance (query throughout and accuracy/relevance of query results) of Biofilm-DIDS is assessed using user curation and system logs.

Biofilm-DIDS stores reference collections and other data needed to validate the biofilm hypotheses generated as a query result and returns the biofilm phenotypes as a function of 2D material properties. Biofilm-DIDS will use partially available datasets (e.g., defect density of Gr coating), biofilm genomics (GSE83516), and images (DA-G20 filaments on Cu/Gr) from literature to guide experimental design aspects of the 2D material synthesis (Area 2) and phenotype tests (Area 3). Table 8.1 outlines an overview of Biofilm-DIDS subtasks (materials, biofilm, and material biofilm interaction categories) to investigate genome and gene regulatory networks that trigger copper stress resistance and biocorrosion in DA-G20 biofilms, in response to the surface properties of Cu/Gr and Cu/hBN.

TABLE 8.1

Sub-Goal/Task (Repositories) for DA-G20 Biofilm Reference Case

Materials	**Identify Materials and the Surface Properties that Impact DA-G20 Biofilm Phenotypes**
UM1	Develop a list of materials (Cu, Cu/Gr, and Cu/h-BN) and relevant surface properties
UM2	Narrow down surface properties: crystallographic orientation, defect concentration, hydrophilicity, charge, accessible area, barrier properties, electrical conductivity
UM3	Develop a complete set of preexisting datasets of material properties for simulations
UM4	Synthesize new 2D material properties with well-characterized nanostructure and predict biofilm phenotypes (biocorrosion). This step fills knowledge gaps in the literature
UM5	Create a test dataset to model and predict the biological mechanism (e.g., peptide interaction) in response to a given material property
UM6	Create a dataset to assess other biofilm phenotypes (biocompatibility and bacteria attachment) in response to crystallographic orientation, defect density, hydrophilicity, and charge of 2D material on a copper surface
UM7	Generate a new dataset to fill knowledge gaps (e.g., nanostructure characterization with and without biofilms)
Biofilms	**Collect Biofilm Properties/Configuration Based on Existing Knowledge for Prediction**
UB1	Estimate each collection for coverage and completeness
UB2	Create a test dataset to build the machine learning model for material property prediction from copper toxicity
UB3	Create a gene collection involved in the biofilm stress response and enrich it with OMICS data to create a protein collection involved in the biofilm stress response and then enrich it with OMICS data
UB4	Create a test dataset to build a model for the gene of interest and phenotype of interest prediction from the dataset of known genes
UB5	Extend the gene list using gene regulatory network analysis and protein network analysis
UB6	Unknown genes and proteins
UB7	Create the dataset to identify conserved patterns regulating the stress response using pattern detection
Biofilm-Material	**Create Biofilm Phenotype Response Dataset on a Given Material (Graphene, hBN). Correlate Material Properties with the Biological Information (e.g., Gene, Protein, Metabolite, Compound)**
UBM1	Create dataset collection to profile material data for a given biofilm's genomic landscape
UBM2	Create an integrated dataset to predict biofilm genomics profile based on the material properties

An effective approach is required to collect datasets and meta-datasets for materials of interest (e.g., Gr and hBN) and their surface properties from disparate sources. We identified about 50 repositories of interest including six literature repositories for TM, 12 material property databases, 15 biofilm databases, and 14 related repositories. Some of these sources include NCBI, Pubmed, PMC, IHS Markit materials database, Materials Project for computed information on known

TABLE 8.2
Sub-Goal/Task (Repositories) for DA-G20 Biofilm Reference Case

	NCBI	Pubmed	PMC	Biofilm-DIDS	HIS Markit	Materials Project	Polymerizer	DANA	MatMatch	WoM	SciCrunch	BioNumbers	aBioFilm	BaAMPs	BiofOmics	Dryad	BacDive DB	GenBank
Materials	X	X	X	X														
UM1	X	X	X		X	X		X	X									
UM2	X	X	X		X	X		X	X									
UM3	X	X	X			X	X		X					X				
UM4	X	X	X			X	X	X										
UM5	X	X	X		X		X						X	X				
UM6	X	X	X							X		X	X	X				
UM7	X	X	X															
Biofilm	X	X	X	X				X		X	X	X	X					
UB1	X	X	X					X		X		X	X	X			X	
UB2	X	X	X					X		X		X	X	X			X	
UB3	X	X	X					X		X		X	X	X				
UB4	X	X	X					X		X		X	X					X
UB5	X	X	X	X						X		X	X	X				X
UB6	X	X	X							X			X	X				
UB7	X	X	X										X					
Biofilm-Material	X	X	X	X									X	X				

and predicted materials, Polymerizer, DANA information, Bionumbers, aBiofilm, and BaAmps (Table 8.2).

Table 8.2 maps each of the sub-questions (shown in Table 8.1) to the relevant data sources that are used to achieve it. For example, consider the sub-question UM6: "Create dataset to assess other biofilm phenotypes in response to crystallographic orientation, defect density, hydrophilicity, and charge of 2D material on a copper surface." Biofilm-DIDS yields a matrix that correlates 2D material property publication and dataset with the biocompatibility of DA-G20. We identified seven repositories that provide chemical, structural, and biological information of 2D materials, quorum quenchers, anti-biofilm agents, and the link to PubChem or Chemspider (e.g., C143H230N42O37S7). They also provided the organism involved (e.g., *Pseudomonas aeruginosa*, strain ATCC 9027) with the link to NCBI taxonomy, the biofilm development stage (e.g., biofilm formation), the biological event, and reference linked to a PubMed paper. Currently, the data collected from these repositories only provides partial information to complete sub-question UM4, but it contains relevant information on materials of interest. This information is fused with the text dataset obtained from the PubMed literature to complete the matrix. A simple search on PubMed with "*Pseudomonas aeruginosa* biofilm nanotube" returns 14 results. Pantenale et al. provide a relevant dataset to update our test set matrix with multimodal imaging and adhesion datasets.

8.4 USING BIOFILM-DIDS TO EXTRACT BIOCORROSION GENE OF INTEREST FROM THE LITERATURE AND MATERIAL DIMENSION PREDICTION

Biofilm-DIDS has been used to solve over 20 biointerface problems in collaboration with lab scientists, with five publications on diverse use cases including TM for biocorrosion gene marker identification, biofilm formation studies, gene name entity resolution for SRB organism collection, essential gene prediction, and deep learning strategies for addressing issues with small datasets in 2D material research in microbial corrosion.[34–38] Here, we present how one can use Biofilm-DIDS to extract biocorrosion datasets for downstream knowledge discovery more accurately than current repositories. Working with expert scientists, we resolved as follows the use case "Develop the list of materials (Cu, Cu/GR, and Cu/h-BN) and relevant surface properties."

8.4.1 Expert Informed Relevant Dataset Extraction from User Free Text Question

The development of Biofilm-DIDS started with six research problems and expanded to over 50 sub-problems relevant to hypothesis-driven experimental validation. These problems include one problem in biofilm engineering (Dr. Sani's Lab), in material engineering (Dr. Jasthi's Lab), and in biointerface engineering (Dr. Gadhamshetty's Lab), all experts from the South Dakota School of Mines and Technology. The implementation of these use cases helped us test and use toolkits such as TM modules. At this stage, Biofilm-DIDS used over 15 data extraction modules or packages to retrieve datasets from published data sources and five annotation tools to allow expert user curation of our dataset. These datasets integrated into our repository are currently undergoing the continuous curation process for quality improvement. We also integrated custom datasets from our collaborator's lab projects (e.g., SEM SRB biofilm dataset analyzer with our tools). We are using different architectures to make these tools discoverable including an API that will make our tool Software as a Service (SaaS) accessible at the HTTP level for any secured application.

To demonstrate some of the functionality of Biofilm-DIDS, consider the simple query: "Develop the list of material (Cu, Cu/G.R. and Cu/h-BN and relevant surface properties." If this search is executed directly in PubMed, no results are returned as of today (Figure 8.4).

A successful search in Biofilm-DIDS returns a results page (Figure 8.5) containing a summary of resources that meet the query requirements. Resources include datasets, tools, and analysis. The "View Details/Download Document" functionality is offered so that the user can obtain additional information regarding a specific item in the result set (Figure 8.6). The downloaded document lists PubMed IDs that are associated with the query, for example, the ID 33784559 entered in the PubMed search returns the article (Figure 8.7).

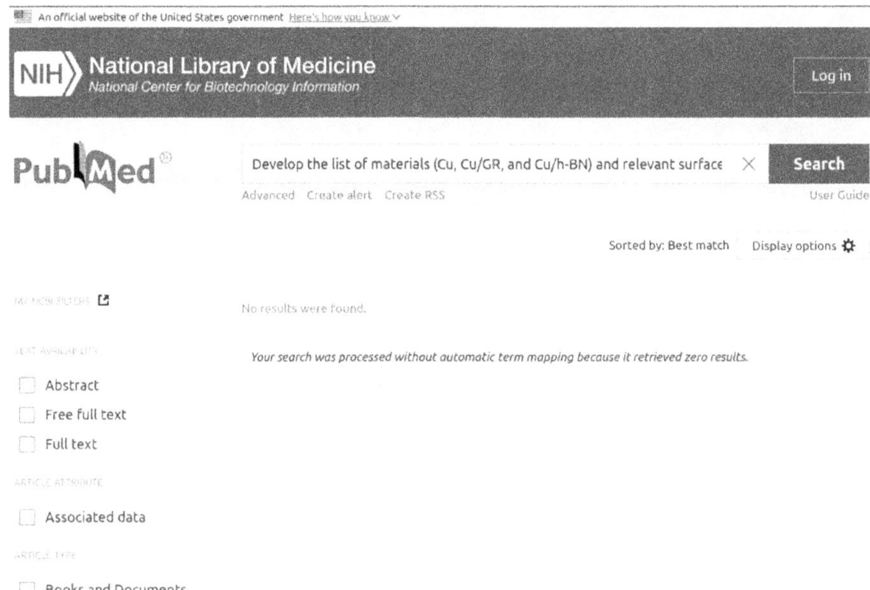

FIGURE 8.4 Test free text query use case on PubMed.

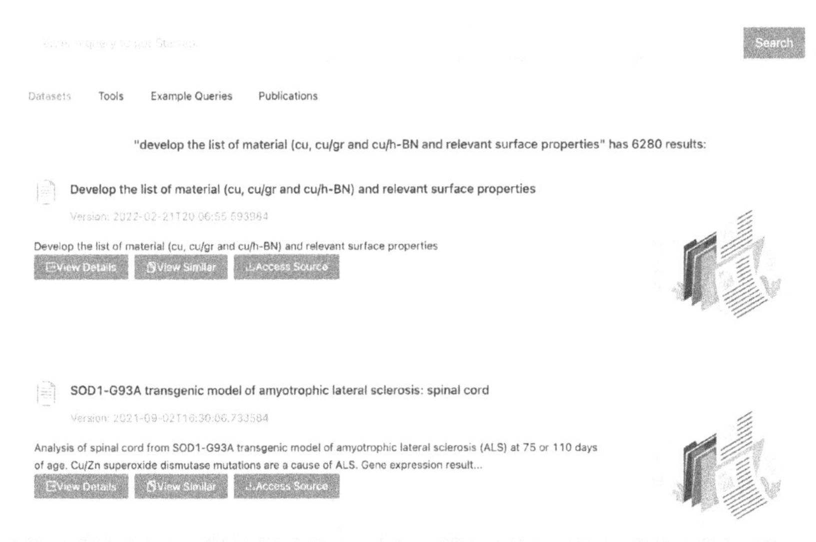

FIGURE 8.5 Biofilm-DIDS search results based on query: list of material (cu, cu/gr, and cu/h-BN and relevant surface properties.

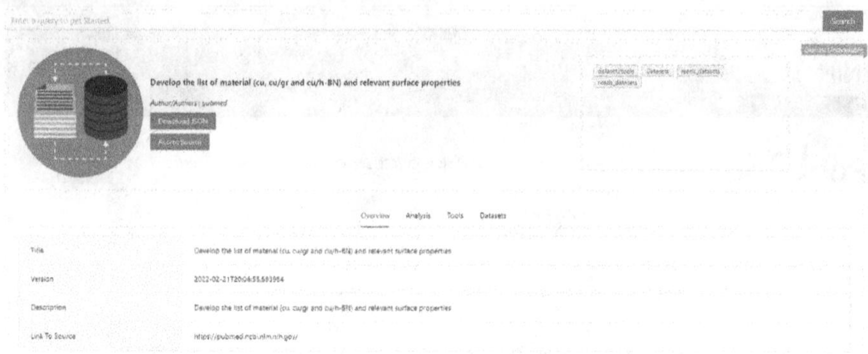

FIGURE 8.6 Biofilm-DIDS details for item: Develop the list of material (cu, cu/gr, and cu/h-BN) and relevant surface properties.

FIGURE 8.7 The list of PubMed IDs. (a) Retrieved from Biofilm-DIDS query allows users to obtain details information. (b) Related to a list of materials and relevant surface properties.

8.4.2 Downstream Analysis for Material Dimension Prediction

Instead of retrieving the publication relevant to their query, the researcher may want to find datasets or tools to predict material dimensions from our tool collections.[39] This question is built on the ability of Biofilm-DIDS to extract knowledge from over eight repositories of materials. With this dataset, a user can discover a new computational material and predict its dimension, and if this is a 2D dimension, they can validate it in the lab and publish it as a new 2D material discovered from computational methods. Using the previous query, if you select the publication tab, you have relevant papers as described in the previous section. If you select the dataset tab with a query involving "material dimension," you will have the dataset on material dimension and a tool for the prediction of the material dimension (Figures 8.8–8.11).

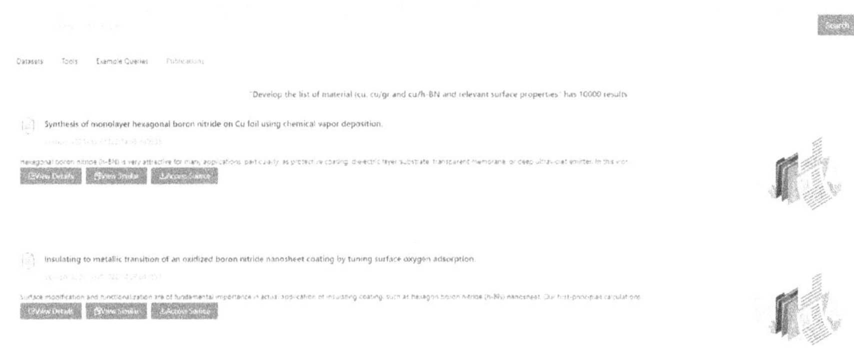

FIGURE 8.8 Publication result using free text query.

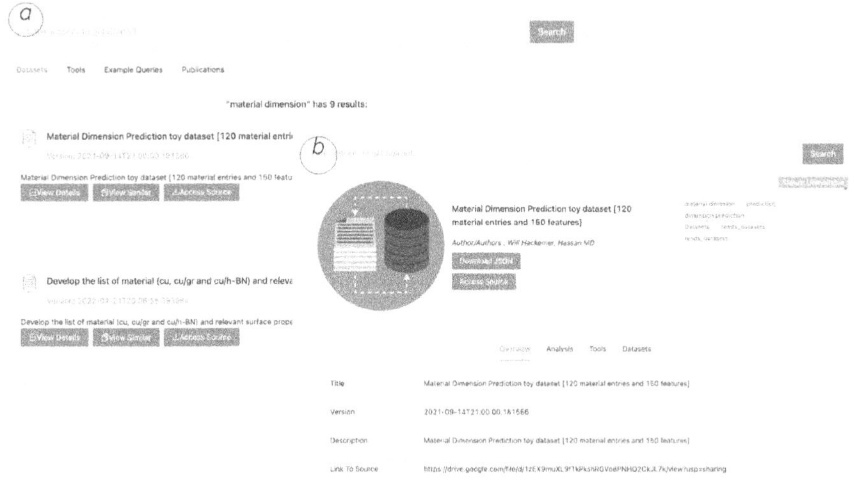

FIGURE 8.9 Dataset results (a) and detail of the first result (b) from material dimension free text query.

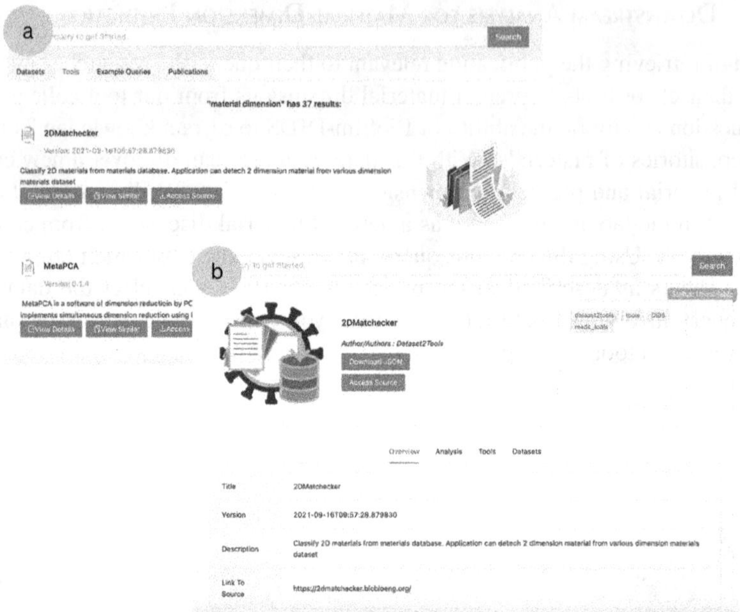

FIGURE 8.10 Analytic tool results (a) and detail of the first result (b) relevant to the dataset retrieved for material dimension analysis.

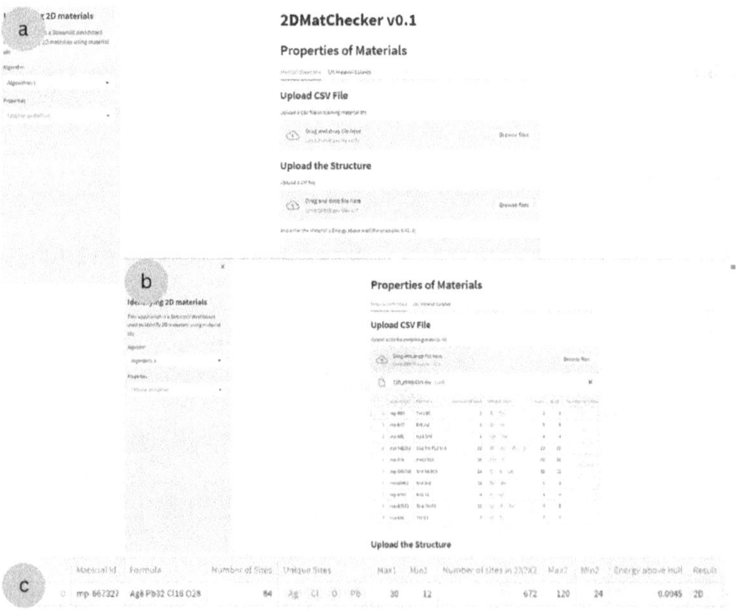

FIGURE 8.11 Downstream analysis to predict the material dimension from single (c) or multiple (b) material ID using 2DMatChecker (a) predictor (https://2dmatchecker.bicbioeng.org/). Used with permission from BicBioEng Lab.

8.5 CONCLUSIONS

Discovering new material in the age of big data science is a big challenge added to an already complex system such as a biointerface. Biointerface science aims to connect the non-living (material) and the living things (microbe here or biofilm). We proposed in this chapter how we can use Biofilm-DIDS as a one-stop shop data integrator to increase access to relevant multimodal datasets. Following the dataset discovery, Biofilm-DIDS allows the scientist to perform a downstream analysis when applicable using relevant analytic tools such as material dimension prediction using 2DMatChecker. The current version of Biofilm-DIDS contains datasets of around 50 expert informed use cases from over 40 data sources and their metadata. The future development of Biofilm-DIDS will rely on the current advances in AI power query analysis and community engagement to enrich the knowledge base with more use cases.

ACKNOWLEDGMENTS

We acknowledge the funding support from the National Science Foundation (NSF) RII T-1 FEC award (#1849206), NSF RII T-2 FEC award (# 1920954), and the Institutional Development Award (IDeA) from the National Institute of General Medical Sciences of the National Institutes of Health P20GM103443.

REFERENCES

[1] Walsh, T. R.; Knecht, M. R. Biointerface Structural Effects on the Properties and Applications of Bioinspired Peptide-Based Nanomaterials. *Chem Rev*, **2017**, *117* (20), 12641–12704. https://doi.org/10.1021/ACS.CHEMREV.7B00139/ASSET/IMAGES/MEDIUM/CR-2017-00139P_0046.GIF.

[2] Aderem, A. Systems Biology: Its Practice and Challenges. *Cell*, **2005**, *121* (4), 511–513. https://doi.org/10.1016/j.cell.2005.04.020.

[3] Donlan, R. M. Biofilms: Microbial Life on Surfaces. *Emerg Infect Dis*, **2002**, *8* (9), 881–890. https://doi.org/10.3201/eid0809.020063.

[4] Koo, H.; Allan, R. N.; Howlin, R. P.; Stoodley, P.; Hall-Stoodley, L. Targeting Microbial Biofilms: Current and Prospective Therapeutic Strategies. *Nat Rev Microbiol*, **2017**, *15* (12), 740–755. https://doi.org/10.1038/nrmicro.2017.99.

[5] Stewart, E. J.; Ganesan, M.; Younger, J. G.; Solomon, M. J. Artificial Biofilms Establish the Role of Matrix Interactions in Staphylococcal Biofilm Assembly and Disassembly. *Sci Rep*, **2015**, *5* (February), 1–14. https://doi.org/10.1038/srep13081.

[6] Beyenal, H.; Sani, R. K.; Peyton, B. M.; Dohnalkova, A. C.; Amonette, J. E.; Lewandowski, Z. Uranium Immobilization by Sulfate-Reducing Biofilms. *Environ Sci Technol*, *38* (7), 2067–2074. https://doi.org/10.1021/es0348703.

[7] Davey, M. E.; O'Toole G, A. Microbial Biofilms: From Ecology to Molecular Genetics. *Microbiol Mol Biol Rev*, **2000**, *64* (4), 847–867.

[8] Römling, U.; Kjelleberg, S.; Normark, S.; Nyman, L.; Uhlin, B. E.; Åkerlund, B. Microbial Biofilm Formation: A Need to Act. *J Intern Med*, **2014**, *276* (2), 98–110. https://doi.org/10.1111/joim.12242.

[9] Barrier, T. Systems Analysis. In *Encyclopedia of Information Systems*, Bidgoli, H. (Ed.), Elsevier; **2003**, 345–349. https://doi.org/10.1016/B0-12-227240-4/00177-5.

[10] Dubinsky, Y.; Hazzan, O.; Talby, D.; Keren, A. System Analysis and Design in a Large-Scale Software Project: The Case of Transition to Agile Development. In *International Conference on Enterprise Information Systems*; **2006**. https://www.scitepress.org/Link.aspx?; doi:10.5220/0002451900110018

[11] Dingsøyr, T.; Nerur, S.; Balijepally, V.; Moe, N. B. A Decade of Agile Methodologies: Towards Explaining Agile Software Development. *J Syst Softw*, **2012**, *85* (6), 1213–1221. https://doi.org/10.1016/J.JSS.2012.02.033.

[12] Lu, W.; Xiao, R.; Yang, J.; Li, H.; Zhang, W. Data Mining-Aided Materials Discovery and Optimization. *J Mater*, **2017**, *3* (3), 191–201. https://doi.org/10.1016/j.jmat.2017.08.003.

[13] Che, D.; Safran, M.; Peng, Z. From Big Data to Big Data Mining: Challenges, Issues, and Opportunities. In *International Conference on Database Systems for Advanced Applications*, Database Systems for Advanced Applications. DASFAA 2013. Lecture Notes in Computer Science, Hong, B.; Meng, X.; Chen, L.; Winiwarter, W.; Song, W. (Eds.), Springer, Berlin, Heidelberg; **2013**, *7827*, 1–15. https://doi.org/10.1007/978-3-642-40270-8_1

[14] Wang, S.; Cao, J.; Yu, P. S. Deep Learning for Spatio-Temporal Data Mining: A Survey. *IEEE Trans Knowl Data Eng*, **2022**, *34* (8), 3681–3700. https://doi.org/10.1109/TKDE.2020.3025580.

[15] Subramaniam, M.; Chundi, P.; Goedert, J. D. Data-Driven Adaptive Learning Environment for Project-Based Construction Engineering. *Technol Educ Learn*, **2015**, *3* (3). https://doi.org/10.2316/JOURNAL.209.2015.3.209-0036.

[16] Wigan, M. R.; Clarke, R. Big Data's Big Unintended Consequences. *Comput (Long Beach Calif)*, **2013**, *46* (6), 46–53.

[17] Gudipati, M.; Rao, S.; Mohan, N. D.; Gajja, N. K. Big Data: Testing Approach to Overcome Quality Challenges. *Big Data: Challeng Opport*, **2013**, *11* (1), 65–72.

[18] Lee, J.; Yoon, W.; Kim, S.; Kim, D.; Kim, S.; So, C. H.; Kang, J. BioBERT: A Pre-Trained Biomedical Language Representation Model for Biomedical Text Mining. *Bioinformatics*, **2019**, *36* (4), 1234–1240. https://doi.org/10.1093/bioinformatics/btz682.

[19] Otter, D. W.; Medina, J. R.; Kalita, J. K. A Survey of the Usages of Deep Learning for Natural Language Processing. *IEEE Trans Neural Netw Learn Syst*, **2019**, *32*(2), 604–624. doi: 10.1109/TNNLS.2020.2979670

[20] Peng, B.; Galley, M.; He, P.; Cheng, H.; Xie, Y.; Hu, Y.; Huang, Q.; Liden, L.; Yu, Z.; Chen, W.; et al. Check Your Facts and Try Again: Improving Large Language Models with External Knowledge and Automated Feedback. **2023**. ArXiv, abs/2302.12813

[21] Kassa, G.; Liu, J.; Hartman, T. W.; Dhiman, S.; Gadhamshetty, V.; Gnimpieba, E. Artificial Intelligence Based Organic Synthesis Planning for Material and Bio-Interface Discovery. In *Microbial Stress Response: Mechanisms and Data Science*, ACS Symposium Series; **2023**, *1434*, 93–111. https://doi.org/10.1021/BK-2023-1434.CH006.

[22] Zanfack, D. R. G.; Bellaïche, A.; Etchebest, C.; Dhiman, S.; Gadhamshetty, V.; Bomgni, A. B.; Gnimpieba, E. Z. Data Mining and Machine Learning over HPC Approach Enhancing Antibody Conformations Prediction. In *Microbial Stress Response: Mechanisms and Data Science*, ACS Symposium Series; **2023**, *1434*, 75–92. https://doi.org/10.1021/BK-2023-1434.CH005.

[23] Burgos-Morales, O.; Gueye, M.; Lacombe, L.; Nowak, C.; Schmachtenberg, R.; Hörner, M.; Jerez-Longres, C.; Mohsenin, H.; Wagner, H. J.; Weber, W. Synthetic Biology as Driver for the Biologization of Materials Sciences. *Mater Today Bio*, **2021**, *11*, 100115. https://doi.org/10.1016/J.MTBIO.2021.100115.

[24] Niu, Z.; Hua, G.; Wang, L.; Gao, X. Knowledge-Based Topic Model for Unsupervised Object Discovery and Localization. *IEEE Trans Image Process*, **2018**, *27* (1), 50–63. https://doi.org/10.1109/TIP.2017.2718667.

[25] Sabbir, A.; Jimeno-Yepes, A.; Kavuluru, R. Knowledge-Based Biomedical Word Sense Disambiguation with Neural Concept Embeddings. *Proc IEEE Int Symp Bioinform Bioeng*, **2017**, *2017*, 163–170. https://doi.org/10.1109/BIBE.2017.00-61.

[26] Brough, D.B., Wheeler, D.; Kalidindi, S.R. Materials Knowledge Systems in Python - A Data Science Framework for Accelerated Development of Hierarchical Materials. *Integr Mater Manuf Innov*, **2017**, *6* (1), 36–53. doi: 10.1007/s40192-017-0089-0.

[27] Zhao, Q.; Yang, H.; Liu, J.; Zhou, H.; Wang, H.; Yang, W. Machine Learning-Assisted Discovery of Strong and Conductive Cu Alloys: Data Mining from Discarded Experiments and Physical Features. *Mater Des*, **2021**, *197*. https://doi.org/10.1016/j.matdes.2020.109248.

[28] Chattopadhyay, I.; J, R. B.; Usman, T. M. M.; Varjani, S. Exploring the Role of Microbial Biofilm for Industrial Effluents Treatment. *Bioengineered*, **2022**, *13* (3), 6420–6440. https://doi.org/10.1080/21655979.2022.2044250.

[29] Agany, D. D. M.; Pietri, J. E.; Gnimpieba, E. Z. Assessment of Vector-Host-Pathogen Relationships Using Data Mining and Machine Learning. *Comput Struct Biotechnol J*, **2020**, *18*, 1704–1721. https://doi.org/10.1016/J.CSBJ.2020.06.031.

[30] Wang, C.; Fu, H.; Jiang, L.; Xue, D.; Xie, J. A Property-Oriented Design Strategy for High Performance Copper Alloys via Machine Learning. *NPJ Comput Mater*, **2019**, *5* (1). https://doi.org/10.1038/s41524-019-0227-7.

[31] Guyon, I.; De, A. M. An Introduction to Variable and Feature Selection André Elisseeff. *J Mach Learn Res*, **2003**, *3*, 1157–1182.

[32] Sanderson, K. GPT-4 Is Here: What Scientists Think. *Nature*, **2023**. https://doi.org/10.1038/D41586-023-00816-5.

[33] Moy, K.; Tae, C.; Wang, Y.; Henri, G.; Bambos, N.; Rajagopal, R. An OpenAI-OpenDSS Framework for Reinforcement Learning on Distribution-Level Microgrids. In *IEEE Power and Energy Society General Meeting*, Washington, DC, July 26–29 2021; **2021**, 1–5. https://doi.org/10.1109/PESGM46819.2021.9638106.

[34] Thakur, P.; Rauniyar, S.; Tripathi, A. K.; Saxena, P.; Gopalakrishnan, V.; Singh, R. N.; Olakunle, M. A.; Gnimpieba, E. Z.; Sani, R. K. Identifying Genes Involved in Biocorrosion from the Literature Using Text-Mining. In *2021 IEEE International Conference on Bioinformatics and Biomedicine, BIBM 2021*, Houston, TX; **2021**, 3586–3588. https://doi.org/10.1109/BIBM52615.2021.9669354.

[35] Thakur, P.; Alaba, M. O.; Rauniyar, S.; Singh, R. N.; Saxena, P.; Bomgni, A.; Gnimpieba, E. Z.; Lushbough, C.; Goh, K. M.; Sani, R. K. Text-Mining to Identify Gene Sets Involved in Biocorrosion by Sulfate-Reducing Bacteria: A Semi-Automated Workflow. *Microorganisms*, **2023**, *11* (1). https://doi.org/10.3390/MICROORGANISMS11010119.

[36] Saxena, P.; Tripathi, A. K.; Thakur, P.; Rauniyar, S.; Gopalakrishnan, V.; Singh, R. N.; Olakunle, M. A.; Gnimpieba, E. Z.; Sani, R. K. Integration of Text Mining and Biological Network Analysis to Access Essential Genes in Desulfovibrio Alaskensis G20. In *2021 IEEE International Conference on Bioinformatics and Biomedicine, BIBM 2021*, Houston, TX; **2021**, 3583–3585. https://doi.org/10.1109/BIBM52615.2021.9669712.

[37] Tripathi, A. K.; Saxena, P.; Thakur, P.; Rauniyar, S.; Gopalakrishnan, V.; Singh, R. N.; Olakunle, M. A.; Gnimpieba, E. Z.; Sani, R. K. Discovery of Genes Associated with Sulfate-Reducing Bacteria Biofilm Using Text Mining and Biological Network Analysis. In *2021 IEEE International Conference on Bioinformatics and Biomedicine, BIBM 2021*, Houston, TX; **2021**, 3589–3591. https://doi.org/10.1109/BIBM52615.2021.9669374.

[38] Allen, C.; Aryal, S.; Do, T.; Gautum, R.; Hasan, M. M.; Jasthi, B. K.; Gnimpieba, E.; Gadhamshetty, V. Deep Learning Strategies for Addressing Issues with Small Datasets in 2D Materials Research: Microbial Corrosion. *Front Microbiol*, **2022**, *13*. https://doi.org/10.3389/FMICB.2022.1059123.

[39] Gnimpieba, E. Z.; VanDiermen, M. S.; Gustafson, S. M.; Conn, B.; Lushbough, C. M. Bio-TDS: Bioscience Query Tool Discovery System. *Nucleic Acids Res*, **2017**, *45* (D1), D1117–D1122. https://doi.org/10.1093/NAR/GKW940.

9 Machine Learning-Guided Optical and Raman Spectroscopy Characterization of 2D Materials

Md Hasan-Ur Rahman, Manoj Tripathi, Alan Dalton, Mahadevan Subramaniam, Suvarna N.L. Talluri, Bharat K. Jasthi, and Venkataramana Gadhamshetty

9.1 INTRODUCTION

Machine learning (ML) enables exciting tools to extract novel information from vast datasets and organize the data efficiently. It is considered a sub-field of artificial intelligence where statistical algorithms are performed in a systematic manner to improve data interpretation. It is much more likely digitally standardizing the protocols with continuous improvements and learnings. In the modern world, the accumulation of big data and its processing have a direct impact; thus, ML-based techniques are referred to as the "fourth industrial revolution" [1]. ML tools can assist researchers in redefining scientific models and designs and optimizing the process parameters, which could not be tackled with a conventional approach from the discovery of new materials to their final deployment (Figure 9.1a). In several disciplines of science and technology, engineers and researchers use ML to address complex research questions and to predict the design, synthesis, and characterization of molecules and materials [2–5]. Figure 9.1b presents the generic ML framework for predicting material property from the feature engineering of the material to the final trained model for structural prediction. Along this route, the chosen material has to go through several stages of model training (i.e., mapping relationships with conditional factors and decisional attributes), and model evaluation (such as property-labeled materials fragments) [6].

In the last decade, ML has been broadening its applicability in quantum dots, nanoscopic materials (ranging from 1 to 100 nm in thickness), thin films, and a broad family of two-dimensional (2D) materials to characterize structure-property relationship. Peculiar to thin films and 2D materials, surface characterization techniques,

DOI: 10.1201/9781003132981-9

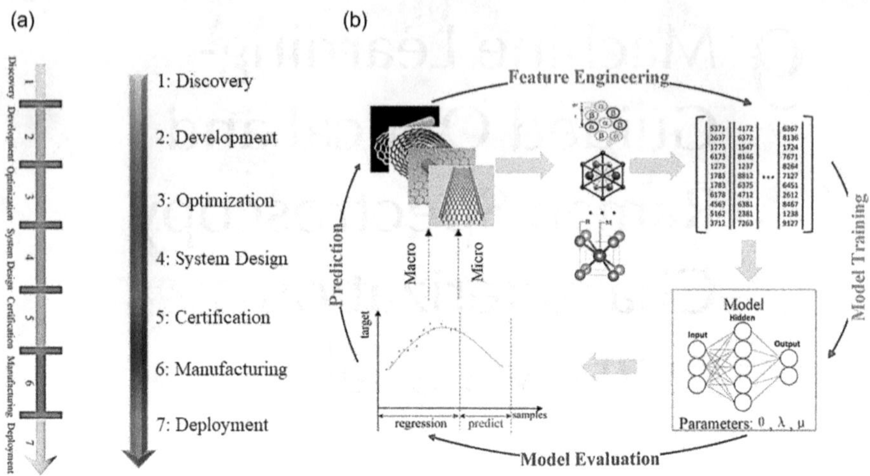

FIGURE 9.1 A general comparison between (a) traditional approach and (b) machine learning approach in materials science for crystals, thin films, and 2D material characterization. (Reproduced from Liu, Y. et al., J. Mater., 3, 159, 2017. With permission from Elsevier.)

including high-resolution scanning probe microscopy and optical and electron spectroscopy, are commonly used for surface evaluation and assessment. Nevertheless, these methods generate myriad datasets, non-linear relationships between variables and parameters, which are extraordinarily complex for high-throughput screening and interpretation. Researchers spend significant time analyzing the data and need domain expertise to create a meaningful relationship between all the different variables.

In the present section, we will focus on a few common practices in 2D material characterizations using ML tools: machine-learning optical identification (MOI) using examples of graphene and MoS_2, random forest regression (RFR), kernel ridge regression (KRR), and Gaussian mixture model (GMM) utilized in Raman spectroscopy to extract invaluable insights of graphene and molybdenum disulfide (MoS_2). We will briefly discuss the challenges and opportunities of ML algorithms for 2D material characterization to enable wide-ranging impact.

9.2 ESTABLISHED SURFACE CHARACTERIZATION TECHNIQUES

The elementary surface characterization of 2D materials entails optical microscopy for the initial assessment of 2D materials, followed by other spectroscopic and probe techniques for quantitative outcomes, such as thickness, uniformity, and defects. One of the typical examples is visualizing the optical contrast of transferred graphene layer/s over Si/SiO_2 substrate (usually 300 nm thick oxide) fabricated through mechanical exfoliation (ME) and chemical vapor deposition (CVD), respectively (Figure 9.2a–c) [8–10]. The thicker graphene (i.e., bulk) over the oxide layer of silica substrate absorbs more visible light than an atomically thin layer (1L). A similar

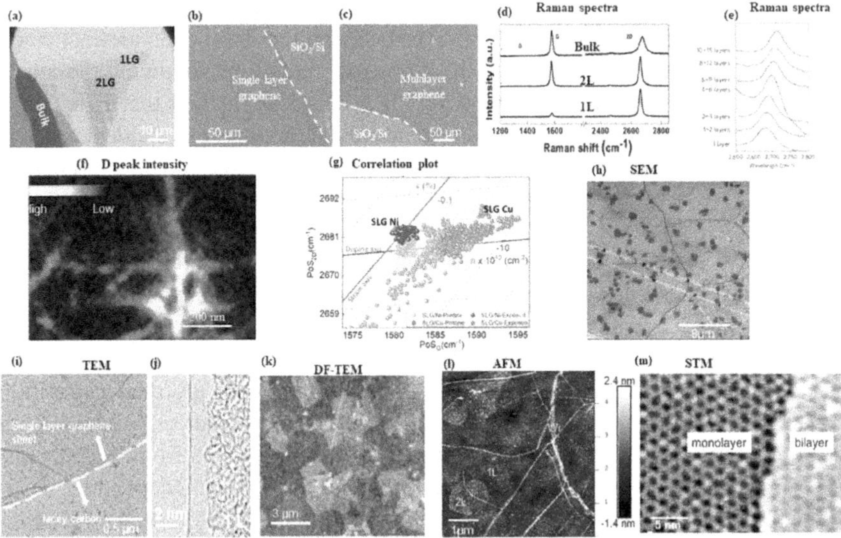

FIGURE 9.2 Optical images of graphene fabricated by (a) mechanical exfoliation. (Reprinted from Tripathi, M. et al., ACS Omega, 3, 17000, 2018. Copyright 2018, American Chemical Society); (b) CVD fabricated and transferred on Si/SiO_2. (Reproduced from Chilkoor, G. et al., Encycl Water, 1, 2019. With permission from John Wiley and Sons); (c) CVD fabricated and transferred over Si/SiO_2. (Reproduced from Chilkoor, G. et al., Encycl. Water, 1, 2019. With permission from John Wiley and Sons); (d) Raman spectra of single-layer to few-layer graphene showing D, G, and 2D peak positions. (Reprinted from Yavari, F. et al., Sci. Rep., 1, 1, 2011. Copyright 2011); (e) Raman spectra of graphene 2D peak shift with varying layers. (Reproduced from Hwangbo, Y. et al., Carbon N Y, 77, 454, 2014. With permission from Elsevier); (f) Raman map showing the distribution of graphene wrinkles through Raman active D peak intensity. (Reproduced from Tripathi, M. et al., ACS Nano, 15, 2520, 2021. Copyright 2021, American Chemical Society); (g) The correlation plot for graphene over Cu and Ni sensing corrosion from H_2SO_4. (Reprinted with the permission from Chilkoor, G. et al., ACS Nano, 15, 447, 2021. Copyright 2021, American Chemical Society); (h) SEM images of CVD graphene on silica substrate. (Reprinted with the permission from Tripathi, M. et al., ACS Appl. Mater. Interfaces, 10, 51, 44614, 2018. Copyright 2018, American Chemical Society); (i) Typical TEM images of single-layer graphene on lacey carbon. (Reproduced from Chilkoor, G. et al., Encycl Water, 1, 2019. With permission from John Wiley and Sons); (j) High-resolution TEM image of single-layer graphene sheet. (Reprinted with the permission from Reina, A. et al., Nano Lett., 9, 30, 2009, Copyright 2009, American Chemical Society); (k) False-color DF-TEM image of graphene. (Reproduced from Lee, G.H. et al., Science, 340, 1074, 2013. With permission from The American Association for the Advancement of Science); (l) AFM topography of CVD graphene. (Reprinted with the permission from Tripathi, M. et al., ACS Appl. Mater. Interfaces, 10, 51, 44614, 2018, Copyright 2018, American Chemical Society); and (m) STM image of Gr/SiC at −1.5 V. (Reproduced from Premlal, B. et al., Appl. Phys. Lett., 94, 263115, 2009. With permission from AIP Publishing.)

optical contrast has been observed in polycrystalline graphene layers that indicate the wide-scale thickness distribution. Nevertheless, under laser excitation with controlled energy, one can quantify the thickness distribution by following the atomic vibrational modes related to Raman peaks. It is based on the fact that light can be scattered inelastically, leading to the difference in the frequency of incident and scattered photons, which strongly relates to the properties of solid materials. The phenomenon of inelastic light scattering is called the Raman effect, which is used to study fundamental excitations of solid-state matter and molecules [11]. Thus, Raman spectroscopy is another crucial non-destructive tool based on optical characterization to get the signature of graphene, and it also deploys extensively to other 2D material characterization. The fingerprint Raman features of graphene are D peak (~1355 cm^{-1}), G peak (~1580cm^{-1}), and 2D peak (~ 2700 cm^{-1}) (shown in Figure 9.2d) [12]. The salient attributes of G and 2D Raman modes are their capabilities to change the position, shape, and intensity based on the number of layers and their interaction with local surroundings. Raman signals of graphene from monolayer to a few layers stacked in the Bernal (AB) configuration will vary depending on the number of layers [13,14]. Figure 9.2e portrays the 2D peak shift to a higher frequency region along with specific peak width (full-width half maximum, FWHM) with the increased number of graphene layers (utilizing $\lambda = 514$ nm wavelength excitation energy). The Raman active disordered peak "D" is useful to monitor structural defects through absolute intensity (I_D) or relative to ratio (I_D/I_G) to reveal wrinkles (see, e.g., Figure 9.2f), edges (zigzag), and bubbles [15]. Additionally, the frequency shift (cm^{-1}) of Raman modes and their correlation, such as 2D vs G peak positions, can reveal underlying strain and doping effects in graphene [16]. It is carried out using a reference coordinate from a suspended specimen (O, intersect of strain and doping axis in Figure 9.2g), which is assumed to be the minimum influence from strain and doping. The distribution of Raman modes (G, 2D) deviated from the reference coordinate through external stimulation (e.g., temperature, impurities, see Figure 9.2g) indicates the extent of carrier concentration and extension/compression of carbon lattice. Thus, Raman spectroscopy can detect subtle changes in the graphene host materials to monitor the corrosion dynamics of underlying metals such as copper (Cu) and nickel (Ni).

The results of Raman spectra can be complemented with other surface characterization techniques, such as scanning electron microscopy (SEM) and transmission electron microscopy (TEM), for an in-depth investigation of real space images and crystallographic information. For the opaque substrate, the localized electron beam interacts with the substrate to generate secondary electrons to develop the image as an impression of topography. This technique based on probing electrons is known as SEM. SEM is useful for scanning large regions to evaluate graphene grain size, growth rates, nucleation density, structural defects, and coating uniformity. Furthermore, the SEM is integrated with additional detectors like energy-dispersive spectroscopy (EDS) and electron backscatter diffraction (EBSD) to obtain insights into the elemental chemical composition and crystallographic orientation of graphene and other 2D materials. Figure 9.2h illustrates a typical SEM image of single-layered graphene (bright area), bilayered graphene (dark patches) and distribution of wrinkles (dark lines). The contrast in SEM micrograph is due to the relationship of accelerating voltage with the number of layers and substrate [17]. It relates to the availability of secondary electrons generated at the topmost layer of the graphene surface. Also, the metallic subsurface (e.g., Cu, Ni) yields higher secondary electrons than the lighter carbon atoms revealing a brighter region.

For the transparent samples, TEM enables atomic-scale characterization to investigate the layers number, in situ growth, and transformation of graphene. Figure 9.2i and j illustrates a typical single-layer graphene micrograph and its high-resolution around the edge region. The number of layers is delved by counting the contrast line along the backfolded edge of a graphene sheet [18].

Dark-field TEM (DF-TEM) is a handy tool that rapidly detects local structures (such as grain sizes) over a large area. Figure 9.2k demonstrates the mapping of the grain structure of graphene where the individual false-color area represents distinct crystal orientation. The probing of the graphene surface with a sharp physical object (usually doped Si) illustrates three-dimensional topography generated due to physical interaction through the technique atomic force microscopy (AFM) [19]. AFM is employed to realize the graphene layer thickness and surface roughness, which is difficult to detect through optical contrast and electron imaging. The topological facets of graphene are sensitive to the underlying substrate. Consequently, it can provide the atomic structure and nanoscale morphology [20]. Figure 9.2l depicts the AFM image of polycrystalline graphene over Si/SiO_2 wafer showing single-layer (1L), bilayer (2L), and wrinkled (Wr) regions. The probing of the conducting graphene through the metallic tip apex (Au, tungsten, pt-Ir etc.) is useful for providing atomic-scale resolution through scanning tunneling microscopy (STM). It is carried out by applying a potential bias across the tip and the substrate and monitoring the tunneling current between them, separated by a few nanometers (1–10 nm). The atomic resolution of graphene single and bilayers results in triangular patterns with hexagonal symmetry, as shown in Figure 9.2m.

Among all the above characterization methods, optical imaging and Raman spectroscopy remain the quick and most viable methods of capturing signatures of 2D materials. Hence, the ML methods discussed in this chapter will focus primarily on these methods.

9.3 ML-GUIDED OPTICAL DETECTION OF 2D MATERIALS

One of the initial requirements of 2D materials and thin films is their optical detection, which involves a great deal of human effort and domain expertise to pinpoint the fingerprint features accurately. The optical detection of 2D materials relies on the experience of the researchers and even seasoned professionals struggle to deal with sophisticated 2D heterostructures. The integration of ML with an optical microscope surpasses some of the crucial factors of detection, especially in fundamental research of 2D following a classical strategy:

$$\text{Goal}(\text{given problem}) + \text{Sample}(\text{raw data}) + \text{Algorithm}(\text{data processing}) = \text{Model}\ [7].$$

ML algorithms can work with thousands of optical images simultaneously and reduce the excessive time required for the detection of 2D material aspects. For instance, graphene exfoliation via the scotch tape method onto Si/SiO_2 substrate will generate different thicknesses and shapes of graphene flakes, which are stochastically distributed on the substrate [29]. All these variations in thicknesses of graphene layers will

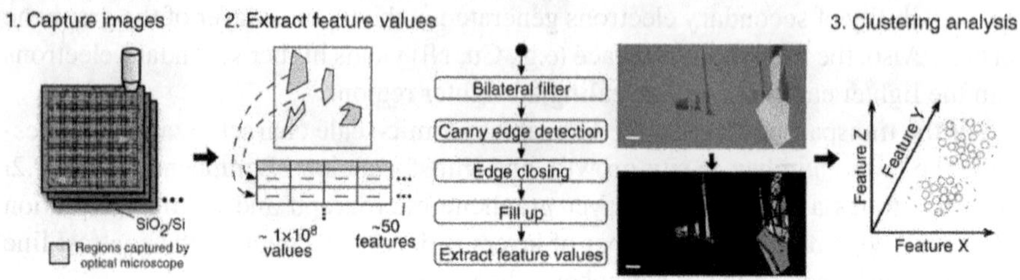

FIGURE 9.3 Stages of data-driven analysis system from a collection of optical microscope photographs to ML model features extraction. (Reprinted from Masubuchi, S. and Machida, T. npj 2D Mater. Appl., 3, 1, 2019. Copyright 2019.)

lead to collection of many datasets that will require relevant graphene characterization. One can input the parameters of graphene in data-driven analysis algorithms and extract the fingerprint features. This procedure integrates the optical microscopy with a ML unsupervised algorithm based on the Bayesian Gaussian mixture model (Figure 9.3). The automatic identification of graphene entails the following steps: (1) First, graphene was exfoliated in Si/SiO_2 substrate using scotch tape method (2) Then, several optical images (70000) were taken by automatic microscope (3) The images were loaded in algorithm and decomposed to HSV (Hue Saturation Value) images and clustered (4) From HSV color images, a scatter plot was established, and the feature values were represented in three-dimensional format to extract the key features (4) Finally, the feature values were analyzed by open-source data platforms (e.g., Python, Jupyter and notebook).

The MOI method is further expanded for the identification of other class of 2D materials such as MoS_2, tungsten disulfide (WS_2) and other transition metal dichalcogenides (TMDs). Unlike graphene, TMD layers comprise a tri-layered configuration with metal at the center. MOI method utilizes a supervised ML model, a support vector machine (SVM) algorithm (Figure 9.4). The SVM analyzes the red, green, and blue (RGB) color insights from the optical images of 2D nanostructures and extracts pivotal aspects based on the number of layers, defects, impurities, and stacking faults. The MOI work in two steps: the training process and the testing process. Initially, the microscope will collect 2D materials (for example graphene or MoS_2) images at different magnifications, and the software will sort them according to pre-established datasets. Then, the images are inserted into the training process to establish a dataset of fingerprint features based on the SVM analysis of RGB channel intensity. The RGB model is linked with the 2D materials' pre-existing datasets based on AFM and Raman spectroscopy. During the testing step, the optical information of the 2D materials is sorted by algorithm in distinct categories. Finally, the RGB information is translated into false-color images distinguishing the substrate, 2D material, and impurities.

MOI optical detection can also be integrated with scanning probe techniques, like AFM. The combined characterization techniques lead to the intelligent identification of $MoS_{2,}$ as shown in Figure 9.4. The MOI system utilized the optical images of MoS_2 in the training process as input variables (Figure 9.4a and c). Furthermore, the

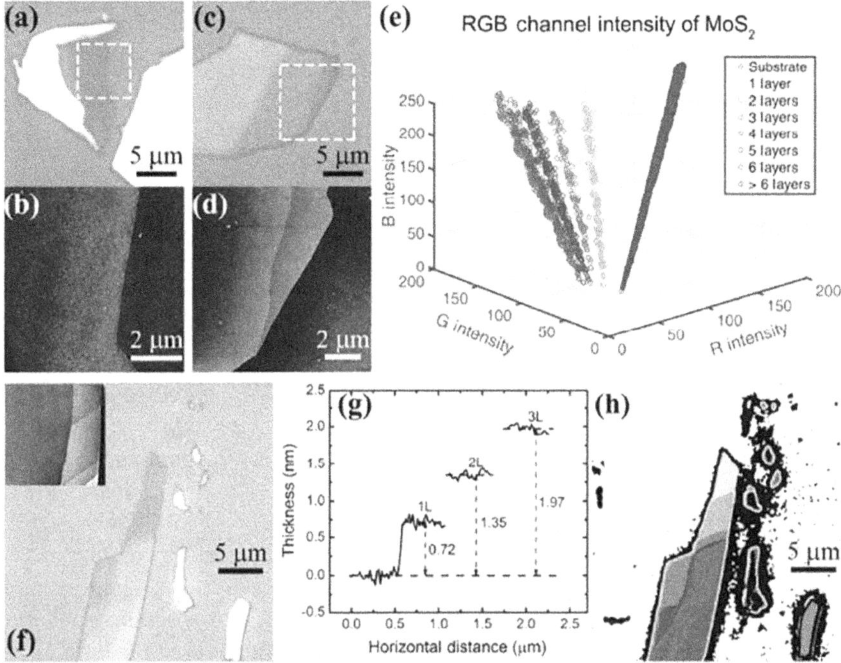

FIGURE 9.4 MOI of MoS_2 (a) and (c) Optical images of MoS_2 for training purposes; (b) and (d) Corresponding AFM images; (e) Training result of different layers over a fixed substrate; (f) Optical images of mixed layer MoS_2 for testing purpose; (g) Corresponding AFM datasets; (h) Testing result showing colored map based on thickness distribution. (Reprinted from Lin, X. et al., Nano Res., 11, 6316, 2018. Copyright 2018, Springer Nature.)

SVM algorithm processed AFM images (Figure 9.4b and d) and established a database of RGB channel intensity versus thickness (i.e., number of layers). The resulting SVM model (Figure 9.4e) represents the MoS_2 RGB features corresponding to the substrate. Subsequently, in the testing step (Figure 9.4f and g), the MOI analyzes the RGB information of optical images along with AFM thickness to link them with SVM model. Finally, the model creates a false-color image of MoS_2 sample based on the number of layers with distinct color for different regions (Figure 9.4h), which will allow the quantification of number of MoS_2 layers present in large areas in considerably less time. One of the additional advantages of MOI-based detection is its ability to identify the contaminated regions in MoS_2 and transfer process residues, as shown in Figure 9.4h (black regions).

Thus, MOI can work on a large family of 2D materials where the model can extract RGB features from optical images and provide useful information about the number of layers present in large number of optical image datasets collected. This procedure minimizes the utilization of multiple sophisticated instruments and ML-guided optical detection method will reduce the cost of 2D material characterization. Additionally, the MOI tools can be deployed to realize the elusive aspects of

2D materials based heterostructure and accelerate the commercial applications of 2D materials.

ML-guided optical detection technique is helpful for the initial screening of 2D materials for thickness distribution. Nevertheless, it limits accuracy, composition, structure, and precision for determining the number of layers and impurities, which is crucial for the electronic and optoelectronic industry.

9.4 ML-GUIDED RAMAN SPECTROSCOPY DETECTION OF 2D MATERIALS

Raman spectroscopy is a viable tool to analyze the molecular structure, layer number, functionalization [32], strain, and structural defects/disorders of 2D materials [15]. Nevertheless, the Raman spectra contain innumerable datasets to examine and establish a meaningful correlation which is complex to decipher. ML-guided Raman detection can improve the efficacy of 2D material characterization and reduce the significant burden in fundamental and applied science. The integration of Raman spectroscopy with scalable production techniques such as CVD [25] will be useful to monitor the uniformity, cracks, adlayers, and applicable for quality control.

One of the ML algorithms deployed for Raman spectral analysis is the random forest regressor, which does not require extensive statistics for processing and interpretation [33]. In RFR, the raw data from Raman spectra are used for training and generating new datasets. Sequentially, a decision tree is generated, and the unused data are implemented to test the model's efficacy. In the later stages, RFR will search for fingerprint features from Raman spectral datasets and the model will take a decision based on majority voting. In the broader perspective, RFR is a learning algorithm consisting of multiple tree structures with several branches. Each tree is set-up based on training sample sets and a random variable, and every tree can cast a single vote for decision-making [34]. Figure 9.5 represents the RFR learning flowchart for extracting Raman fingerprint features of TMD (MoS_2) onto Si/SiO_2 substrate. Generally, MoS_2 has two major active Raman modes, including E^1_{2g} at $388 cm^{-1}$ and A^1_g at $407 cm^{-1}$, that are associated with first-order in-plane and out-of-plane Raman bands, respectively [35]. The Raman frequency exhibits significant differences depending on the number of layers. As the number of MoS_2 layers increased, the frequency of the E^1_{2g} peak decreases and the A^1_g peak increases. For the crack regions, the model is taking silicon Raman mode at $\sim 520 cm^{-1}$ since the film is assumed to be continuous and the unexposed regions will be a substrate (Si/SiO_2) only. Hence, the RFR model input these active Raman modes and extracts the number of layers as output parameters. There are five input variables: (1) intensity and (2) frequency of the E^1_{2g} peak; (3) intensity and (4) frequency of the A^1_g peak; (5) Raman frequency difference between these two peaks, which are designated as α, β, γ, δ, and $\mathcal{E}$, respectively, in the RFR training sample sets. The algorithm will extract features of monolayer, bilayer, and crack regions of MoS_2 as output variables.

Another potential ML algorithm employed in Raman spectroscopy interpretation is kernel ridge regression to solve an inverse problem (a large set of observations that is responsible for its generation) of 2D materials. The KRR-based ML-guided

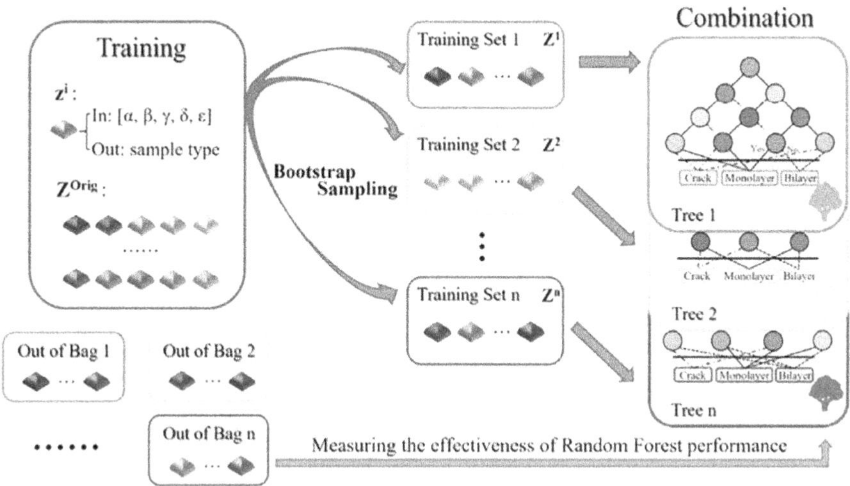

FIGURE 9.5 Process flowchart of RFR learning mechanism for MoS_2 features extraction. (Reprinted from Mao, Y. et al., Nanomaterials, 10, 1–13, 2020. Copyright 2020.)

algorithm is helpful in addressing sophistication in the important fingerprint features from vast datasets. A Gaussian kernel (radial basis function - rbf) is executed to establish a non-linear fitting for addressing the complexity of Raman spectra. Gaussian kernel efficiently extracts meaningful interpretation from non-linear datasets [37]. The analysis consists of the following steps: (1) visual inspection of spectral datasets to establish a general trend, (2) principal component analysis (PCA) for extracting small input variables from the spectral data, (3) model performance evaluation by cross-validation and overfitting test, and (4) features significance.

When graphene layers are stacked on top of each other, with or without twisted angles between interlayers, defects and structural disorders produce intriguing properties, such as superconductivity and magnetism [38,39]. The relationship between Raman spectroscopic details and the twisted angle of bilayer graphene (tBLG) has been investigated using graphene stacked layers as an example. It is carried out by monitoring the key attributes of graphene Raman modes of D, G and 2D peaks, and their intensity, Raman shifts and FWHM as a standard to realize the quality and number of layers [40]. From visual inspection and PCA analysis of tBLG, the most significant parameter was found G band at 1672 cm^{-1}. Hence, the KRR-rbf will take the Raman shift of G band (cm^{-1}) as an input variable and extract the meaningful relationship between the twist angle (Θ°) and G band (cm^{-1}). At the initial stage, the KRR go through training, and the unseen data will be utilized for testing purposes. Then, the algorithm will process the cross-validation test to avoid overfitting. Finally, the model will plot the features as a function of twisted angle to extract the interesting patterns (Figure 9.6).

ML-based KRR prediction is limited to generating training models focused on particular Raman mode (G band) only; nevertheless, it lacks the capability for

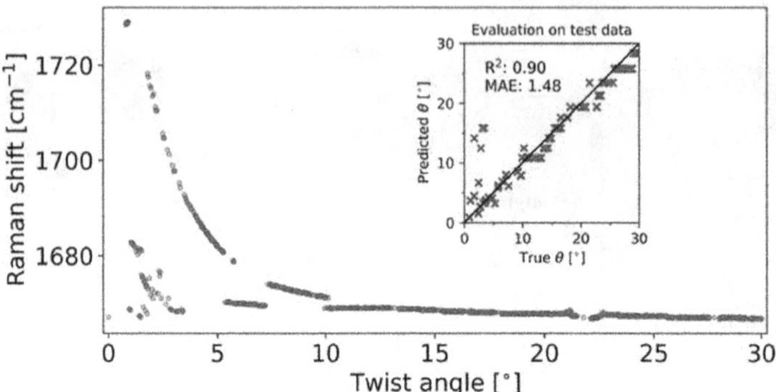

FIGURE 9.6 KRR prediction of Raman spectrum at different twist angles (Θ°) between interlayer graphene. (Reproduced from Sheremetyeva, N. et al., Carbon N Y, 169, 455, 2020. With permission from Elsevier.)

complex spectral analysis such as spectral broadening (FWHM), 2D peaks interpretation and resonant Raman processes pertinent to twist angles of tBLG [42]. Therefore, an improved ML algorithm is proposed as a Gaussian mixture model combined with Raman spectroscopy to overcome such intricacies. GMM is a data clustering technique, which assumes the datasets are generated from a finite number of Gaussian distributions with unknown parameters. The model utilizes the expectation-maximization algorithm, covariance matrices and weightings of the N-Gaussian probability distributions to find the relationship of the datasets [42]. There is a plethora of information in Raman spectra, mapping and manual peak fitting will reduce the dimensionality of the parameters to train the model. At the initial stage, the model utilizes the significant attributes (G and 2D) extracted from Raman peak fitting as an input variable. The algorithm compares the distance between the points in a finite-dimensional space in a selected area input features (G and 2D peaks) to create a scatterplot (Figure 9.7a). Figure 9.7b and c represent the Raman map of G and 2D peaks positions (i.e., frequency) respectively, for the same region. Then, the GMM labeled the clusters based on the similarities and created eight distinguished regions. The clusters were assigned numbers relating to the population density from the most (cluster 1) to the least (cluster 8) population. The shape of the cluster is drawn from 2σ away from the mean values, where "σ" is the direction-dependent standard deviation. The output of the clusters result is shown in the inset of Figure 9.7a. Clusters (2, 3, and 6) represent the distinct types of tBLG, clusters (1 and 5) represent the single layer of graphene, cluster (4) accounts for adlayer regions, and clusters (7 and 8) are not fitting with any particular trends due to the low weighing of the population density to the background. This pre-trained data clustering model can be applied to any other regions of tBLG to extract fundamental aspects.

9.5 COMMON CHALLENGES TO ML IN RAMAN SPECTROSCOPY

Like other emerging techniques, the ML approach in material science, especially in 2D material detection, is rapidly gaining ground. Nevertheless, there is a long

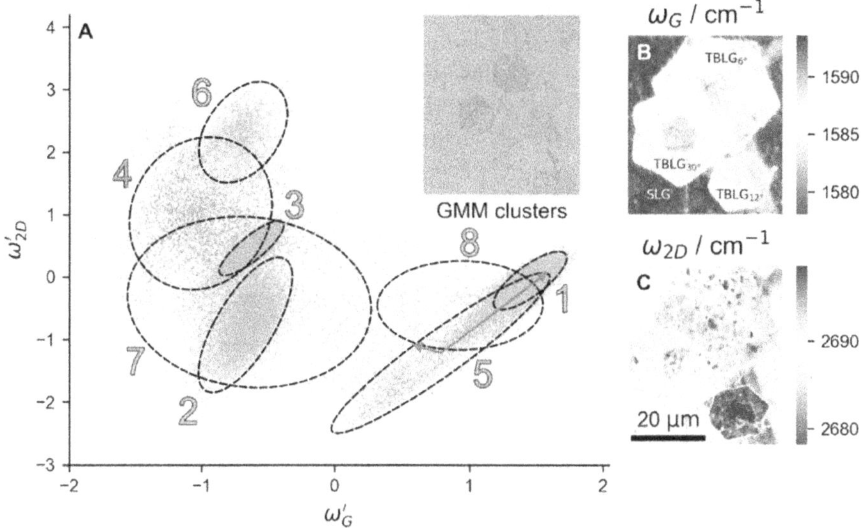

FIGURE 9.7 (a) GMM learning mechanism for tBLG features extraction from 1 to 8, inset shows the output in color configuration. (b and c) Raman mapping of G and 2D peak positions (frequency, cm^{-1}). (Reproduced from Vincent, T. et al., Carbon N Y, 201, 141, 2021. With permission from Elsevier.)

way ahead before it accepts as an impeccable technique. In a few instances, several analytical models for ML-guided algorithms such as RFR, KRR, and GMM in Raman spectroscopy are in practice to extract the fingerprint features of 2D materials. Although the validation of scientific theories is one of the common challenges of ML-based algorithms, these ML models performed well in synthetic data mostly captured in ideal conditions but are incapable of capturing the experimental conditions which are influenced from sample preparation, local surroundings, defects, impurities and other factors. Hence, these models cannot be readily deployed to characterize the experimental Raman spectra. For instance, KRR predicts the relationship between the Raman spectra and the twist angles of tBLG by overestimating the G bands. Furthermore, additional sidebands are occurring near the G band [43], which the ML is not considering for prediction. The experimental resonance that changes due to laser excitation energy is also neglected in establishing the model database, along with other crucial factors such as 2D band position, intensity, and width. In experimental conditions, these features of 2D peaks significantly varied with twist angles. Most scientific theories for Raman interpretation of graphene/graphite come in conjunction with D, G, and 2D bands, similar to the other family of 2D materials (e.g., TMDs, E^1_{2g}, and A_{1g}). Hence, the KRR model cannot be readily deployed to realize the properties of tBLG. Further refinement of the computational Raman spectral dataset is required to correlate with the experimental observations. As the computational complexity increases, the time and cost associated with the ML significantly increase and nullify ML sole benefit. Although the GMM-guided Raman spectra study included the additional band (2D peak) properties, the model can handle a low amount of input variables and the dimensionality reduction requires peak fitting by manual human intervention to realize the significant input parameters. The GMM can integrate with unsupervised ML model like PCA to reduce the dimensionality, which allows the entire Raman spectra as input variables without expert intervention and automatically identifies the features within short span of time (in seconds). However, the resulting clusters are harder to interpret. Moreover, the ML algorithms efficacy depends on the training datasets and the appropriate labeling of the data by professionals. The ML-based algorithm needs to be mature for interpretation in experimental conditions; in the present scenario, the 2D material dataset is insufficient to provide conclusive decisions on the relationship between multiple properties.

9.6 FUTURE PROSPECTS

The ML is the fastest growing field because of the emergence of new learning algorithms and theories, the continuous refinement and availability of online datasets, and the computational cost reduction [44]. As the Raman spectral database is enriched over time, the ML learns efficiently and predicts the designed outcomes. Furthermore, the integration of hybrid database: computational + experimental in the Raman equipment will lead to the translation of the model in real-life applications. This in-built database will be beneficial for more sophisticated 2D heterostructure characterization and applications. The above-mentioned algorithm can easily be applied to other classes of 2D material characterization of complex structure-property relationship like doping, mechanical, and oxidation straining effects and enhance to unleashing new properties. Moreover, the ML-guided Raman model and the database can easily be transferred to other surface characterization techniques like AFM, SEM, TEM for fundamental research in the exploration of novel 2D materials. The high-throughput screening of ML integrated with Raman spectroscopy will accelerate the fabrication of 2D material devices in various industries. Supervised regression-based ML algorithms and unsupervised algorithms are effective in exploring 2D materials based on multiple imaging modalities. A multi-modal ML approach using an ensemble of ML models could provide a comprehensive view of 2D materials and provide new insights about their properties.

9.7 SUMMARY

The presented chapter provides an overview of ML algorithms application and advancement in the detection of 2D materials. A classical comparison between the traditional technique and ML approach is demonstrated for the interpretation of a wide variety of materials. Several ML approaches are discussed in detail, and their interpretation from optical characterization and Raman spectroscopy is associated to 2D materials using the example of graphene and TMDs. Several advantages and limitations have been highlighted of ML tools for interpreting 2D materials. Despite the shortcomings of ML algorithms in Raman characterization, the ML can perform efficiently and better than humans in terms of time and cost. Nevertheless, the ML tools are lacking to emulate the experimental conditions to unleash comprehensive information. The Raman spectra contain a vast set of information, and the experts spend several hours extracting the meaningful relationship. In modern research, ML algorithms are continuously improving through automation, screening of quality data, implementation, and interpretation. The ML-guided Raman analysis integration with other surface characterization techniques will continue to improve the conventional characterization of 2D materials.

REFERENCES

[1] "The Fourth Industrial Revolution | Foreign Affairs." https://www.foreignaffairs.com/world/fourth-industrial-revolution (Accessed Feb. 01, 2023).

[2] A. P. Bartók *et al.*, "Machine learning unifies the modeling of materials and molecules," *Sci Adv*, vol. 3, no. 12, Dec. 2017, doi: 10.1126/SCIADV.1701816/SUPPL_FILE/1701816_SM.PDF.

[3] K. T. Butler, D. W. Davies, H. Cartwright, O. Isayev, and A. Walsh, "Machine learning for molecular and materials science," *Nature* vol. 559, no. 7715, pp. 547–555, Jul. 2018, doi: 10.1038/s41586-018-0337-2.

[4] C. Allen *et al.*, "Deep learning strategies for addressing issues with small datasets in 2D materials research: Microbial Corrosion," *Front Microbiol*, vol. 13, p. 5095, Dec. 2022, doi: 10.3389/FMICB.2022.1059123/BIBTEX.

[5] R. Sikder, T. Zhang, and T. Ye, "Predicting THM formation and revealing its contributors in drinking water treatment using machine learning," *ACS ES&T Water*, 2023, doi: 10.1021/ACSESTWATER.3C00020/ASSET/IMAGES/LARGE/EW3C00020_0006.JPEG.

[6] O. Isayev, C. Oses, C. Toher, E. Gossett, S. Curtarolo, and A. Tropsha, "Universal fragment descriptors for predicting properties of inorganic crystals," *Nat Commun*, vol. 8, no. 1, pp. 1–12, Jun. 2017, doi: 10.1038/ncomms15679.

[7] Y. Liu, T. Zhao, W. Ju, and S. Shi, "Materials discovery and design using machine learning," *J Mater*, vol. 3, no. 3, pp. 159–177, Sep. 2017, doi: 10.1016/J.JMAT.2017.08.002.

[8] S. Roddaro, P. Pingue, V. Piazza, V. Pellegrini, and F. Beltram, "The optical visibility of graphene: Interference colors of ultrathin graphite on SiO(2)," *Nano Lett*, vol. 7, no. 9, pp. 2707–2710, Sep. 2007, doi: 10.1021/NL071158L.

[9] J. H. Warner, F. Schäffel, A. Bachmatiuk, and M.H. Rümmeli, Chapter 5-characterisation techniques . In J. H. Warner, F. Schäffel, A. Bachmatiuk, and M.H. Rümmeli (Eds), *Graphene: Fundamentals and Emergent Applications*, Elsevier, pp. 229–332, Jan. 2013, doi: 10.1016/B978-0-12-394593-8.00005-9.

[10] M. Tripathi *et al.*, "Laser-based texturing of graphene to locally tune electrical potential and surface chemistry," *ACS Omega*, vol. 3, no. 12, pp. 17000–17009, Dec. 2018, doi: 10.1021/ACSOMEGA.8B02815/ASSET/IMAGES/LARGE/AO-2018-02815R_0001.JPEG.

[11] A. Jorio and M. S. Dresselhaus, "Raman Spectroscopy in Nanoscience and Nanometrology: Carbon Nanotubes, Nanographite and Graphene," p. 354, 2010 (Accessed Mar. 07, 2023. https://www.wiley.com/en-us/Raman+Spectroscopy+in+Graphene+Related+Systems-p-9783527408115

[12] A. Ferrari and J. Robertson, "Interpretation of Raman spectra of disordered and amorphous carbon," *Phys Rev B*, vol. 61, no. 20, p. 14095, May 2000, doi: 10.1103/PhysRevB.61.14095.

[13] L. M. Malard, M. A. Pimenta, G. Dresselhaus, and M. S. Dresselhaus, "Raman spectroscopy in graphene," *Phys Rep*, vol. 473, no. 5–6, pp. 51–87, Apr. 2009, doi: 10.1016/J.PHYSREP.2009.02.003.

[14] A. C. Ferrari *et al.*, "Raman spectrum of graphene and graphene layers," *Phys Rev Lett*, vol. 97, no. 18, p. 187401, Oct. 2006, doi: 10.1103/PHYSREVLETT.97.187401/FIGURES/3/MEDIUM.

[15] M. Tripathi *et al.*, "Structural defects modulate electronic and nanomechanical properties of 2D materials," *ACS Nano*, vol. 15, no. 2, pp. 2520–2531, Feb. 2021, doi: 10.1021/ACSNANO.0C06701/ASSET/IMAGES/LARGE/NN0C06701_0005.JPEG.

[16] J. E. Lee, G. Ahn, J. Shim, Y. S. Lee, and S. Ryu, "Optical separation of mechanical strain from charge doping in graphene," *Nat Commun*, vol. 3, no. 1, pp. 1–8, Aug. 2012, doi: 10.1038/ncomms2022.

[17] H. Hiura, H. Miyazaki, and K. Tsukagoshi, "Determination of the number of graphene layers: Discrete distribution of the secondary electron intensity stemming from individual graphene layers," *Appl Phys Exp*, vol. 3, no. 9, p. 095101, Sep. 2010, doi: 10.1143/APEX.3.095101/XML.

[18] A. Reina *et al.*, "Large area, few-layer graphene films on arbitrary substrates by chemical vapor deposition," *Nano Lett*, vol. 9, no. 1, pp. 30–35, Jan. 2009, doi: 10.1021/NL801827V/SUPPL_FILE/NL801827V_SI_003.PDF.

[19] M. Tripathi, G. Paolicelli, S. Daddato, and S. Valeri, "Controlled AFM detachments and movement of nanoparticles: Gold clusters on HOPG at different temperatures," *Nanotechnology*, vol. 23, no. 24, p. 245706, May 2012, doi: 10.1088/0957-4484/23/24/245706.

[20] M. Ishigami, J. H. Chen, W. G. Cullen, M. S. Fuhrer, and E. D. Williams, "Atomic structure of graphene on SiO2," *Nano Lett*, vol. 7, no. 6, pp. 1643–1648, Jun. 2007, doi: 10.1021/NL070613A/ASSET/IMAGES/LARGE/NL070613AF00004.JPEG.

[21] M. Tripathi *et al.*, "Laser-based texturing of graphene to locally tune electrical potential and surface chemistry," *ACS Omega*, vol. 3, no. 12, pp. 17000–17009, Dec. 2018, doi: 10.1021/ACSOMEGA.8B02815/ASSET/IMAGES/LARGE/AO-2018-02815R_0001.JPEG.

[22] G. Chilkoor, N. Shrestha, S. P. Karanam, V. K. K. Upadhyayula, and V. Gadhamshetty, "Graphene coatings for microbial corrosion applications," *Encycl Water*, pp. 1–25, Dec. 2019, doi: 10.1002/9781119300762.WSTS0125.

[23] F. Yavari, Z. Chen, A. v. Thomas, W. Ren, H. M. Cheng, and N. Koratkar, "High sensitivity gas detection using a macroscopic three-dimensional graphene foam network," *Sci Rep*, vol. 1, no. 1, pp. 1–5, Nov. 2011, doi: 10.1038/srep00166.

[24] Y. Hwangbo *et al.*, "Interlayer non-coupled optical properties for determining the number of layers in arbitrarily stacked multilayer graphenes," *Carbon N Y*, vol. 77, pp. 454–461, Oct. 2014, doi: 10.1016/J.CARBON.2014.05.050.

[25] G. Chilkoor *et al.*, "Atomic layers of graphene for microbial corrosion prevention," *ACS Nano*, vol. 15, no. 1, pp. 447–454, Jan. 2021, doi: 10.1021/ACSNANO.0C03987/ASSET/IMAGES/LARGE/NN0C03987_0006.JPEG.

[26] M. Tripathi *et al.*, "Friction and adhesion of different structural defects of graphene," *ACS Appl Mater Interfaces*, vol. 10, no. 51, pp. 44614–44623, Dec. 2018, doi: 10.1021/ACSAMI.8B10294/SUPPL_FILE/AM8B10294_SI_003.AVI.

[27] G. H. Lee *et al.*, "High-strength chemical-vapor-deposited graphene and grain boundaries," *Science*, vol. 340, no. 6136, pp. 1074–1076, May 2013, doi: 10.1126/SCIENCE.1235126/SUPPL_FILE/LEE-SM.PDF.

[28] B. Premlal *et al.*, "Surface intercalation of gold underneath a graphene monolayer on SiC(0001) studied by scanning tunneling microscopy and spectroscopy," *Appl Phys Lett*, vol. 94, no. 26, p. 263115, Jul. 2009, doi: 10.1063/1.3168502.

[29] K. S. Novoselov *et al.*, "Electric field in atomically thin carbon films," *Science*, vol. 306, no. 5696, pp. 666–669, Oct. 2004, doi: 10.1126/SCIENCE.1102896/SUPPL_FILE/NOVOSELOV.SOM.PDF.

[30] S. Masubuchi and T. Machida, "Classifying optical microscope images of exfoliated graphene flakes by data-driven machine learning," *npj 2D Mater Appl*, vol. 3, no. 1, pp. 1–7, Jan. 2019, doi: 10.1038/s41699-018-0084-0.

[31] X. Lin *et al.*, "Intelligent identification of two-dimensional nanostructures by machine-learning optical microscopy," *Nano Res*, vol. 11, no. 12, pp. 6316–6324, Dec. 2018, doi: 10.1007/s12274-018-2155-0.

[32] U. Schnabel, C. Schmidt, J. Stachowiak, A. Bösel, M. Andrasch, and J. Ehlbeck, "Cover picture: Plasma process. polym. 2/2018," *Plasma Process Polym* , vol. 15, no. 2, p. 1870004, Feb. 2018, doi: 10.1002/PPAP.201870004.

[33] J. Yang and H. Yao, "Automated identification and characterization of two-dimensional materials via machine learning-based processing of optical microscope images," *Extreme Mech Lett*, vol. 39, Sep. 2020, doi: 10.1016/J.EML.2020.100771.

[34] Y. Liu, Y. Wang, and J. Zhang, "New machine learning algorithm: Random forest," In *Lecture Notes in Computer Science (including subseries Lecture Notes in Artificial Intelligence and Lecture Notes in Bioinformatics)*, Springer, vol. 7473, LNCS, pp. 246–252, 2012, doi: 10.1007/978-3-642-34062-8_32.

[35] B. R. Carvalho, L. M. Malard, J. M. Alves, C. Fantini, and M. A. Pimenta, "Symmetry-dependent exciton-phonon coupling in 2D and bulk MoS2 observed by resonance Raman scattering," *Phys Rev Lett*, vol. 114, no. 13, Apr. 2015, doi: 10.1103/PHYSREVLETT.114.136403.

[36] Y. Mao *et al.*, "Machine learning analysis of Raman spectra of MoS2," *Nanomaterials*, vol. 10, no. 11, pp. 1–13, Nov. 2020, doi: 10.3390/NANO10112223.

[37] A. K. Bishwas, A. Mani, and V. Palade, "Gaussian kernel in quantum learning," *Int J Quantum Inform*, vol. 18, no. 3, May 2020, doi: 10.1142/S0219749920500069.

[38] U. Mogera and G. U. Kulkarni, "A new twist in graphene research: Twisted graphene," *Carbon N Y*, vol. 156, pp. 470–487, Jan. 2020, doi: 10.1016/J.CARBON.2019.09.053.

[39] Y. Cao *et al.*, "Unconventional superconductivity in magic-angle graphene superlattices," *Nature*, vol. 556, no. 7699, pp. 43–50, Mar. 2018, doi: 10.1038/nature26160.

[40] A. C. Ferrari and D. M. Basko, "Raman spectroscopy as a versatile tool for studying the properties of graphene," *Nat Nanotechnol*, vol. 8, no. 4, pp. 235–246, Apr. 2013, doi: 10.1038/nnano.2013.46.

[41] N. Sheremetyeva, M. Lamparski, C. Daniels, B. van Troeye, and V. Meunier, "Machine-learning models for Raman spectra analysis of twisted bilayer graphene," *Carbon N Y*, vol. 169, pp. 455–464, Jul. 2020, doi: 10.1016/j.carbon.2020.06.077.

[42] T. Vincent, K. Kawahara, V. Antonov, H. Ago, and O. Kazakova, "Data cluster analysis and machine learning for classification of twisted bilayer graphene," *Carbon N Y*, vol. 201, pp. 141–149, Jul. 2021, doi: 10.1016/j.carbon.2022.09.021.

[43] K. Kim *et al.*, "Raman spectroscopy study of rotated double-layer graphene: Misorientation-angle dependence of electronic structure," *Phys Rev Lett*, vol. 108, no. 24, p. 246103, Jun. 2012, doi: 10.1103/PHYSREVLETT.108.246103/FIGURES/4/MEDIUM.

[44] M. I. Jordan and T. M. Mitchell, "Machine learning: Trends, perspectives, and prospects," *Science*, vol. 349, no. 6245, pp. 255–260, Jul. 2015, doi: 10.1126/SCIENCE.AAA8415.

10 Atomistic Experiments for Discovery of 2D Coatings

Biological Applications

Sourav Verma, Rabbi Sikder, Dipayan Samanta, Lan Tong, and Kenneth M. Benjamin

10.1 INTRODUCTION

Two dimensional (2D) materials such as graphene and its derivatives, termed graphene-family nanomaterials (GFNs), have gained considerable traction in research as novel materials due to their unique physical and chemical properties. Some potential applications include biomedical device sensors and coatings to inhibit biofilm production on metal surfaces.[1,2] Computational modeling and simulation methods, more specifically, classical mechanics approaches such as molecular dynamics (MD) simulations can be used to predict interactions between biomolecules and 2D materials at atomic levels to study their interfacial chemistry and physics. Over the last few decades, the development of MD techniques has rendered itself as a powerful biophysics tool and has led to significant advancements in the field of 2D material discovery and studying their intrinsic properties as well as understanding their interaction with biomolecules, allowing us to investigate intriguing questions on the nature of biomolecule-2D system mechanisms.[3] Furthermore, the results generated from such studies and these interactions can be fed into machine learning algorithms as training data sets to predict and extrapolate biomolecular behavior near various functionalized and defective 2D surfaces. Machine learning has also been adopted by the computational research community for forcefield parameterization and development for use in modeling these unique and complex 2D-biomolecule interfacial systems.[4,5] This chapter aims to provide an overview of the emerging area of computational interfacial biology, chemistry, and physics: atomistic and coarse-grained molecular dynamics simulations of chemically complex models of biomolecule-2D interface by focusing on key methodology present in current literatures; as well as introduce bioinformatics and machine learning tools used to handle and process the data generated. More specifically, recent advances in the field of computational techniques, bioinformatics, and machine learning for tasks such as force field development and free energy methodologies to study the free energy profiles; highlighting key challenges and prospects will be discussed.

While various levels of theories are used to study 2D materials and their interactions with biomolecules in the computational space, the choice of approach usually depends on the time and length scale of the system and problem in focus (Figure 10.1).

DOI: 10.1201/9781003132981-10

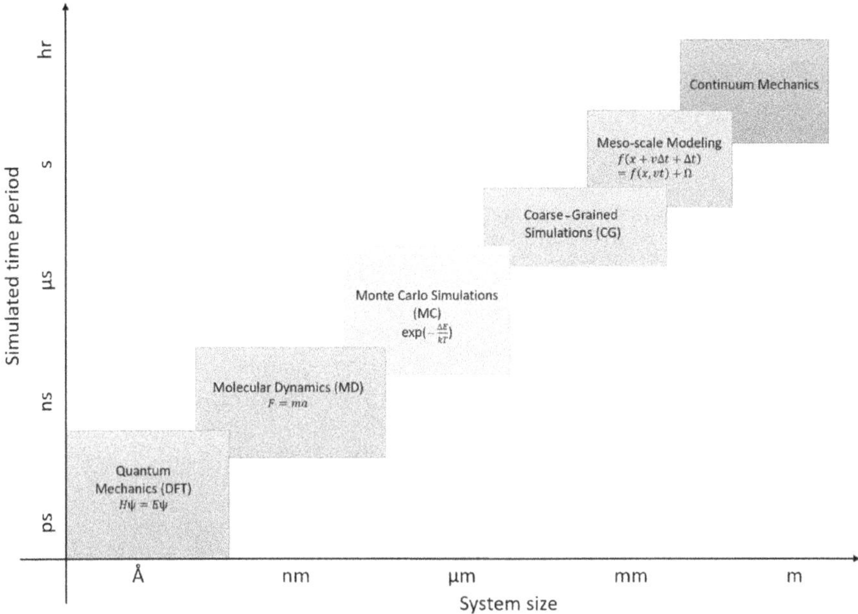

FIGURE 10.1 Time vs length scale representation of various simulation models.

MD is the most prominently used approach to capture 2D-biomolecue interfacial phenomena and study it's dynamics.

10.2 MOLECULAR DYNAMICS (ALGORITHMS AND METHODS)

Molecular dynamics (MD) is used to calculate a system's select properties through time propagation (evolution of system). In MD, all particles in a given system obey Newton's equation of motion, i.e., the classical second order differential equation of motion; that is given by:

$$F_i = m_i \frac{d^2 x_i}{dt^2} = -\nabla V\left(x_{ij}\right),\ i = 1,2,\ldots,N \tag{10.1}$$

Here, $V\left(x_{ij}\right)$ w.r.t position of each particle in the system is a time independent interatomic potential energy function. The Eq. 10.1 is solved for all atoms in a system with 3N position and velocity coordinates to study its dynamics. This fundamental equation for MD gives out the position and velocity of particles in a system and its evolution with time.

Simulation of any given system in MD requires us to solve this second-order differential equation of motion by implementing some numerical integration technique since an analytical solution for the same is impossible to achieve. The Verlet algorithm[6] is one of the most widely used integration scheme in MD. In a given system, the updated position and velocity of the particle following a time evolution according to the Verlet algorithm is given by Eqs 10.2 and 10.3.

$$x(t+\partial t) = 2x(t) - x(t-\partial t) + a(t)\partial t^2 \tag{10.2}$$

Here, the position of any given particle in the system at time t is given by $x(t)$; at a later time $(t+\partial t)$ where ∂t is the time step, a particle's updated position can be written as $x(t+\partial t)$. Similarly, $v(t)$ and $a(t)$ are velocity and acceleration of the particle at time t. Equations 10.2 and 10.3 are derived by neglecting the higher order terms of Taylor series expansion of the Verlet algorithm.

$$v(t) = \frac{x(t+\partial t) - x(t-\partial t)}{2\partial t} \tag{10.3}$$

These equations can be used iteratively to solve the equations of motion and obtain the next set of updated positions and velocities for every particle in a system it progresses toward equilibrium in a simulation.

Modeled empirical forcefields are used to describe the interactions in a system and the nature of the force acting on each particle in that given system. There are various forcefields developed and derived in the research community that can be found in literature. Depending on the uniqueness of the system, specific interactions can be modeled.

10.2.1 Empirical Forcefields

Empirical forcefields are broadly categorized into two forms: reactive and non-reactive (polynomial) forcefields. The key difference between these two subcategories is that in the case of a polynomial forcefield, the standard equation (Eq. 10.4) usually consists of terms to capture long-range, angle-bend, torsion energetics of a system but does not consist of a many-body term in the equation that may account for bond formation or breakage, i.e., no chemistry is captured in the system throughout the simulation. While a reactive forcefield (Eqs 10.5 and 10.7) consists of a bond order term which makes it possible for the system to undergo reaction (bond breaking/formation), these models are widely used to study reaction kinetics of a system. Some of the widely used reactive forcefield developed in the community include ReaxFF[7] and AIREBO[8]. ReaxFF is a state of the art 'reactive' forcefield developed by the van Duin group, capable of capturing chemical reactions in MD simulations. ReaxFF has been widely adapted and parameterized for complex 2D materials as well as biological systems in recent years[9] including but not limited to studying growth mechanisms of 2D transition metal carbides[10] and development of ReaxFF protein reactive forcefield (protein-2013)[11] used to simulate biomolecules and membrane fuel cells. Some of the developed forcefields used in literature to study biomolecule-2D interactions have been discussed in this section.

A number of empirical force fields have been developed to accurately model conformational energies and intermolecular interactions involving proteins, nucleic acids, and other molecules with related functional groups which are of interest in organic and biological chemistry. Currently, the most widely used all-atom polynomial force fields for proteins are OPLS/AA[12], CHARMM22[13] and AMBER.[14] For the scope of this chapter, AMBER forcefield is briefly discussed as a prime example of all-atom polynomial model implemented to study biomolecules computationally. The energy terms for Assisted Model Building with Energy Refinement (**AMBER**) force field are as follows:

$$E_{\text{system}} = \sum_{\text{bonds}} K_r\left(r - r_{eq}\right)^2 + \sum_{\text{angles}} K_\theta\left(\theta - \theta_{eq}\right)^2 + \sum_{\text{dihedrals}} \frac{V_n}{2}\left[1 + \cos\left(n\varphi - \gamma\right)\right]$$
$$+ \sum_{i<j}\left[\frac{A_{ij}}{R_{ij}^{12}} - \frac{B_{ij}}{R_{ij}^{6}} + \frac{q_i q_j}{\epsilon R_{ij}}\right] \tag{10.4}$$

The last term for Eq. 10.4 consists of non-bonded van der Waals interaction given by 12–6 Lennard-Jones (LJ) potential and the electrostatic interaction given by Coulomb's law. The rest of the terms in Eq. 10.4 define the bonded or intramolecular interactions which deal with forces present within a given molecule, i.e., energetics due to bond stretching, angle bending and torsional forces.

The Adaptive Intermolecular Reactive Empirical Bond Order potential (**AIREBO**) developed for hydrocarbons has been widely used in literature to study graphene based 2D materials and has proven to accurately capture their structural and thermodynamic properties.[15] The AIREBO potential consists of three terms, covalent bonding interactions, LJ term and torsion interaction; given by:

$$E_{\text{system}} = \frac{1}{2}\sum_i \sum_{j \neq i}\left[E_{ij}^{\text{REBO}} + E_{ij}^{LJ} + \sum_{k \neq i,j} \sum_{l \neq i,j,k} E_{kijl}^{\text{TORSION}}\right] \tag{10.5}$$

The E_{ij}^{REBO} term describes short-ranged C–C, C–H and H–H interactions and this reactive term is described as:

$$E_{ij}^{\text{REBO}} = V_{ij}^R\left(r_{ij}\right) + b_{ij} V_{ij}^A\left(r_{ij}\right) \tag{10.6}$$

The E_{ij}^{REBO}term is similar in its functional form to the REBO potential. For atoms i and j, V_{ij}^R, and V_{ij}^A are pairwise repulsive and attractive potential for atom types C and H, which are function of distance r_{ij} between the two atoms and the many-body bond order terms b_{ij}. A distance-dependent switching function that switches off the E_{ij}^{REBO} interactions when the atom pairs exceed the bonding distances.

For **ReaxFF**, the current model form of the potential consists of both reactive and non-reactive interactions between atoms, which allows the potential to accurately model both covalent and electrostatic interactions for a vast range of systems.

$$E_{\text{system}} = E_{\text{bond}} + E_{\text{over}} + E_{\text{angle}} + E_{\text{tor}} + E_{\text{vdWaals}} + E_{\text{Coulomb}} + E_{\text{Specific}} \tag{10.7}$$

The total energy of the system, E_{system} is divided into bond order dependent and independent contributions. E_{bond} is a function of interatomic distance and described the energy due to bond formation between atoms, this is calculated as:

$$\mathrm{BO}_{ij} = \mathrm{BO}_{ij}^{\sigma} + \mathrm{BO}_{ij}^{\pi} + \mathrm{BO}_{ij}^{\pi\pi} \tag{10.8}$$

$$= \exp\left[p_{\mathrm{bo1}}\left(\frac{r_{ij}}{r_o^{\sigma}}\right)^{p_{\mathrm{bo2}}}\right] + \exp\left[p_{\mathrm{bo3}}\left(\frac{r_{ij}}{r_o^{\pi}}\right)^{p_{\mathrm{bo4}}}\right] + \exp\left[p_{\mathrm{bo5}}\left(\frac{r_{ij}}{r_o^{\pi\pi}}\right)^{p_{\mathrm{bo6}}}\right]$$

where, BO_{ij} is the bond order term between atoms i and j which is a function of interatomic distance r_{ij} and the equilibrium bond length r_{o}; p_{bo} terms are empirical parameters; σ, π and $\pi\pi$ are the bond characteristics. Equation 10.8 takes into consideration covalent interactions in transition state structures which allows Reaxff to accurately predict reaction barriers for specific systems.

Understanding and studying the solvation effect is essential for exploring the structural dynamics of biomolecules in aqueous solution near 2D surfaces. Many implicit (continuum) and explicit (all atom) models are proposed and used to model solvents in a system. TIP3P and SPC/E are some of the widely used water models to simulate aqueous solutions in both the computational biomolecular and interface communities.[16–18]

The full functional form of the potential functions discussed in this section and their description can be found in their respective original literature. Due to the complexity of these models, specifically for biomolecule-2D systems, it is advisable to select the correct forcefield to model the interactions in your system and validate the forcefield and its implementations against the data present in the literature.

10.2.2 Periodic Boundary Conditions

Simulating a large number of particles in a given system and observing its evolution via MD is restricted by the computational power at hand. Often researchers focus on simulating a box with reduced volume and number of particles compared to the actual system's size, this in turn reduces the computational cost on both time and length scale, while still mimicking and capturing the key chemistry and physics of the system without comprising the accuracy. This is achieved by implementing periodic boundary conditions (PBC).

For PBC, the boundaries of the simulation box are considered to be continuous along all axes and periodic in nature, i.e., The simulation consists of an infinitely large system with the 'original' box repeating in each direction. Implementing PBC itself can be computational expensive and infeasible, as solving the long-range interactions term in various forcefields such as the one mentioned in Eqs. 10.4 and 10.5 in Section 10.2.1, for an infinite periodic system is impossible to compute. The concept of cutoff is introduced, for a single particle in PBC, instead of interacting with all the $(N-1)$ particles in the system, a sphere with radius r_{cutoff} around the particle is defined and the particle is allowed to interact with other particles within this defined sphere only (see Figure 10.2).

While using a cutoff with PBC significantly reduces the computational cost, sudden truncation of potentials introduces discontinuities in a system, which violates energy conservation along with producing incorrect thermodynamic properties of systems.

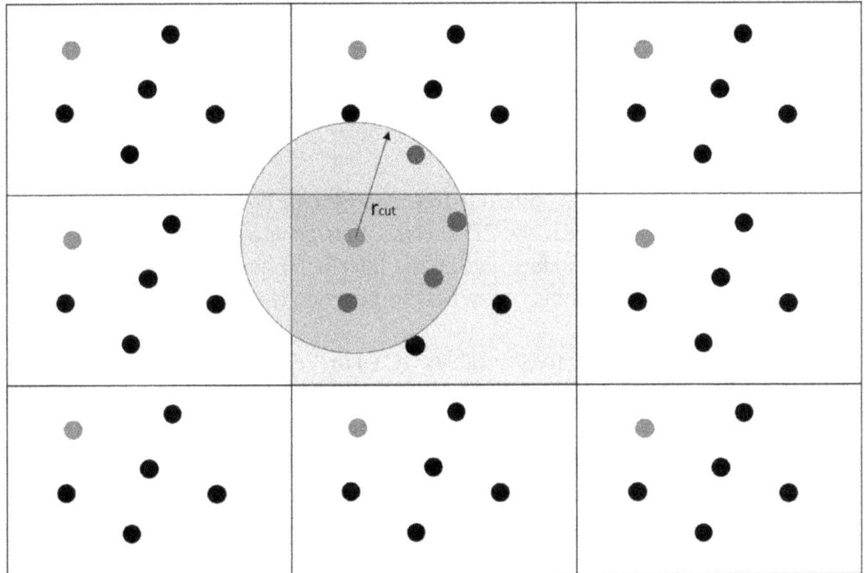

FIGURE 10.2 Periodic boundary conditions representation in 2D. The simulation box in the center is replicated in every direction to produce an infinite system. The *light-grey* colored particle interacts with every particle within a defined sphere of radius r_{cutoff}.

To tackle this challenge, long-range corrections and soft 'fading' of potentials are applied at the cutoffs.[19]

When implementing PBC within a MD simulation, one should take care that the size of the box has an appropriate length scale for the given system and the property being computed. This will help to avoid finite-size effects and related errors in computational results.

10.2.3 Binding Energy

While adsorption of biomolecules, which are complex and large in nature, are difficult to study due to computational limitations, literature studies have probed the interaction between the interaction sites and functional groups of amino acid side chain at the outermost surface of proteins on adsorption surfaces, giving an insight into protein adsorption mechanism on the surface. As such, small molecule amino acids serve as suitable model compounds to mimic protein-surface adsorption.[20–22] Dragneva *et al.*[22] studied and presented the adsorption of 20 proteogenic amino acids on a graphene surface. Furthermore, the effect of solvation on adsorption behavior of amino acids in presence of water was investigated. The binding energy E_{binding} is defined as:

$$E_{\text{binding}} = E_{\text{system}} - \left(E_{\text{slab}} + E_{\text{adsorbate}}\right) \tag{10.9}$$

where E_{system} is the total energy for the adsorbate-slab system; E_{slab} is the total energy of the pristine or defective 2D slab without any adsorbate in the system; and $E_{adsorbate}$ is the total energy of the adsorbate in the bulk phase.

10.2.4 Free Energy

For a biomolecule-2D system, binding free energy can be used to determine the affinity of the biomolecule near the 2D surface, thus generating a free energy landscape is an essential step in understanding the interfacial phenomena. Various techniques and methodologies employed in MD to generate the free energy profiles are discussed below.

In MD simulations, 'collective variables' (CV) or 'reaction coordinates' or 'variable sets' are used to reduce the degrees of freedom of a system into few parameters, which can be analyzed individually via ensemble averaging. Here, collective variables are any set of differentiable function of atomic cartesian coordinates, x_i, with i between 1 and N, the total number of atoms:

$$\xi(t) = \xi(X(t)) = \xi(x_i,(t),\ x_j(t),\ x_k(t)\ldots),\ 1 \le i,\ j,\ k\ \ldots \le N \qquad (10.10)$$

The restraints or biasing potentials can be applied to multiple variables or set $\xi(X)$ to calculate the potential mean force (PMF) on the system using different enhanced sampling methods, such as metadynamics, adaptive biasing force (ABF) and umbrella sampling.[23–25]

10.2.5 Umbrella Sampling

Umbrella sampling method is utilized to probe and generate the free energy landscape of a given system, as a function of a single reaction coordinate. Specifically, a biased harmonic potential is induced in the system to overcome the energy barrier separating any two regions of configuration space. The biased harmonic potential added to the system is simply defined as:

$$V(\xi) = \frac{1}{2}k\left(\frac{\xi - \xi_0}{w_\xi}\right) \qquad (10.11)$$

Here, ξ is centered at ξ_0 and is scaled by its characteristic length scale $w_\xi \cdot k$ is chosen equal to $\kappa_B T$ (thermal energy), the resulting probability distribution of $z = \xi_i - \xi_0 / w_\xi$ (dimensionless) is approximately a Gaussian with mean of 0 and standard deviation of 1.

Implementing the umbrella sampling methodology to obtain the free energy landscape of a system in a stepwise manner can be shown using Figure 10.3. The biasing potential is added to the natural unbiased system followed by division of the landscape into bins according to the reaction coordinate(s) (1D in the case of umbrella sampling), and the biased potential is allowed to act on the system distributed in bins

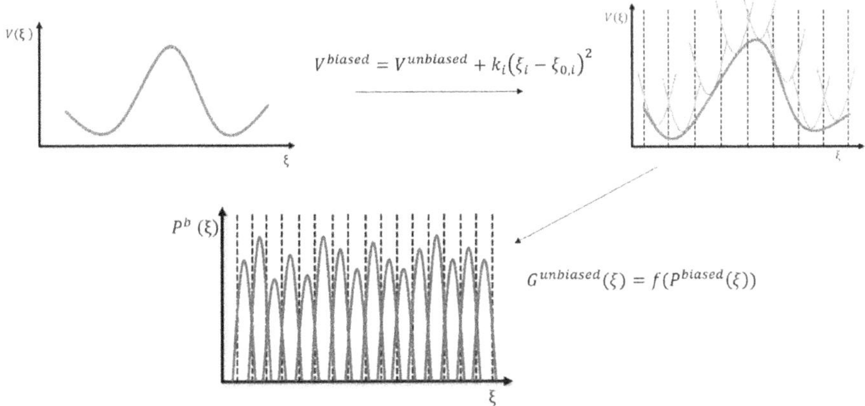

FIGURE 10.3 Schematic illustration of the umbrella sampling method. The small U-curves represent the harmonic bias potentials that are added to the unbiased system at different CV points (windows) along the CV space to generate the probability distribution.

to obtain the probability distribution. The free energy of the natural unbiased system is a function of the biased probability distribution.[26]

The concept of umbrella sampling and evaluating the free energy landscape of a system using MD can be better understood with an example. Here we study the interaction of a small biomolecule (phenylalanine) and its interaction with pristine graphene (2D surface). To calculate the free energy of adsorption of the capped AC-Amino acid-NHMe to the graphene surface, we defined X (the reaction coordinate) as the distance along the z-axis from the center of mass (COM) of AC-Amino acid-NHMe to the center of mass of an atom from the atomic layer of graphene (illustrated in Figure 10.4a) and calculated the potential mean force as a function of this distance, PMF(X). The z-axis is orthogonal to the plane of the graphene. Umbrella sampling combined with the weighted histogram analysis method, WHAM[27](post processing software/method to generate PMF curves from probability distributions), was used to calculate the PMF using a single window on the domain $X \in [3, 10]$ Å, with a bin width of $\Delta X = 0.5$ Å. The resulting histogram for the free energies study is shown in Figure 10.4b. The PMF curves are generated for the amino acid-graphene system is presented in Figure 10.4c.

Dasetty *et al.*[28] employed similar methodology to study the free energy of all 20 proteinogenic amino acids onto pristine graphene surface as a function of Z direction using umbrella sampling. The amino acid-graphene system was modeled using the force fields—Amberff99SB-ILDN/TIP3P, CHARMM36/modified-TIP3P, OPLS-AA/M/TIP3P, and Amber03w/TIP4P/2005, providing a comparative assessment on these forcefield and their ability to correctly capture the adsorbed state and free energy landscape of amino acids on graphene. Here, the systems were equilibrated at 300 K for 1 ns using NVT ensemble and steepest descent energy minimization algorithm. A time step of 2 fs was used for all MD simulations. Final production simulations were performed at 300 K for 10 ns. A spring constant (k) of 8000 kJ/mol/nm^2 for

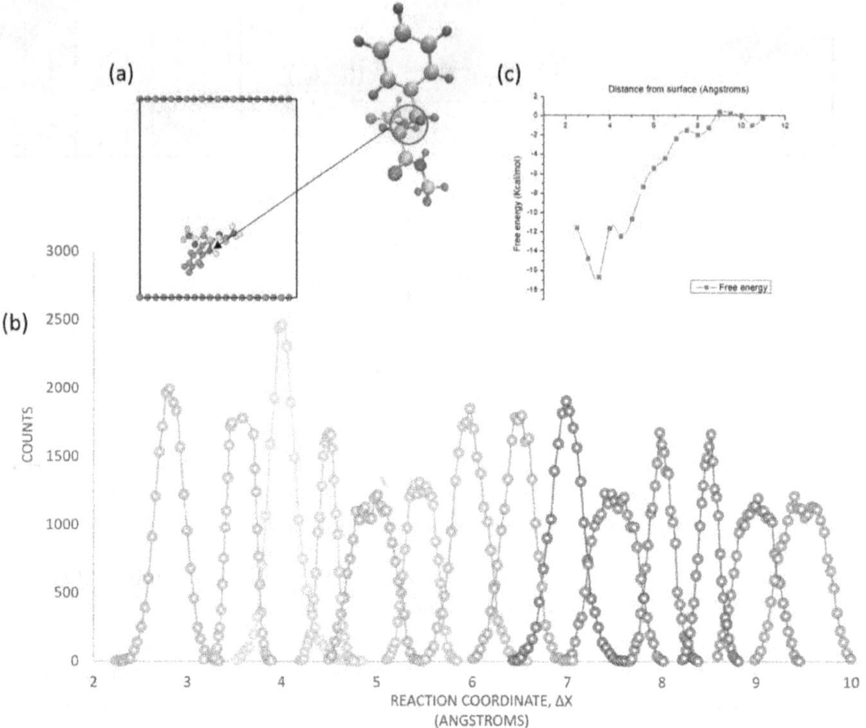

FIGURE 10.4 (a) Illustration of the harmonic spring on the center of mass of the biomolecule through reaction co-ordinate Z. (b) The histograms of the configurations within the umbrella sampling windows and the harmonic spring against the distance of graphene and phenylalanine. (c) PMF of phenylalanine on graphene generated, similar to PMF generated for various forcefields by Dasetty et al.[28]

0.4 nm $\leq \xi \leq$ 0.8 nm and k = 4000 kJ/mol/nm^2 for 0.9 nm $\leq \xi \leq$ 2.0 nm was employed with a spacing of 0.05 nm for 0.4 nm $< \xi <$ 0.8 and 0.1 nm for 0.9 nm $< \xi <$ 2.0 nm for good overlap between the distribution of neighboring windows. The PMF curves are then generated for amino acid-graphene system using various forcefield.

Similarly, Zheng *et al.*[29] studied the conformation change and aggregation of HIV-1 Vpr13–33 on graphene oxide (GO) by employing the umbrella sampling method. Water was represented by the TIP3P model. The system was energy minimized followed by equilibration for 500ps, 1 bar constant pressure and 298K temperature, followed by NVT simulation at 298K for 500ns for umbrella sampling. Thirty (30) configurations were generated along the z-axis. Here, z coordinates of COM distance between Vpr13–33 and GO in each configuration differed by 0.1nm. Each window was equilibrated for 5 ns and a production run of 5 ns was continued for sampling. The PMF curve was obtained using WHAM. Both single peptide on GO and double peptide aggregation in water PMFs were generated. Unfolding of peptide and loss of secondary structure near GO surface was observed with highly stable $\pi - \pi$ interactions; electrostatic interactions prevent the peptide from folding further. Interactions between single peptide and GO are much stronger than the inter-peptide interactions.

While not in the scope of this chapter, other thermodynamic integration methods such as Metadynamics, ABF and well-tempered Metadynamics, etc. have been implemented to study conformational free energies of various systems in MD. Metadynamics in particular uses gaussian hills and history dependent CV to explore phase space. Metadynamics simulations can be used to reveal the binding affinities and transition pathways of biomolecules near 2D and metal surfaces.[30,31]

10.2.6 Coarse-Grained Modeling

For studying binding energies and/or exploring conformational changes (protein folding near 2D surface) in a 2D-biomolecule system, computational length and time scale challenges are a major blockade to be faced. While trying to simulate and study interfacial science in a given system, coarse-grained modeling techniques employed tackle these challenges in a system by reducing the overall complexity of the system. Coarse-grained biomolecular systems are less computationally expensive than their all-atomistic system counterparts because coarse-grained models reduce the number of interaction sites and heavy atoms in a system (see Figure 10.5). Coarse-grained models are developed to contain fewer degrees of freedom (e.g., removal of the carbon–hydrogen bond vibrational modes), and are parameterized with smoother potential energy surfaces. This in turn leads to a smoother potential energy surface which reduces the challenges associated with overcoming energy barriers while exploring free energy landscapes, thereby leading to more efficient sampling. The MARTINI coarse-grained forcefield, which employs a four-to-one mapping (a single interaction site/bead is used to represent four heavy atoms), has been successfully implemented for simulating wide range of biomolecular systems such as DNA fragments, CG version of standard and polarizable water molecules, lipids and polysaccharide fragments, etc.[32–35]

In conclusion, coarse-grained simulations can access and evolve systems to length and time scales far beyond those that are practically achievable by all atomistic molecular dynamics simulations. Due to complexity and large number of heavy atoms in biological systems, coarse-grained modeling methods are the subject of considerable current interest in this community. However, since they do not represent molecules as all atomistic models, coarse-grained models lose out on finer details and inaccurately depict some important chemical features within a system.

10.3 EMPLOYMENT OF MD ON FUNCTIONAL 2D MATERIALS

2D materials are structures with a thickness on the order of 1–2 atomic layers. Many 2D materials of research interest are categorized in classes or treated as a special material like graphene, with the most relevant being transition metal dichalcogenides (TMDCs; WTe_2, MoS_2, etc.), atom thin layers of elements like tin and bismuth (stanene and bismuthene), and hexagonal boron nitride (h-BN), which is somewhat graphene-like in geometry and electron configuration but composed of covalently

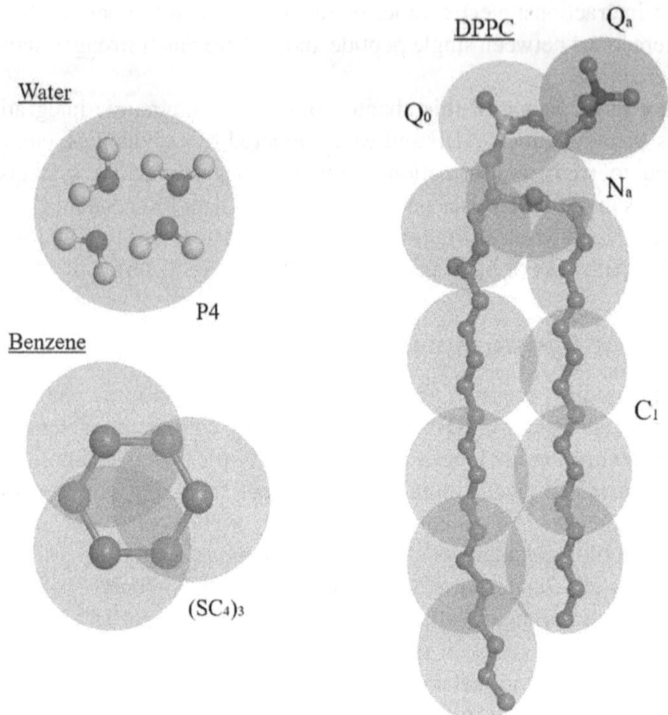

FIGURE 10.5 Mapping between the all-atom and the coarse-grained MARTINI models for water, benzene and DPPC membrane lipid molecules. Eg: one P4 bead/interaction site with specific LJ parameters is used to define four all-atom water molecules. Here the CG beads are shown as transparent vdW spheres and the hydrogens are only shown for atomistic water molecules.

bonded boron and nitrogen. Grown from single layer, heterostructures or multilayer 2D materials (e.g., multi-layer graphene) has been of utmost importance recently in a wide range of application. Multi-layer 2D materials or hetero 2D structures are vertically-stacked mechanically-assembled monolayer flakes and are held together with long range forces such as van der Waals forces.[36] There are several wide range applications of 2D materials in fields of (1) electrical and electronic (e.g., battery storage system and semi-conductors), (2) biomedical (e.g., drug carriers), and (3) microbiology (e.g., prevention of corrosion in steel pipes).[37–39] Computational techniques have been widely used for 2D materials modeling and studying their properties such as electronic structure and mechanical properties, from ab initio level of theory to classical and semiclassical approaches.[40]

10.3.1 Graphene and Its Structural Defects

Of all materials known, graphene has the highest tensile strength, the highest electron mobility, and the highest thermal conductivity. The term graphene-family nanomaterials (GFNs) refer to many different graphene-like materials which can be classified by their number of layers (few-layer graphene (FLG) to graphene nanosheets (GNSs)) and/or their degree of oxidation, with graphene oxide (GO) being oxidized and reduced GO (rGO) being oxidized and then reduced. Carbon nanotubes could be considered a member of GFNs as they are essentially a rolled layer of graphene with universal sp^2 bonding in a cylindrical geometry and are the subject of their own intensive research.[41–45]

All carbon-carbon bond lengths in graphene are 0.142 nm, and graphene's unit cell is a rhombus with edge lengths (lattice constant) of 0.246 nm and two central basis carbon atoms.[46] This means that the shortest linear distance between two non-adjacent carbon atoms within the same hexagon is 0.246 nm, and the distance between any hexagonal center to an adjacent one is also 0.246 nm.

Defects in graphene sheets can alter its mechanical, chemical, and electronic properties. These defects are either undesired and generated during the manufacturing stage or can be engineered for use in important applications.[47] Point defects such as single-vacancy (SV) and Stone-Wales(SW) defects are widely studied in literature[48]; here the SV defect refers to a missing lattice atom, while the SW defect means one of the C-C bonds is rotated by 90 degrees resulting in four hexagons in a pristine graphene sheet transforming into two pentagons and two heptagons (see Figure 10.6).

Yoon *et al.*[49] studied the generation of defects in graphene using irradiation simulations in MD. Carbon atoms in graphene were modeled using ReaxFF and the interaction parameters between graphene and ions were optimized using density functional theory (DFT) calculations and universal repulsive potential. The system was equilibrated at 300K, ions with impact energy of 25keV were irradiated on the center area of graphene. Extremely small-time steps or 0.005-0.02 fs were used during irradiation. He^+-irradiated graphene exhibited SW defects most frequently (~65%) while Ne^+-, Ar^+-, and Kr^+- irradiated graphene exhibited SV defects most frequently (~73%)

Generating these point defects can also affect the binding and absorption of key biomolecules onto graphene surface. The surface defects lead to enhanced charge

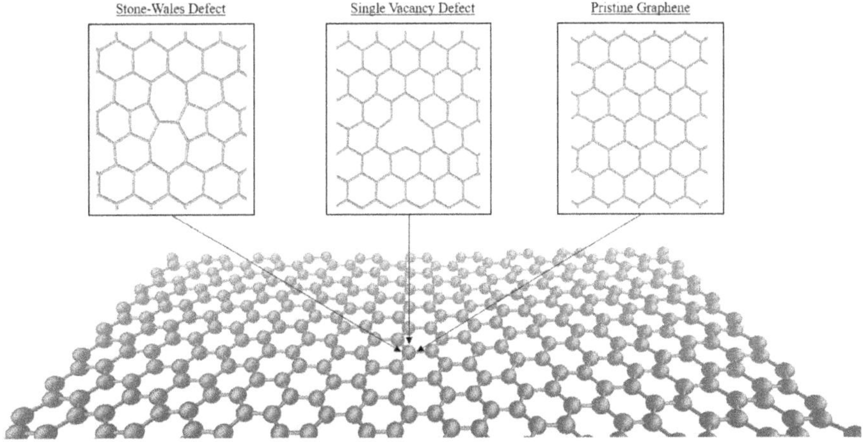

FIGURE 10.6 Pristine graphene and two most common point defects of graphene.

transfer and exposed active sites and has attained great attention toward development of non-enzymatic electrochemical biosensors.[50] The defects in 2D materials results in unexpected changes in protein behavior due to denaturation and unfolding. Gu *et al.*, investigated the impact of defective graphene and ideal graphene on a model protein (chicken villin headpiece subdomain, HP35) using molecular dynamics studies. The report suggested that the protein has undergone severe denaturation, while the protein was observed to be undamaged on ideal graphene.[51] The contacted amino acid residues was observed to be tightly anchored near the defects owing to favorable electrostatic interactions, however, at the interface, the residues are highly restrained. Therefore, the thermal movements of the remaining residues led the protein to denature or unfolding.[51,52] A biofilm associated study showed that graphene monolayer on copper surface enhanced the biogenic sulfide attack by 5-folds as compared to the bare copper, but multilayered graphene, when used, inhibited the biofilm formation.[53] Dong *et al.*, showed that the effectiveness of graphene coated copper surface toward microbial corrosion is time-dependent and the prolonged exposure to ionic environments results in defective graphene coatings.[54] Furthermore, Chilkoor *et al.*, demonstrated that anaerobic microbial corrosion due to *Oleidesulfovirbio alaskensis* G20 can be inhibited with a monolayer of h-BN. The impermeable nature of the monolayer prevents the diffusion of corrosive metabolites toward the metallic substrate.[55] Therefore, h-BN coatings are promisingly effective at minimizing galvanic effects as the local defects act as a cathodic site for anchoring and reducing terminal electron acceptors, which resulted in inhibition of microbial corrosion.[55,56]

10.3.2 The Emergence of Bioinformatics: Applications and Methodologies

Open-source MD codes such as LAMMPS[57] are widely used to model and simulate complex 2D-biomolecule systems by employing the various algorithms and techniques discussed in previous sections. Bioinformatics online and offline software are powerful tools that can aid in performing this thorough investigation of the interaction between the 2D materials and biomolecules, more specifically bioinformatics can aid in predicting protein structures of interest and propose optimized conformation of the said protein that can be used in the 2D-biomolecule system.[58,59] Bioinformatics is a multidisciplinary field which can be defined as an amalgamation between biological science and information technology to retrieve and solve "big data" problems using biological databases and programming algorithms.[60] Bioinformatics has broad range applications in all three genomics- structural, functional, and comparative genomics.[61,62] The prediction of protein structures at the secondary and tertiary level is the structural genomics, whereas assigning the functionality to an unknown protein using already available data is defined as functional genomics.[63] Comparative genomics, however, tells the evolutionary nature of a species and its environmental origin.[64,65]

The offline python-based command line interface, Modeler v10.3 can be employed to predict the structure of unknown protein with amino acid sequence as an input and template protein structure.[66,67] The template protein structure is an experimental (X-ray or nuclear magnetic resonance) protein structure and can be retrieved from Protein DataBank.[68] The interface generates the desired number of poses or

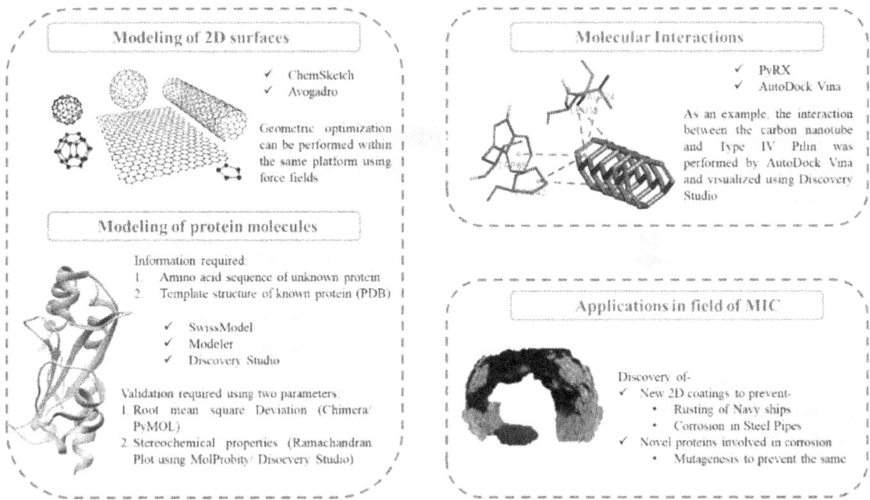

FIGURE 10.7 Computational tools and strategy to discover novel 2D coatings and their application as MIC inhibitors.

conformers; the most stable being the least discrete optimized potential energy score. The validation of the modeled protein structure can be performed using MolProbity (online server) and PyMOL (offline software) for stereochemical correctness, and superimposed root mean square deviation (RMSD), respectively.[69–71] The less the RMSD, the more correct are the coordinates and torsions of the α- and β-carbons.[72] Molecular docking is one of the most powerful strategies to calculate binding energy between the surfaces and visualize the most crucial amino acids that participates during the course of interaction.[73] There are several tools available as online server or offline installation packages that can be used directly for rigid or flexible docking, most popularly, AutoDock (The Scripps Research Institute) and PyRX (incorporates vina, genetic and Lamarckian genetic algorithm).[74,75] The platform returns a number of interaction-poses with different ligand RMSDs (represents various conformers of the same ligand at all possible hydrophobic pocket sites in a protein). The most negative binding energy is the most favorable interaction and provides the most stable complex. As an example, Figure 10.7 denotes the overall workflow and application of protein modeling and molecular interaction to study microbial induced corrosion (MIC).

10.3.3 Current Trends in Biomolecular Simulation and Modeling

Several studies had been performed using inhibitory material and the computational simulation helped to understand the inhibition and interaction mechanism at the surface atomistic level. Khadom *et al.*, theoretically simulated the *Citrus Aurantium* leaf extract as a bio-inhibitor for biochemical corrosion of mild steel in acidic solution using homology modeling and molecular docking.[76] Hussein *et al.*, studied the inhibition of *Acidithiobacillus Ferrooxidans* bacteria, which is responsible for the corrosion using the compound 1-Isoquinolinyl phenyl ketone by employing AutoDock vina algorithm and density functional theory (DFT).[77] Ahmed *et al.*, studied the interactions between dietary fibers and meat proteins to understand the textural changes within the protein with estimation of parameters, such as the formation of hydrogen bonds, the free energy of binding, and Van der Waals and desolvation energy.[78] These provides an understanding about evaluations of an extent of parameters from interaction studies. Furthermore, Kinghorn *et al.*, provided insights on the progress of aptamer (a small stretch of nucleic sequences that are known for its specificity and high-binding affinity) bioinformatics, and how the computational simulations with regard to fragment-based aptamer design, and identification of lead aptamers from high-throughput sequencing data have progressed over years.[79] These procedural aptamer studies may be crucial for the identification of protein domains that are promising to have higher affinity toward 2D surfaces. Zhao *et al.*, used graphitic carbon nitride as an analog of graphene, and performed interaction studies with 20 amino acids using DFT, and reported that graphitic carbon nitride attaches to amino acids using the amino group (-NH2).[80] Unal *et al.*, reported promising antimicrobial effect of graphene oxide nanosheets on the SARS-CoV-2 Surface Proteins and Cell Receptors using interaction study as a basis.[81] Therefore, bioinformatics provides a broader opportunity to elucidate 2D materials and determine their antimicrobial effect with consideration of unknown target proteins and chemically modified surfaces as inhibitors.

10.4 MACHINE LEARNING

Since the development of Machine Learning (ML), it has been used in many fields including bioinformatics, materials science, data mining, and computer vision.[82] ML models can be applied to predict the fundamental properties (e.g., mechanical, and elastic properties), and thermal stability of 2D materials, which have the potential to advance the process of designing new 2D materials.[83–85] Moreover, ML optimization algorithms (for example, Bayesian optimization) can be used to design and predict new 2D materials with desired properties.[86,87] ML methods and optimization techniques can also be applied to develop forcefields for molecular simulation. Since the early 2010s, ML has also been used extensively to predict protein structural information such as secondary structures, accessible surface areas, and torsional angles.[88]

10.4.1 ML Methods for 2D Materials

Kastuar *et al.* developed ML-based predictive models using temperature-dependent elastic and mechanical properties of 2D crystals.[83] The worked implemented XGBoost and LightGBM to predict the lattice constant using temperature, space group, vacuum size, C_{11}, and C_{12} as input features. XGBoost is a scalable machine learning algorithm that creates a weak learner at each step and improves prediction accuracy by building a set of decision trees. After summing all models, it creates the final tree model defined as,

$$\widehat{y_i}^{(t)} = \sum_{n=1}^{t} f_n(x_i) \tag{10.12}$$

Where, $\widehat{y_i}^{(t)}$ is the final model, $f_n(x_i)$ are the generated tree models. The optimum algorithm is obtained by minimizing the following loss function:

$$\text{Objective function} = \sum_{n=1}^{t}\left[g_n f_t(x_n) + \frac{1}{2}h_n f_t^2(x_n)\right] + \varphi(f_t) \qquad (10.13)$$

where g_i and h_i are the first and second order gradient statistics on the loss function, $\varphi(f_t)$ is the regularization term. By using regularized objective model, XGBoost is able to prevent the overfitting problem. All decision tree-based ML models include model-based feature importance analysis techniques that provide a rank of feature importance. The XGBoost model outperformed the LightGBM in terms of the following two loss functions: R^2 (R-squared) and MSE (mean squared error).[89] Finally, the most influential parameters for the lattice constant were determined using model-based feature importance analysis. Tawfik *et al.* employed four ML models including random forest, support vector machine, relevance vector machine, and neural network with a combination of DFT to predict the interlayer distance and the band gap of hybrid 2D materials.[84,90] For the models input 1529 descriptors were calculated using the property-labeled materials fragments, which has excellent performance for ML application of crystals.[91]

To create new 2D materials, a variety of techniques, including defect engineering and atom or molecule adsorption, can be used.[87] Among these methods, defect engineering is an appealing option for identifying new 2D materials and application of ML methods can speed up the process of predicting defect properties in 2D materials. Frey *et al.* employed ML methods in designing ideal defect structures.[92] First, deep neural networks (DNNs) were used to predict material properties in order to find optimal host 2D materials. Since the DNNs need large number of training data, this work implemented 'transfer learning'[93] for the prediction of formation energy, band gap, and Fermi energy. After removing 8 compounds containing heavy elements, the process yielded the identification of 150 wide band gap 2D materials. More than 10,000 defect structures were produced by combining 150 wide band gaps and 70 defect structures containing all possible vacancies, divacancies, antisites, and common dopants. To identify potential defects, one classification model capable of identifying the deep center defect and one regressor mode capable of predicting defect formation energy were developed. For ML model input, the structural and chemical properties of the host materials and defects were used. For both the classification and regressor problems, a random forest algorithm was used. To find the best defect candidate, a defect score function was defined that expresses fitness as a possible deep center for quantum emission. The goal is to maximize the function value, which indicates the best defect candidates.

10.4.2 ML for Force Field Development and Parameterization

As discussed earlier in this chapter, forcefields in MD expresses atomic interactions via parametrized analytical functional forms. A forcefield is heavily reliant on experimental data for calibration, raising the question of whether ML approaches can be used to evaluate the nonlinear associations between atomic configurations and forces using benchmark data (e.g., quantum mechanics based materials simulations).[94]

Li *et al.* applied the genetic algorithm to optimize force fields parameters.[95] Genetic algorithm is based on biological evolution and mimics the process of natural selection and greatly use for optimization purposes.[96] The study focused on the nonbonded electrostatic and van der Waals (vdW) parameters because other parameters (e.g., bond, dihedral, and torsional parameters) have trivial effect on condensed liquid phase systems modeled using the Atomic Multipole Optimized Energetics for Biomolecular Applications (AMOEBA) forcefield.[97] The work considered the AMOEBA functional form due to its relative simplicity. The AMOEBA force field includes 44 independent parameters for computing electrostatic energy and 10 parameters for computing vdW energy. For optimizing electrostatic parameters, at the first step, atomic multipoles were obtained from quantum mechanics electrostatic potential on the Connolly surface of a single isolated methanol molecule. The second step involved the use of a large dataset containing 4943 methanol dimers to optimize 44 electrostatic parameters that minimize a predefined optimization function. Following electrostatic parameters optimization, vdW interactions parameters were optimized to find the best match the interaction between a central methanol molecule and its closet super-molecule. The optimized electrostatic parameters were remained unchanged during vdW optimization. An objective function was defined that looks for vdW parameters which will minimize the objective function.

Moreover, ML can be utilized to develop forcefields from atomic configurations and forces, which has the potential to greatly speed up atomistic materials modeling processes. However, atomic configurations need be converted to numeric representation, which is commonly referred to as fingerprints. The next step is to select the appropriate set of fingerprint features for a training model. The generation of fingerprints frequently produced high-dimensional data. High-dimensional data is defined as the number of features close to or higher than the sample size[98] and can degrade the accuracy and computational speed of ML models.[99] Principal component analysis (PCA) can be employed to reduce the dimensionality of large datasets. PCA generates a linear combination of variables from a large number of variables to reduce the dimension of data while retaining most of the variation in the dataset.[100] Finally, ML models such as DNNs and nonlinear regression can be implemented for developing forcefields.

10.4.3 ML for Protein Structure Prediction

Protein structure prediction (PSP) is a central problem in structural bioinformatics. The goal of protein bioinformatics is to reveal the relationships between amino acids and its function. These insights can be used to identify and design proteins that can bind specific targets, act as catalysts in reactions, or guide biotechnology advances. In recent years, ML, particularly deep neural networks, has been extensively used for PSP. PSP models can take input of protein sequence in variety of formats, including multiple sequence alignment (MSA) and position-specific scoring matrices (PSSMs) and can return the results in a variety of formats (e.g., 1-D, 2-D, and 3-D prediction).[101–104]

PSP systems using ML consist of three components: (1) the inputs which contain protein sequences in different forms, (2) a 'ML algorithm' typically deep neural networks are widely used for PSP, (3) the outputs that can be represented by 1-D prediction (e.g., solvent accessibility prediction), 2-D prediction (e.g., contact maps), and 3-D prediction (e.g., tertiary structure of a protein).

10.5 SUMMARY

Over the years, with an emerging appeal to solve complex interactions, computational and/or bioinformatic techniques that involve molecular dynamics and atomistic simulation are of utmost importance among today's researchers. Within this chapter we have presented the fundamental concept of molecular simulations focusing particularly on the molecular dynamics technique and its algorithms. Here, molecular dynamics approach can be employed to study interactions at 2D-biomolecule interfaces. We have highlighted the use of system energy to calculate the binding energies of biomolecule to 2D surface as well as addition of biasing potential to the system to implement umbrella sampling method, to map out free energies for the system.

While defect-free graphene growth is still under investigation and these defects remain largely undesired, studies have shown defective graphene surfaces have properties distinct from the pristine layers. What remains to be known is what effect these surface defects have on biomolecule adhesion, since very little foundational atomistic-level information is available on whether these defects produce positive or negative surface adsorption characteristics relevant to biomolecule adhesion. Recent advances and findings on defect engineering and adhesion of biomolecules on defective graphene using MD has been discussed.

A brief introduction to bioinformatics has been presented, as a promising field to deal with proteins; thereafter to determine the role of unknown proteins and evaluate the chemical bonding nature of interactions using molecular modeling and molecular docking, respectively. To interpret and process the data available and generated while studying biomolecule-2D interface, some key concepts related to use of machine learning (ML) to aid computational methods have also been discussed. The emerging algorithms of machine learning such as XGBoost and LightGBM are in extensive use to determine properties of 2D materials and to develop decision tress for discovery of potential new 2D materials. ML can also be incorporated for forcefield parameterization and development along with aiding in protein structure prediction.

The amalgamation of computational and bioinformatics strategies with machine learning is a thought-provoking approach to put a deep insight into the atomistic interactions of biomolecules at 2D surfaces.

REFERENCES

1. Cacaci, M. *et al.* Graphene oxide coatings as tools to prevent microbial biofilm formation on medical device. *Adv Exp Med Biol* **1282**, 21–35 (2020).
2. Agarwalla, S. V. *et al.* Hydrophobicity of graphene as a driving force for inhibiting biofilm formation of pathogenic bacteria and fungi. *Dent Mater* **35**, 403–413 (2019).

3. Wang, Q. *et al.* Computer simulation of biomolecule-biomaterial interactions at surfaces and interfaces. *Biomed Mater (Bristol)* **10**. Preprint at https://doi.org/10.1088/1748-6041/10/3/032001 (2015).
4. Avery, C., Patterson, J., Grear, T., Frater, T. & Jacobs, D. J. Protein function analysis through machine learning. *Biomolecules* **12**, 1246 (2022).
5. Kaptan, S. & Vattulainen, I. Machine learning in the analysis of biomolecular simulations. *Adv Phys: X* **7**. Preprint at https://doi.org/10.1080/23746149.2021.2006080 (2022).
6. Verlet, L. Computer 'experiments' on classical fluids. I. Thermodynamical properties of Lennard-Jones molecules. *Phys Rev* **159**, 98 (1967).
7. Van Duin, A. C. T., Dasgupta, S., Lorant, F. & Goddard, W. A. ReaxFF: A reactive force field for hydrocarbons. *J Phys Chem A* **105**, 9396–9409 (2001).
8. Stuart, S. J., Tutein, A. B. & Harrison, J. A. A reactive potential for hydrocarbons with intermolecular interactions. *J Chem Phys* **112**, 6472–6486 (2000).
9. Senftle, T. P. *et al.* The ReaxFF reactive force-field: Development, applications and future directions. *npj Comput Mater* **2**, 1–14 (2016).
10. Sang, X. *et al.* In situ atomistic insight into the growth mechanisms of single layer 2D transition metal carbides. *Nat Commun* **9**, 1–9 (2018).
11. Zhang, W. & Van Duin, A. C. T. Improvement of the ReaxFF description for functionalized hydrocarbon/water weak interactions in the condensed phase. *J Phys Chem B* **122**, 4083–4092 (2018).
12. Jorgensen, W. L., Maxwell, D. S. & Tirado-Rives, J. Development and testing of the OPLS all-atom force field on conformational energetics and properties of organic liquids. *J Am Chem Soc* **118**, 11225–11236 (1996).
13. MacKerell, A. D. *et al.* All-atom empirical potential for molecular modeling and dynamics studies of proteins. *J Phys Chem B* **102**, 3586–3616 (1998).
14. Cornell, W. D. *et al.* A second generation force field for the simulation of proteins, nucleic acids, and organic molecules. *J Am Chem Soc* **117**, 5179–5197 (1995).
15. Qian, C., Mclean, B., Hedman, D. & Ding, F. A comprehensive assessment of empirical potentials for carbon materials COLLECTIONS A reactive potential for hydrocarbons with intermolecular interactions Machine learning for interatomic potential models A comprehensive assessment of empirical potentials for carbon materials. *J Chem Phys* **153**, 61102 (2020).
16. Mark, P. & Nilsson, L. Structure and dynamics of the TIP3P, SPC, and SPC/E water models at 298 K. *J Phys Chem A* **105**, 9954–9960 (2001).
17. Kleinjung, J. & Fraternali, F. Design and application of implicit solvent models in biomolecular simulations. *Curr Opin Struct Biol* **25**, 126 (2014).
18. Ren, P. *et al.* Biomolecular electrostatics and solvation: A computational perspective. *Q Rev Biophys* **45**, 427 (2012).
19. Siperstein, F., Myers, A. L. & Talu, O. Long range corrections for computer simulations of adsorption. *Mol Phys* **100**, 2025–2030 (2002).
20. Welch, C. M. *et al.* Computation of the binding free energy of peptides to graphene in explicit water. *J Chem Phys* **143**, (2015).
21. Hughes, Z. E. & Walsh, T. R. What makes a good graphene-binding peptide? Adsorption of amino acids and peptides at aqueous graphene interfaces. *J Mater Chem B* **3**, 3211–3221 (2015).
22. Dragneva, N. *et al.* Favorable adsorption of capped amino acids on graphene substrate driven by desolvation effect. *J Chem Phys* **139**, 174711 (2013).
23. Hénin, J., Fiorin, G., Chipot, C. & Klein, M. L. Exploring multidimensional free energy landscapes using time-dependent biases on collective variables. *J Chem Theory Comput* **6**, 35–47 (2009).
24. Fiorin, G., Klein, M. L. & Hénin, J. Using collective variables to drive molecular dynamics simulations. *Mol Phys* **111**, 3345–3362 (2013).

25. Bernardin, A. *et al. Collective Variables Module Reference Manual for LAMMPS.* Albuquerque, NM: Sandia National Laboratories (2013). https://docs.lammps.org/PDF/colvars-refman-lammps.pdf
26. You, W., Tang, Z. & Chang, C. E. A. Potential mean force from umbrella sampling simulations: What can we learn and what is missed? *J Chem Theory Comput* **15**, 2433–2443 (2019).
27. Grossfield, A. An implementation of WHAM: The Weighted Histogram Analysis Method Version 2.0.10.
28. Dasetty, S., Barrows, J. K. & Sarupria, S. Adsorption of amino acids on graphene: Assessment of current force fields. *Soft Matt* **15**, 2359–2372.
29. Zeng, S., Zhou, G., Guo, J., Zhou, F. & Chen, J. Molecular simulations of conformation change and aggregation of HIV-1 Vpr13-33 on graphene oxide. *Sci Rep* **6**, 1–7 (2016).
30. Laio, A. & Parrinello, M. Escaping free-energy minima. *Proc Natl Acad Sci USA* **99**, 12562–12566 (2002).
31. Sampath, J., Kullman, A., Gebhart, R., Drobny, G. & Pfaendtner, J. Molecular recognition and specificity of biomolecules to titanium dioxide from molecular dynamics simulations. *npj Comput Mater* **6**, 1–8 (2020).
32. De Jong, D. H. *et al.* Improved parameters for the MARTINI coarse-grained protein force field. *J Chem Theory Comput* **9**, 687–697 (2013).
33. Roel-Touris, J. & Bonvin, A. M. J. J. Coarse-grained (hybrid) integrative modeling of biomolecular interactions. *Comput Struct Biotechnol J* **18**, 1182–1190 (2020).
34. Singh, N. & Li, W. Recent advances in coarse-grained models for biomolecules and their applications. *Int J Mol Sci* **20**, 3774 (2019).
35. Monticelli, L. *et al.* The MARTINI coarse-grained force field: Extension to proteins. *J Chem Theory Comput* **4**, 819–834 (2008).
36. Androulidakis, C., Zhang, K., Robertson, M. & Tawfick, S. Tailoring the mechanical properties of 2D materials and heterostructures. *2D Mater* **5**. Preprint at https://doi.org/10.1088/2053-1583/aac764 (2018).
37. Mujib, S. bin, Ren, Z., Mukherjee, S., Soares, D. M. & Singh, G. Design, characterization, and application of elemental 2D materials for electrochemical energy storage, sensing, and catalysis. *Mater Adv* **1**. Preprint at https://doi.org/10.1039/d0ma00428f (2020).
38. Kurapati, R., Kostarelos, K., Prato, M. & Bianco, A. Biomedical uses for 2D materials beyond graphene: Current advances and challenges ahead. *Adv Mater* **28**. Preprint at https://doi.org/10.1002/adma.201506306 (2016).
39. Ebrahimi, A. (Invited) Tailoring 2D Materials As Artificial Enzymes for Developing Electrochemical Biosensors. *ECS Meeting Abstracts, MA2021-01*, IOP Publishing (2021). https://iopscience.iop.org/article/10.1149/MA2021-01551409mtgabs
40. Carvalho, A., Trevisanutto, P. E., Taioli, S. & Neto, A. H. C. Computational methods for 2D materials modelling. *Rep Prog Phys* **84**, 106501 (2021) doi:10.1088/1361-6633/ac2356.
41. Balandin, A. A. *et al.* Superior thermal conductivity of single-layer graphene. *Nano Lett* **8**, 902–907 (2008).
42. Lee, C., Wei, X., Kysar, J. W. & Hone, J. Measurement of the elastic properties and intrinsic strength of monolayer graphene. *Science* **321**, 385–388 (2008).
43. Geim, A. K. & Novoselov, K. S. The rise of graphene. *Nat Mater 2007 6:3* **6**, 183–191 (2007).
44. Sanchez, V. C., Jachak, A., Hurt, R. H. & Kane, A. B. Biological interactions of graphene-family nanomaterials: An interdisciplinary review. *Chem Res Toxicol* **25**, 15–34 (2012).
45. Ou, L. *et al.* Toxicity of graphene-family nanoparticles: A general review of the origins and mechanisms. *Part Fibre Toxicol* **13**. Preprint at https://doi.org/10.1186/s12989-016-0168-y (2016).

46. Yang, G., Li, L., Lee, W. B. & Ng, M. C. Structure of graphene and its disorders: a review. *Sci Technol Adv Mater* **19**, 613–648. Preprint at https://doi.org/10.1080/14686996.2018.1494493 (2018).
47. Achtyl, J. L. *et al.* Aqueous proton transfer across single-layer graphene. *Nat Commun* **6**, 6539 (2015).
48. Banhart, F., Kotakoski, J. & Krasheninnikov, A. V. Structural defects in graphene. *ACS Nano* **5**, 26–41 (2011).
49. Yoon, K. *et al.* Atomistic-scale simulations of defect formation in graphene under noble gas ion irradiation. *ACS Nano* **10**, 8376–8384 (2016).
50. Briggs, N. *et al.* A roadmap for electronic grade 2D materials. *2D Mater* **6**. Preprint at https://doi.org/10.1088/2053-1583/aaf836 (2019).
51. Gu, Z. *et al.* Defect-assisted protein HP35 denaturation on graphene. *Nanoscale* **11**, 19362–19639 (2019).
52. Li, B., Bell, D. R., Gu, Z., Li, W. & Zhou, R. Protein WW domain denaturation on defected graphene reveals the significance of nanomaterial defects in nanotoxicity. *Carbon N Y* **146**, 257–264 (2019).
53. Chilkoor, G. *et al.* Atomic layers of graphene for microbial corrosion prevention. *ACS Nano* **15**, 447–454 (2021).
54. Dong, Y., Liu, Q. & Zhou, Q. Time-dependent protection of ground and polished Cu using graphene film. *Corros Sci* **90**, 69–75 (2015).
55. Chilkoor, G. *et al.* Hexagonal Boron Nitride: The Thinnest Insulating Barrier to Microbial Corrosion. *ACS Nano* **12**, 2242–2252 (2018).
56. Santos, J., Moschetta, M., Rodrigues, J., Alpuim, P. & Capasso, A. Interactions between 2D materials and living matter: A review on graphene and hexagonal boron nitride coatings. *Front Bioeng Biotechnol* **9**. Preprint at https://doi.org/10.3389/fbioe.2021.612669 (2021).
57. Plimpton, S. Short-range molecular dynamics. *J Comput Phys* **117**, 1–42 (1997).
58. Navanietha Krishnaraj, R., Samanta, D. & Sani, R. K. Computational nanotechnology: A tool for screening therapeutic nanomaterials against Alzheimer's disease. In *Neuromethods* vol. 132, 613–635. New York: Springer (2018). https://link.springer.com/protocol/10.1007/978-1-4939-7404-7_21
59. Krishnaraj, R. N., Samanta, D., Kumar, A. & Sani, R. Bioprospecting of thermostable cellulolytic enzymes through modeling and virtual screening method. *Can J Biotechnol* **1**, 19 (2017).
60. Tripathi, A. K. *et al.* Transcriptomics and functional analysis of copper stress response in the sulfate-reducing bacterium Desulfovibrio alaskensis G20. *Int J Mol Sci* **23**, 1396 (2022).
61. Mahato, D., Samanta, D., Mukhopadhyay, S. S. & Krishnaraj, R. N. A systems biology approach for elucidating the interaction of curcumin with Fanconi anemia FANC G protein and the key disease targets of leukemia. *J Recept Signal Transduct* **37**, 276–282 (2017).
62. Uversky, V. N. Protein intrinsic disorder and structure-function continuum. In *Prog Mol Biol Transl Sci* **166**, 1–17 (2019).
63. Sali, A. *et al.* Comparative protein structure modeling of genes and genomes. *Acta Crystallogr A* **58**, 291–325 (2002).
64. Bansal, A. K. Bioinformatics in microbial biotechnology: A mini review. *Microb Cell Factories* **4**. Preprint at https://doi.org/10.1186/1475-2859-4-19 (2005).
65. Martí-Renom, M. A. *et al.* Comparative protein structure modeling of genes and genomes. *Ann Rev Biophy Biomol Struct* **29**. Preprint at https://doi.org/10.1146/annurev.biophys.29.1.291 (2000).
66. Samanta, D. *et al.* Enhancement of methane catalysis rates in Methylosinus trichosporium OB3b. *Biomolecules* **12**, 560 (2022).

67. Samanta, D. *et al.* Methane Monooxygenases. In R. Chandra and R.C. Sobti (eds) *Microbes for Sustainable Development and Bioremediation*. Boca Raton: CRC Press (2019). doi:10.1201/9780429275876-12.
68. Berman, H. M. *et al.* The protein data bank. *Nucl Acids Res* **28**. Preprint at https://doi.org/10.1093/nar/28.1.235 (2000).
69. Maiorov, V. N. & Crippen, G. M. Significance of root-mean-square deviation in comparing three-dimensional structures of globular proteins. *J Mol Biol* **235**, 625–634 (1994).
70. Yuan, S., Chan, H. C. S. & Hu, Z. Using PyMOL as a platform for computational drug design. *Wiley Interdiscip Rev: Comput Mol Sci* **7**. Preprint at https://doi.org/10.1002/wcms.1298 (2017).
71. Chen, V. B. *et al.* MolProbity: All-atom structure validation for macromolecular crystallography. *Acta Crystallogr D Biol Crystallogr* **66**, 12–21 (2010).
72. Maurice, K. J. SSThread: Template-free protein structure prediction by threading pairs of contacting secondary structures followed by assembly of overlapping pairs. *J Comput Chem* **35**, 644–656 (2014).
73. Patel, R. *et al.* Repurposing the antibacterial drugs for inhibition of SARS-CoV2-PLpro using molecular docking, MD simulation and binding energy calculation. *Mol Divers* **26**, 1–21 (2022).
74. Eberhardt, J., Santos-Martins, D., Tillack, A. F. & Forli, S. AutoDock Vina 1.2.0: New docking methods, expanded force field, and python bindings. *J Chem Inf Model* **61**, 3891–3898 (2021).
75. Dallakyan, S. & Olson, A. J. Small-molecule library screening by docking with PyRx. *Methods Mol Biol* **1263**, 243–250 (2015).
76. Khadom, A. A. *et al.* Theoritical evaluation of Citrus Aurantium leaf extract as green inhibitor for chemical and biological corrosion of mild steel in acidic solution: Statistical, molecular dynamics, docking, and quantum mechanics study. *J Mol Liq* **343**, 116978 (2021).
77. Abdul Hussein, E. *et al.* 1-Isoquinolinyl phenyl ketone as a corrosion inhibitor: A theoretical study. *Mater Today: Proc* **42**, 2241–2246 (2021).
78. Ahmad, S. S., khalid, M. & Younis, K. Interaction study of dietary fibers (pectin and cellulose) with meat proteins using bioinformatics analysis: An in-silico study. *LWT* **119**, 108889 (2020).
79. Kinghorn, A. B., Fraser, L. A., Lang, S., Shiu, S. C. C. & Tanner, J. A. Aptamer bioinformatics. *Int J Mol Sci* **18**. Preprint at https://doi.org/10.3390/ijms18122516 (2017).
80. Zhao, X. F. *et al.* 2D g-C3N4 monolayer for amino acids sequencing. *Appl Surf Sci* **528**, 146609 (2020).
81. Unal, M. A. *et al.* Graphene oxide nanosheets interact and interfere with SARS-CoV-2 surface proteins and cell receptors to inhibit infectivity. *Small* **17**, 2101483 (2021).
82. Wei, J. *et al.* Machine learning in materials science. *InfoMat*, **1**, 338–358. Preprint at https://doi.org/10.1002/inf2.12028 (2019).
83. Kastuar, S. M., Ekuma, C. E. & Liu, Z. L. Efficient prediction of temperature-dependent elastic and mechanical properties of 2D materials. *Sci Rep* **12**, 3776 (2022).
84. Tawfik, S. A. *et al.* Efficient prediction of structural and electronic properties of hybrid 2D materials using complementary DFT and machine learning approaches. *Adv Theory Simul* **2**, 1800128 (2019).
85. Schleder, G. R., Acosta, C. M. & Fazzio, A. Exploring two-dimensional materials thermodynamic stability via machine learning. *ACS Appl Mater Interfaces* **12**, 20149–20157 (2020).
86. Momeni, K. *et al.* Multiscale computational understanding and growth of 2D materials: A review. *npj Comput Mater* **6**. Preprint at https://doi.org/10.1038/s41524-020-0280-2 (2020).

87. Ryu, B., Wang, L., Pu, H., Chan, M. K. Y. & Chen, J. Understanding, discovery, and synthesis of 2D materials enabled by machine learning. *Chem Soc Rev* **51**, 1899–1925. Preprint at https://doi.org/10.1039/d1cs00503k (2022).
88. Hong, Y., Song, J., Ko, J., Lee, J. & Shin, W. H. S-Pred: Protein structural property prediction using MSA transformer. *Sci Reports* **12**, 1–11 (2022).
89. Chicco, D., Warrens, M. J. & Jurman, G. The coefficient of determination R-squared is more informative than SMAPE, MAE, MAPE, MSE and RMSE in regression analysis evaluation. *PeerJ Comput Sci* **7**, 1–24 (2021).
90. Yang, H. *et al.* Predicting heavy metal adsorption on soil with machine learning and mapping global distribution of soil adsorption capacities. *Environ Sci Technol* **55**, 14316–14328 (2021).
91. Isayev, O. *et al.* Universal fragment descriptors for predicting properties of inorganic crystals. *Nat Commun* **8**. Preprint at https://doi.org/10.1038/ncomms15679 (2017).
92. Shenoy, V. B., Frey, N. C., Akinwande, D. & Jariwala, D. Machine learning-enabled design of point defects in 2d materials for quantum and neuromorphic information processing. *ACS Nano* **14**, 13406–13417 (2020).
93. Pan, S. J. & Yang, Q. A survey on transfer learning. *IEEE Trans Knowl Data Eng* **22** 1345–1359. Preprint at https://doi.org/10.1109/TKDE.2009.191 (2010).
94. Botu, V., Batra, R., Chapman, J. & Ramprasad, R. Machine learning force fields: Construction, validation, and outlook. *J Phys Chem C* **121**, 511–522 (2017).
95. Li, Y. *et al.* Machine learning force field parameters from Ab Initio data. *J Chem Theory Comput* **13**, 4492–4503 (2017).
96. Lambora, A., Gupta, K. & Chopra, K. Genetic Algorithm: A Literature Review. In *2019 International Conference on Machine Learning, Big Data, Cloud and Parallel Computing (COMITCon)*, Faridabad: IEEE, 380–384 (2019) doi:10.1109/COMITCON.2019.8862255.
97. Ren, P., Wu, C. & Ponder, J. W. Polarizable atomic multipole-based molecular mechanics for organic molecules. *J Chem Theory Comput* **7**, 3143–3161 (2011).
98. Hastie, T., Tibshirani, R., Friedman, J. & Friedman, J. *The Elements of Statistical Learning: Data Mining, Inference, and Prediction*. New York: Springer (2009).
99. Berisha, V. *et al.* Digital medicine and the curse of dimensionality. *npj Digit Med* **4**, 1–8 (2021).
100. Ringnér, M. What is principal component analysis? *Nat Biotechnol* **26**, 303–304 (2008).
101. Torrisi, M., Pollastri, G. & Le, Q. Deep learning methods in protein structure prediction. *Comput Struct Biotechnol J* **18**, 1301 (2020).
102. AlQuraishi, M. Machine learning in protein structure prediction. *Curr Opin Chem Biol* **65**, 1–8 (2021).
103. Cheng, J., Tegge, A. N. & Baldi, P. Machine learning methods for protein structure prediction. *IEEE Rev Biomed Eng* **1**, 41–49 (2008).
104. Senior, A. W. *et al.* Improved protein structure prediction using potentials from deep learning. *Nature* **577**, 706–710 (2020).

11 Machine Learning for Materials Science
Emerging Research Areas

Bharat K. Jasthi, Venkataramana Gadhamshetty, Grigoriy A. Sereda, and Alexey Lipatov

11.1 INTRODUCTION

Traditional approach for the development of new materials in materials science involves trial-and-error experiments which are often expensive, time-consuming, and are less efficient. Machine learning (ML) approaches can accelerate the discovery of new materials by utilizing the experimental data from various databases to identify the correlations between various experimental variables. ML approaches can develop the models based on the correlations which can be used to predict the properties and accelerate the materials discovery. A detailed information on the ML approaches and methods has been provided in Chapter 3. This chapter provides a summary of research and development efforts in the key areas of metallurgical engineering and materials science where ML approaches have been used. Applying ML approaches to materials science is an interdisciplinary effort where experts from materials science, data science, computer science, and other domain experts related to informatics, biology, and chemistry must work together to address the challenges and remove the barriers to implementation. Working together with experts from various disciplines has a great potential to make accelerated progress in materials discovery and enable new innovations in materials science.

11.2 APPLICATIONS OF ML IN MATERIALS SCIENCE

11.2.1 Additive Manufacturing

Additive manufacturing (AM) is the fabrication of three-dimensional (3D) objects using a computer-aided-design (CAD) model in a layer-by-layer approach to get precise shapes. AM technologies have attracted great interest in recent years because of its ability of make complex-shaped components and also because of its ability to produce personalized and customized components (e.g., prosthetics and biomedical implants). A wide range of materials such as polymers, metals, and composite materials can be additively manufactured and can be used in a wide range of industries and applications (e.g., aerospace, automotive, biomedical, defense, transportation, medical, sensors, and several other applications). Few examples of 3D-printed

DOI: 10.1201/9781003132981-11

FIGURE 11.1 Examples of laser-additive manufactured components for rocket engine applications showing (a) Axially coupled chamber and nozzle, (b) Jacket deposited on chamber with internal channels, (c) Rotating assembly for fuel pump, and (d) Combustion chamber liner with jacket. (Courtesy: NASA [1–3].)

rocket engine components using laser-based additive manufacturing technologies are shown in Figure 11.1.

AM technologies have emerged as a disruptive technology and have several potential applications in a wide range of industries. However, there are still some barriers to overcome such that these technologies can be adopted quickly in the industries. Currently, there are limited available AM materials databases, and there are also inconsistencies in the material properties reported in the literature. The inconsistencies are mainly caused by the defects present in the components and can be correlated to the heat inputs, cooling rates, and the process parameters employed. Recently, ML approaches have gained some traction in the additive manufacturing industry, and the applications of ML are primarily used in five research domains: (1) materials design, (2) materials analytics, (3) in-situ monitoring and defect detection, (4) process modeling and process control, and (5) sustainability of AM process [4,5]. Figure 11.2 shows the ML research domains that are primarily used in AM.

The design and development of materials to achieve desired properties is very important to understand the microstructure of AM products and correlate that with mechanical properties. Several possibilities can exist to achieve the desired properties, and developing these combinations manually by trial-and-error approach is very time-consuming and can be very expensive. The ML approaches can accelerate the discovery and design of new materials and help with the prediction of material properties. A wide range of ML techniques such as support vector machine algorithms, deep learning, decision tree, neural network, linear regression, Bayesian, Gaussian process, and clustering algorithms are commonly used for AM research and applications [5]. These ML techniques have been used for identification of defects [6–10], detection of porosity [11–15], density prediction [16,17], manufacturability [18], stress distribution [19], geometric deviation [20–22], fatigue life [23,24], and other mechanical properties prediction [25–29].

Although ML technologies for AM have been promising, there are still some challenges to be addressed. One of the biggest challenges related to the application of ML

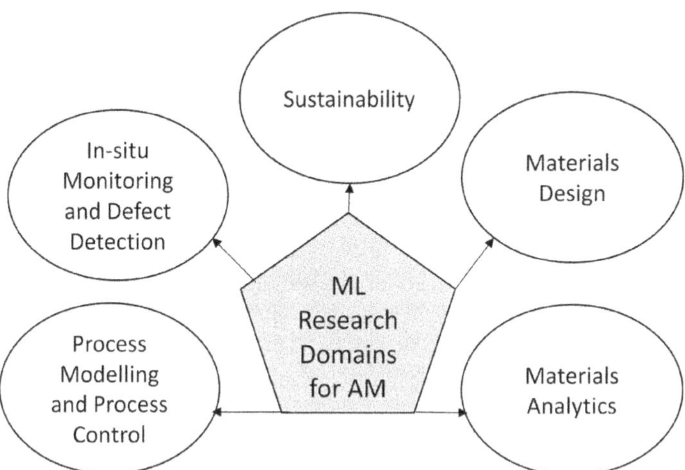

FIGURE 11.2 Machine learning research domains for additive manufacturing. (Adapted from Qin, J. et al., Addit. Manuf., 52, 102691, 2022.)

technologies for AM is the integration and analysis of data from multiple sources. The data from various sources can be challenging to fuse as each of these data may produce conflicting results when ML models are applied [5]. Another biggest challenge is the limited amount data to train, which can influence the performance of ML algorithms. The limited data can lead to inefficient training and may result in failure of models. The interpretation of the ML models for decision-making is another challenge. The ML techniques are typically developed by data scientists and computer science experts, but the AM engineers need interpretable models so that they can understand the significance of the models to optimize the process. Even with several challenges, the ML for AM still provides good opportunities for researchers to design, optimize, and predict properties for various applications.

11.2.2 Combinatorial Synthesis and Machine Learning-Assisted Discovery of Thin Films

Thin films are the atomic layers of materials deposited on a substrate to improve the surface properties of the substrate material. The thickness of thin films can be few atomic layers to few hundreds of nanometers thick. These thin films are deposited onto the substrates to improve surface properties such as the tribological, optical, electrical, chemical, and corrosion properties of the substrate materials. There are a wide range of thin film deposition techniques, but most of the techniques are broadly categorized to either a physical or chemical vapor depositions technique. The discovery of new thin films is often limited by how fast various combinations of materials are created and characterized. Combinatorial synthesis of thin films has been explored to improve the rate at which new thin films and materials are discovered [30–32]. One of the main advantages of combinatorial synthesis is the ability to produce a larger number of samples quickly and at a lower cost. Combinatorial synthesis provides increased flexibility in the discovery of novel materials and surfaces by creating libraries with different compositions, thickness, microstructures, and mechanical properties, which will be useful in screening the materials with desired functional properties. Another benefit in combinatorial synthesis is that all the specimens will be deposited under the same conditions while only changing any one variable (e.g., composition, thickness, and temperature) providing a unique capability to generate a wide range of specimens in a controlled manner. The use of ML approaches along with the combinatorial synthesis has been shown to accelerate the rate of materials discovery compared to traditional serial experimentation techniques.

Physical vapor deposition (PVD) process is the most commonly used technique for the synthesis of combinatorial thin films. There are several approaches on how combinatorial deposition can be performed. One of the simplest approaches is the deposition of the gradient deposition of thin films, where two or more magnetrons are focused onto a stationary substrate. Since the deposition rates of materials change with distance, thin films with multiple combinations of compositions can be deposited on the substrate. Figure 11.3 shows the schematic illustration and an example of gradient composition deposition using the pulsed laser deposition (PLD) process.

The gradient layer deposition is a simple approach and provides an ability to adjust the deposition rates and other variables to produce a wide range of combinations. However, large variations in thickness and stoichiometry can be present in the films, which could be difficult to produce reliable and high-precision materials.

Confocal array deposition is another approach that has been developed to improve the coating uniformity where a metal mask is placed in between the substrate and the magnetrons. The substrate is rotated under the stationary mask which creates discrete test pads at a fixed radial position from the center of the substrate as shown in Figure 11.4a. Compared to gradient deposition, this technique offers some improvement but still has some limitations. One limitation with this approach is that only one radial position of the substrate can be used for the combinatorial experiments and much of the substrate area is left unused. Moreover, the mask can also create asymmetrical exposure and shadowing effects on the test pads which could lead to

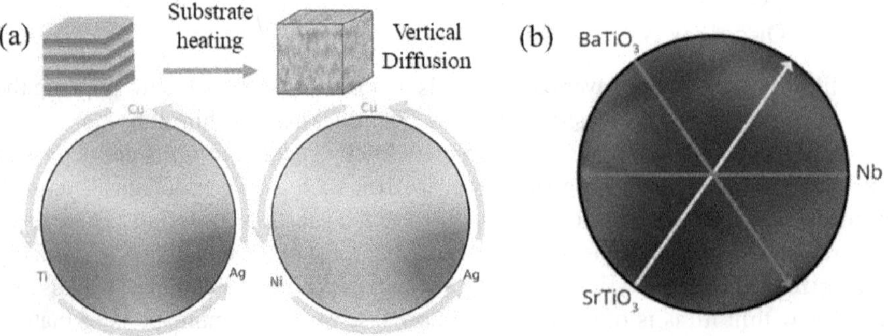

FIGURE 11.3 (a) Schematic illustration of combinatorial synthesis using PLD process [33]. (b) Example gradient thin film of $BaTiO_3$, $SrTiO_3$, and Nb using PLD process. (Courtesy: PVD Products.)

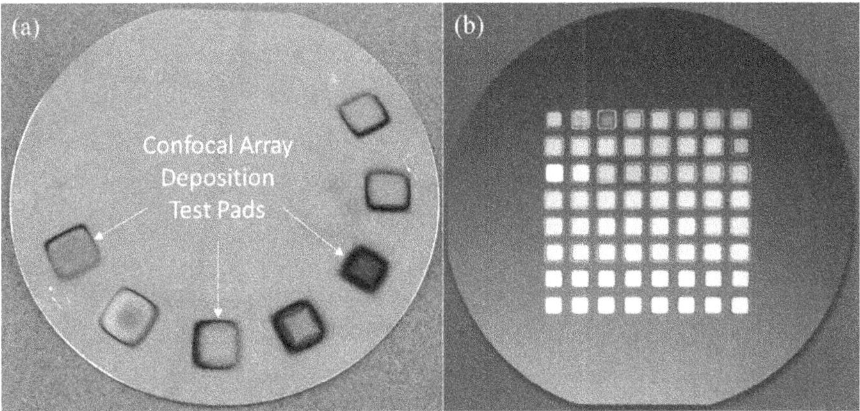

FIGURE 11.4 Combinatorial deposition of test pads deposited using PVD processes showing (a) confocal array deposition and (b) coincidental confocal deposition. (Courtesy: PVD Products.)

variability in coating thickness and composition of test pads. Since a mask will be used for the confocal array deposition, the use of radial frequency (RF) bias would be difficult, and therefore, the confocal array deposition approach may not be able to be used for reactive deposition or in-situ substrate cleaning.

Coincident confocal deposition is an alternate approach where a confocal point of the magnetrons is focused on a stationary mask, and the substrate position will be changed using a X-Y stage to produce combinatorial test pads in a gridded pattern as shown in Figure 11.4b. This approach addresses some of the issues associated with gradient layer and confocal array deposition approaches and produces films with maximum substrate coverage, uniform thickness, and greater control of composition of the test pads. The combinatorial deposition samples can be used to optimize or identify a composition that can provide a specific functional property (e.g., conductivity, dielectric properties, hardness, modulus, and corrosion properties). Figure 11.5 shows hardness modulus and resistivity maps of a ternary Cu-Ni-Ag alloy thin film deposited using gradient combinatorial deposition approach using PLD process.

Figure 11.5 shows the maps of hardness, modulus, and resistivity as a function of chemical composition. These maps can be very helpful in identifying the composition that can provide a higher hardness or electrical conductivity for specific applications. Such information can be very helpful for designing coatings for a wide range of applications. A wide range of other functional properties (e.g., optical, tribological, and electronic properties) can be optimized or tuned for specific applications. The combinatorial thin films can be used in combination with automated high-throughput characterization tools (e.g., X-ray diffraction, scanning electron microscopy, and confocal Raman spectroscopy) to screen and generate a large number of datasets. These datasets help identify the correlations between the structure, properties, and composition of the thin films, which can be used to discover new functional materials.

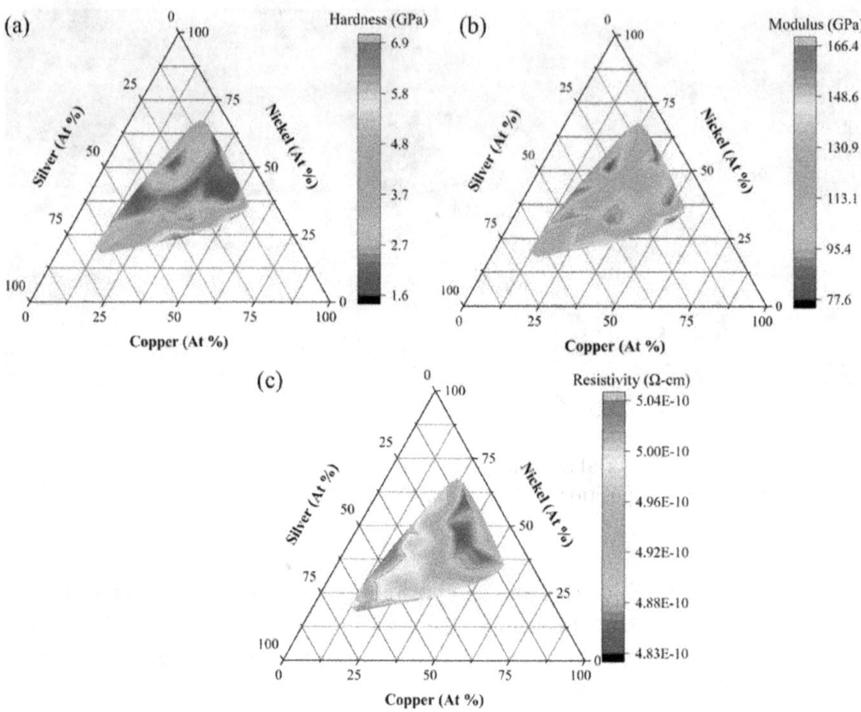

FIGURE 11.5 Gradient combinatorial deposition of Cu-Ni-Ag thin film deposited using PLD process showing (a) hardness, (b) modulus, and (c) resistivity. (Courtesy of Kandadai, V., *Combinatorial Synthesis of Cu-Ni-Ti Thin films using Pulsed Laser Deposition Process*, South Dakota School of Mines and Technology, 2022.)

11.2.3 Machine Learning-Assisted Properties Prediction of Bulk Alloys

Metals and their alloys are used for a wide range of applications from automobiles to rocket engines and for a range of several other applications (e.g., construction, biomaterials, and electronic materials). The design of alloys for specific applications with optimized properties such as the hardness, tensile strength, ductility, toughness, corrosion properties, and fatigue properties are only limited by how many combinations of materials can be made at a given time. The combination of various metals gives unique combination of properties which can be used for specific applications. Depending on the application, a specific alloying element can be added to achieve a specific functional property. For example, addition of Cr and Mo to steel will improve the corrosion resistance. Similarly, addition of Ti, Ta can form carbides (TiC and TaC) in steels which can improve the strength of materials. Likewise, addition of Ni will stabilize the austenite phase and can help with high-temperature stability of the alloys. So, a wide range of elements can be added to design new alloys to achieve specific properties. However, it will be difficult to identify the suitable composition by trial-and-error approach as this can be expensive and will take a significant time to develop. Thermodynamic modeling, molecular dynamics, and density functional theory simulations have been used to understand the stability of phases, solidification behavior,

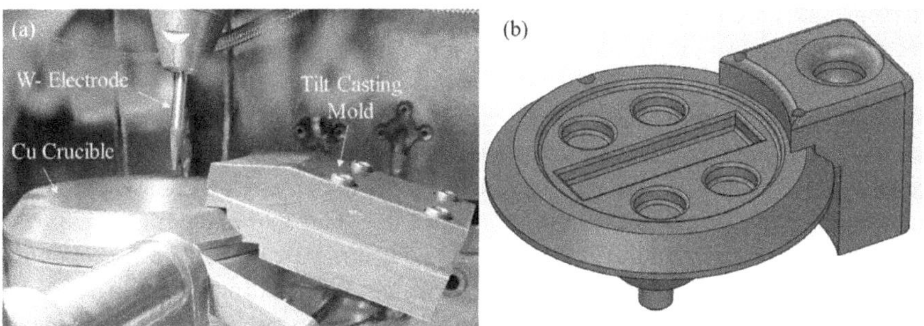

FIGURE 11.6 (a) Vacuum arc melter setup and (b) schematic of button mould.

and precipitation kinetics for various alloys [35–37]. While these techniques are helpful for simple alloys systems, these would be challenging to predict the compositions for multicomponent or compositionally complex alloys. The use of machine learning can enable the prediction of the properties for complex and multialloy systems.

While there are several fabrication techniques available for melting and alloying, arc melting and induction melting are commonly used for fabrication in an industrial setting. Both arc and induction melting techniques are typically used as batch processes for producing metal alloys. However, there are an infinite number of combinations that can be used to create new alloy systems. Combinatory approach melting using arc melting can enable the synthesis of multiple combinations of alloys in significantly less time. Figure 11.6 shows the vacuum arc melter and the schematic of a button mould that can be used for generating multiple melts.

The schematic of mould in Figure 11.6b only shows a few buttons that can be melted to make new alloys. However, moulds up to 32 buttons have been used to produce unique compositionally complex alloys [38]. To train the ML models, several parameters such as lattice constant, configurational enthalpy, atomic radii, melting temperatures, and electronegativity are considered. A wide range of ML models (as described in Chapter 4) were used to optimize the composition of the alloys to achieve a desired property.

11.2.4 Design of Drug-Releasing Materials with Machine Learning

One of the most prominent features of a living organism is a fine-tuned system of regulated biochemical pathways that result in a pattern of concentrations of chemicals released and consumed in the right place and at the right time. An externally administered drug tends to exhibit an unnatural time concentration profile that inevitably limits its therapeutical effectiveness and results in side effects, which is especially critical for cytotoxic anticancer and antimicrobial drugs. The drug-releasing materials help to alleviate this problem by mimicking a natural concentration pattern by releasing a drug in a time-controlled fashion or when triggered by external stimuli. One of the latest examples is a "smart" nanoassembly carrying an anticancer veratridine [39] or eugenol [40] into colon cancer cells and releasing the drug on-target when an MMP-7 enzyme overproduced by cancer cells digests the "gatekeeping" element and unseals the nanoparticle's drug load. However, the bottleneck of the development of drug-releasing materials and especially their clinical translation is a large number of variables to be taken into account, such as type of drug, type of material, type of cancer and its heterogenic environment, genetic heterogeneity of the patient, shape and surface characteristics of the material, drug loading efficiency, penetration through cellular barriers, selectivity, toxicity, drug-release efficiency, nature of the "gatekeeping element," to name a few. This makes the traditional trial-and-error approach prohibitively expensive and technically unrealistic without powerful and iterative data processing methods, which Artificial Intelligence (AI) is.

While AI is now commonly used for drug discovery, its application to drug release is relatively new due to a greater complexity of the underlying mechanisms and is aimed at achieving the efficient trade-off between therapeutic and side effects. The Probabilistic Suffix Annotation (PSA) model making sequential predictions of the drug-cell dynamics was applied to the analysis of the concentration of a drug metronidazole and population of its target bacteria *Giardia Lamblia* [41]. This model adjusts its sensitivity and selectivity based on the threshold level determined by the operator. The future drug-cell dynamics are predicted from current observations of the drug dose and the pathogen population by the Variable Length Markov Model (VLMM) used for increased flexibility, and the Fuzzy C-Mean clustering techniques involving online learning [41]. An Artificial Neural Network (ANN) model is employed in the design of 3D-printed oral devices to achieve the desired dose and drug-release profile based on the surface-to-volume ratio and the combination of fundamental shapes (cylinder, hollow cylinder, and pyramid) [42]. While thousands of oral drug formulations are currently approved, only about 30 long-acting injectables (LAI) have established their safety profiles despite a wide variety of discovered biocompatible polymers [43]. Each drug has unique physicochemical properties, which makes unlikely for one LAI formulation to be ideally suited for all drugs.

The main hurdle in the application of ML to pharmaceutical science is lack of available databases needed to train the models [43]. Of several tested approaches, the random forest (RF) model was the best predictor of drug-release build from the available online Tensor flow [44] and Scikit [45] learn libraries available in *Python*. As opposed to the bulk drug-releasing materials, drug-delivering nanoparticles need to travel through biological barriers bringing another dimension to the realm of ML. The multiple particle-tracking (MPT) analysis considering different modes of diffusion of PEGylated particles through the heterogeneous airway mucus was used to predict the passage times of nanoparticles and the *influenza A* virus, boosting the development of inhalable drug formulations [46]. The model input included fluorescent video microscopy and modeling the interactions with the negatively charged and hydrophobic domain of mucus [46].

After the success of COVID-19 vaccines based on the delivery of mRNA by lipid nanoparticles, this type of materials became the focus of many AI applications. Thus, deep learning models based on the Convolutional Neural Networks (CNN) and Long Short-Term Memory (LSTM) were applied to predict the efficiency of the Green Fluorescent Protein (GFP) mRNA transfer to living cells by lipid nanoparticles based on the time-lapse microscopy data before the GFP expression [47]. Data mining and ML were used to predict drug loading of solid lipid nanoparticles with curcumin

[48]. The authors applied the ANN method as an example of the supervised type of ML, which outperforms its unsupervised counterparts due to its ability to relate the input variables to responses. The input of experimental drug-loading data and results of the molecular dynamics docking simulations help predict the drug-loading of nanoparticles [48]. The key ingredient in the design of drug-delivering nanoparticles is engineering their surface by coating and modification. Application of both linear and nonlinear perturbation theory machine learning (PTNL) algorithms allowed for the prediction of drug-releasing properties of coated metal oxide nanoparticles [49]. The model learned from publicly available datasets generated by preclinical assays and processes the information on the parameters of biological activity, types of proteins, types of coating agents, and the nanoparticle's composition and shape [49].

The physiologically based pharmacokinetic model (PBPK) with the input from the nanotumor database (376 datasets) adequately predicted the drug-delivery efficiency of different nanoparticles to different tumors and outperformed all other ML methods including random forest, support vector machine, linear regression, and bagged model methods [50]. In that work, the cancer type was an important determinant for the deep neural network (DNN), which performed better than linear regression because of better handling of large incomplete datasets and identified new relationships not identified by the user. This model predicted absorption, distribution, metabolism, excretion, and toxicity of nanoparticles based on the input of the type of nanoparticle, its core material, shape, z-potential, hydrodynamic diameter, targeting strategy, tumor model, cancer type, and time profile of the drug-delivery efficiency [50].

A recent 2023 review underscores cancer heterogeneity, patient heterogeneity, interaction with the immune system, and the differences between animal and human physiology and pathology as the major challenges for the targeted drug-delivery systems, which should be capable of sensing biomarkers [51]. The AI methods that are able to analyses large amounts of patient data should be able to help generate personalized treatment plans based on diagnostic. All components of AI (Machine Learning (ML), Deep Learning (DL), Natural Language Processing (NLP), and Computer Vision (CV)) must be deployed in the feed-forward multilayer perception, deep synergy, and other methods to address these challenges and take advantage of the enhanced permeability and retention effect (EPR) to process the data on molecular imaging and drug delivery, to analyze holographic images from the nanoparticle-tracking experiments, and the data of Computed Tomography (CT) and Positron Emission Tomography (PET) scans. For example, variant of the KRAS gene produces unique biomarkers in cancer patients, which can be detected by sulfur fluorescent quantum dots, processed by AI and results in the specific receptors that should be targeted by therapeutics [51].

Finally, AI plays a critical role in devising better nanorobots for drug delivery with effective nanocommunication [52]. The drug-delivering nanorobots can be propelled by the enzymatic decomposition of urea to tiny bubbles of carbon dioxide in the nanomotors made of silica-supported urease and gold nanoparticles [53]. The radiolabelled nanorobots can be tracked *in vivo* by PET scans [54]. AI is critical for integrating "smart" sensors and power supply in the nanorobots [55]. Figure 11.7 summarizes the aspects of the design of drug-delivery materials, which generate the streams of experimental information (characteristics of the material and their drug

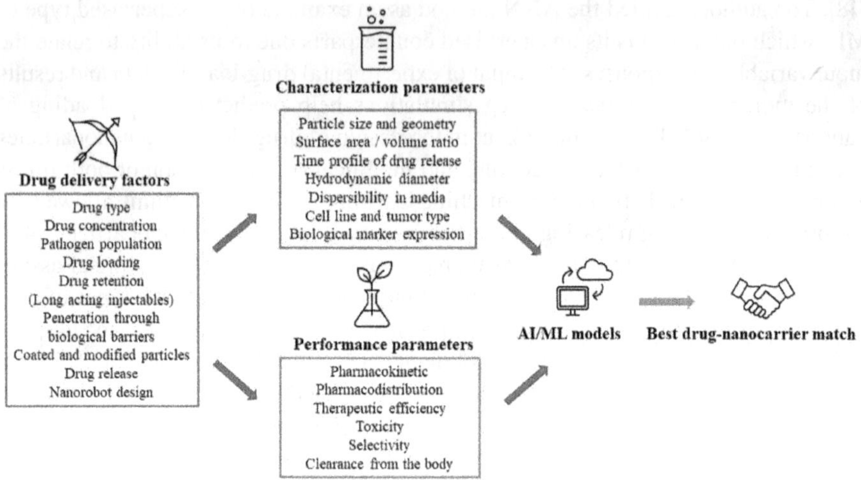

FIGURE 11.7 Input and output streams for the ML—assisted design of drug-delivering materials.

delivery performance) to teach a ML model to predict the best matches between the drug, its delivery system, and their specific biological application.

11.2.5 AI and ML Tools for Search and Discovery of Quantum Materials

In the last decade, quantum materials have become a major topic of condensed matter physics. These include two-dimensional materials [39], topological insulators and superconductors [40], Weyl semimetals [41], and quantum spin liquids [42]. The properties of these materials are determined by the collective behavior of a large number of interacting particles, which cannot be described using single-particle approximation typically used to describe more common characteristics like melting point, band gap, and heat capacity. Quantum materials host various exotic excitations, such as relativistic fermions in Dirac materials [39], Majorana-bound states in topological superconductors [40], and skyrmions in chiral magnets [43]. They hold the promise of transforming high-speed electronics and communication devices, as well as providing a platform for quantum computing [44]. AI-enabled methods are becoming increasingly important for studying quantum materials, due to the complexities and rich physics present in these materials. AI tools have the potential to drastically improve the efficiency of experimental and computational studies and can be used to create, analyze, and visualize high-dimensional heterogeneous data collections [45]. Below we discuss several venues where AI tools can provide a significant boost in the research of new and existing quantum materials.

11.2.5.1 Search and Analysis of Computational Materials Databases

Density functional theory (DFT) remains the workhorse of computational materials science and has been combined with ML algorithms to predict conventional materials properties such as melting temperature, band gap, shear modulus, and heat capacity [46]. Similar approaches have been utilized to search for quantum states in various materials. ML methods provide a systematic way to extract important predictors of materials properties from complex high-dimensional data of DFT calculations, creating ML models that can be used to filter through existing crystal structure databases in search of potential candidates. For example, structural, electronic, and band structure properties data from a materials database were used to create a set of "fingerprints," which were then utilized to create ML models for the critical temperature of hundreds of superconductors [47]. Such models can be used to fast screening of hundreds of thousands of existing and potential materials stored in computational databases created by high-throughput DFT methods.

11.2.5.2 Improve the Accuracy and Throughput of Ab Initio Methods

DFT is the most widely used computational method for simulating materials' properties, but it has several approximations that limit its accuracy for properties relevant to quantum materials. DFT is also resource hungry, and a blanket calculation of all properties of all compounds is not currently feasible. Integrating AI methods improves the approximations in the ab initio calculations and helps produce more accurate results [48,49]. Alternatively, ML can reduce the computational cost of ab initio methods, thus significantly enhancing the throughput of computational materials screening. For example, AI has been used to improve the accuracy of the Allen-Dynes approximation [50], a commonly used formula for predicting the critical temperature of electron-phonon paired superconductors, reducing the number of required DFT calculations and speeding up the discovery of novel superconductors.

11.2.5.3 Search for Stable Phases with Properties Relevant for Quantum Materials

Some types of strongly correlated materials, like superconductors and heavy fermion metal oxides, have been extensively studied for decades and amassed a significant number of systems and experimental data to work with. Other types of quantum materials, like multiferroics and materials with intrinsic topologically nontrivial states, while known for about the same period of time, only recently gained attention from the scientific community thanks to the development of advanced characterization techniques [40]. As such, only a handful of such materials are available for researchers. To facilitate the discovery of new materials, AI tools were developed to search for various indicators, like symmetry and materials chemistry in crystallographic data information, to automatically screen for new topological insulators and semimetals [51,52]. Another fertile ground for finding new quantum material systems is recently discovered 2D materials like graphene, which offer a wealth of potential for technologies such as electronics, sensing, and energy storage. High-throughput DFT has been used to compile publicly available databases of potential 2D materials, while ML models have been used to predict properties important for the synthesis of such materials, i.e., exfoliation energies, formation energies, and to classify them as having low, medium, or high stability [53]. These models have been used to discover materials with specific properties, such as those suitable for photoelectrocatalytic water splitting [53].

11.2.5.4 Making Predictions from Known Materials

Large databases of experimental materials data, such as the phase equilibria diagram and the inorganic crystal structure database (ICSD), are becoming more accessible to researchers. AI approaches are being increasingly applied to such experimental materials data to build models for making predictions. The pioneering work by Villars and Phillips in the 1980s used only three stoichiometric descriptors to cluster the 60 superconductors with $T_c > 10$ K known by then in three distinct groups and made predictions for potential high-temperature superconductors [54]. In more recent work, a neural network was trained using the ICSD database to predict crystal structure information [55] and then group materials according to their similarities in structure and composition, providing a list of potential materials sharing these similarities with known superconductors and topological insulators [55]. Creating an experimental database of structures and properties of quantum materials, however, is a significant challenge due to the manual effort required to extract data points from published articles and the lack of uniformity in experimental characterizations across different groups of researchers. Emerging AI-driven automatic generation of databases can provide an alternative. An example of this is the recently created database of almost 40,000 Curie and Néel phase-transition temperatures of magnetic materials produced from text data in articles using natural language processing (NLP) and related ML methods [56].

11.2.5.5 Extracting "Hidden" Knowledge from Materials Characterization Data

Modern materials characterization instrumentation and computing technology advances have enabled data collection on a much larger scale and with higher precision than ever before, and even a single measurement of one material can generate large volumes of high-dimensional data. This has created the challenge of navigating the vast amounts of data generated in real time while also opening new possibilities for research. ML has been used to augment traditional characterization methods helping to analyze noisy and complex data. For instance, by applying ML methods to the angle-resolved photoemission spectroscopy (ARPES) data of optimally doped cuprates [57], researchers discovered a hidden feature in the spectra, clarifying the role of energy dissipation and quantum entanglement in the superconducting phase. Another work reported the development of a neural network-based classifier trained on X-ray absorption spectroscopy (XAS) data, which was designed to distinguish topological materials from trivial ones [58]. Given that XAS is a widely used characterization technique, such a tool would greatly expedite the experimental identification of topologically nontrivial materials.

11.3 GAPS AND BARRIERS TO IMPLEMENTATION

ML has the great potential, and this chapter only provided a few examples where ML was used to accelerate and enable materials discovery. There are several more applications in the field of materials science where ML's modeling and prediction capabilities can be beneficial. However, there are several gaps and barriers for ML to be implemented in the areas of materials science. Applying ML approaches to materials science is an interdisciplinary effort, but some training is needed to material science engineers and scientists to acquire some ML knowledge and skills. In order to achieve this, some textbooks and other technical resources should be developed at an appropriate level where the material science community can understand and use them for their research. Developing curricula and integrating ML approaches to course modules can help train the students to get basic understanding of ML approaches in materials science. One of the biggest challenges related to the application of ML technologies for materials science is the lack of extensive databases to train, which can influence the performance of ML models. Also, the limited data from multiple sources can also be challenging to integrate as each of these data may produce conflicting results when ML models are applied. The interpretation of the ML models for decision-making is another challenge. The ML techniques are typically developed by data scientists and computer science experts, but the materials scientists need interpretable models so that they can understand the significance of the models to optimize the process. Even with several challenges, the ML for materials science still provides good opportunities for researchers to design, optimize, and predict properties for various applications. While using AI in drug design has become a common approach, its application to the drug-delivering materials is still in its nascent state. The AI/ML methods address the major challenges of drug delivery such as huge diversity of particles, heterogeneity of biological targets and environment, large and incomplete datasets, and integration of "smart" components into nanorobots. The major gap in the practical application of AI models is the lack of user-friendly software for their implementation and the lack of availability of the already trained models ready for further training and making practical predictions. Further, the wide variety of the developed AI models constitute its own parameter of optimization that may be handled by a meta-AI method to help researchers to choose the best AI model for their specific material and target.

REFERENCES

[1] B. Blakey-Milner, P. Gradl, G. Snedden, M. Brooks, J. Pitot, E. Lopez, M. Leary, F. Berto, A. du Plessis, Metal additive manufacturing in aerospace: A review, *Materials & Design* 209 (2021) 110008.

[2] P. Gradl, D.C. Tinker, A. Park, O.R. Mireles, M. Garcia, R. Wilkerson, C. Mckinney, Robust metal additive manufacturing process selection and development for aerospace components, *Journal of Materials Engineering and Performance* 31 (2022) 1–32.

[3] P.R. Gradl, T.W. Teasley, C.S. Protz, M.B. Garcia, D. Ellis, C. Kantzos, Advancing GRCop-based bimetallic additive manufacturing to optimize component design and applications for liquid rocket engines, In *AIAA Propulsion and Energy 2021 Forum*, AIAA, 2021, p. 3231. https://arc.aiaa.org/doi/abs/10.2514/6.2021-3231

[4] C. Wang, X. Tan, S. Tor, C. Lim, Machine learning in additive manufacturing: State-of-the-art and perspectives, *Additive Manufacturing* 36 (2020) 101538.

[5] J. Qin, F. Hu, Y. Liu, P. Witherell, C.C. Wang, D.W. Rosen, T. Simpson, Y. Lu, Q. Tang, Research and application of machine learning for additive manufacturing, *Additive Manufacturing* 52 (2022) 102691.

[6] M. Bugatti, B.M. Colosimo, Towards real-time in-situ monitoring of hot-spot defects in L-PBF: A new classification-based method for fast video-imaging data analysis, *Journal of Intelligent Manufacturing* 33(1) (2022) 293–309.

[7] L. Scime, J. Beuth, A multi-scale convolutional neural network for autonomous anomaly detection and classification in a laser powder bed fusion additive manufacturing process, *Additive Manufacturing* 24 (2018) 273–286.
[8] M. Grasso, B.M. Colosimo, Process defects and in situ monitoring methods in metal powder bed fusion: A review, *Measurement Science and Technology* 28(4) (2017) 044005.
[9] H. Taheri, M.R.B.M. Shoaib, L.W. Koester, T.A. Bigelow, P.C. Collins, L.J. Bond, Powder-based additive manufacturing: A review of types of defects, generation mechanisms, detection, property evaluation and metrology, *International Journal of Additive and Subtractive Materials Manufacturing* 1(2) (2017) 172–209.
[10] L. Chen, X. Yao, P. Xu, S.K. Moon, G. Bi, Rapid surface defect identification for additive manufacturing with in-situ point cloud processing and machine learning, *Virtual and Physical Prototyping* 16(1) (2021) 50–67.
[11] B. Zhang, S. Liu, Y.C. Shin, In-process monitoring of porosity during laser additive manufacturing process, *Additive Manufacturing* 28 (2019) 497–505.
[12] M. Montazeri, A.R. Nassar, A.J. Dunbar, P. Rao, In-process monitoring of porosity in additive manufacturing using optical emission spectroscopy, *IISE Transactions* 52(5) (2020) 500–515.
[13] M. Khanzadeh, S. Chowdhury, M. Marufuzzaman, M.A. Tschopp, L. Bian, Porosity prediction: Supervised-learning of thermal history for direct laser deposition, *Journal of Manufacturing Systems* 47 (2018) 69–82.
[14] Q. Tian, S. Guo, E. Melder, L. Bian, W. Guo, Deep learning-based data fusion method for in situ porosity detection in laser-based additive manufacturing, *Journal of Manufacturing Science and Engineering* 143(4) (2021) 041011.
[15] R. Liu, S. Liu, X. Zhang, A physics-informed machine learning model for porosity analysis in laser powder bed fusion additive manufacturing, *The International Journal of Advanced Manufacturing Technology* 113(7) (2021) 1943–1958.
[16] G.O. Barrionuevo, J.A. Ramos-Grez, M. Walczak, C.A. Betancourt, Comparative evaluation of supervised machine learning algorithms in the prediction of the relative density of 316L stainless steel fabricated by selective laser melting, *The International Journal of Advanced Manufacturing Technology* 113(1) (2021) 419–433.
[17] C.H. Lee, U. Kühn, S.C. Lee, S.J. Park, H. Schwab, S. Scudino, K. Kosiba, Optimizing laser powder bed fusion of Ti-5Al-5V-5Mo-3Cr by artificial intelligence, *Journal of Alloys and Compounds* 862 (2021) 158018.
[18] Y. Zhang, S. Yang, G. Dong, Y.F. Zhao, Predictive manufacturability assessment system for laser powder bed fusion based on a hybrid machine learning model, *Additive Manufacturing* 41 (2021) 101946.
[19] A. Khadilkar, J. Wang, R. Rai, Deep learning-based stress prediction for bottom-up SLA 3D printing process, *The International Journal of Advanced Manufacturing Technology* 102(5) (2019) 2555–2569.
[20] Q. Huang, Y. Wang, M. Lyu, W. Lin, Shape deviation generator-a convolution framework for learning and predicting 3-D printing shape accuracy, *IEEE Transactions on Automation Science and Engineering* 17(3) (2020) 1486–1500.
[21] R.d.S.B. Ferreira, A. Sabbaghi, Q. Huang, Automated geometric shape deviation modeling for additive manufacturing systems via Bayesian neural networks, *IEEE Transactions on Automation Science and Engineering* 17(2) (2019) 584–598.
[22] Z. Zhu, N. Anwer, Q. Huang, L. Mathieu, Machine learning in tolerancing for additive manufacturing, *CIRP Annals* 67(1) (2018) 157–160.
[23] Z. Zhan, H. Li, A novel approach based on the elastoplastic fatigue damage and machine learning models for life prediction of aerospace alloy parts fabricated by additive manufacturing, *International Journal of Fatigue* 145 (2021) 106089.

[24] Z. Zhan, H. Li, Machine learning based fatigue life prediction with effects of additive manufacturing process parameters for printed SS 316L, *International Journal of Fatigue* 142 (2021) 105941.
[25] G.X. Gu, C.-T. Chen, D.J. Richmond, M.J. Buehler, Bioinspired hierarchical composite design using machine learning: Simulation, additive manufacturing, and experiment, *Materials Horizons* 5(5) (2018) 939–945.
[26] C. Herriott, A.D. Spear, Predicting microstructure-dependent mechanical properties in additively manufactured metals with machine-and deep-learning methods, *Computational Materials Science* 175 (2020) 109599.
[27] F. Yan, Y.-C. Chan, A. Saboo, J. Shah, G.B. Olson, W. Chen, Data-driven prediction of mechanical properties in support of rapid certification of additively manufactured alloys, *Computer Modeling in Engineering & Sciences* 117(3) (2018) 343–366.
[28] I. Baturynska, Application of machine learning techniques to predict the mechanical properties of polyamide 2200 (PA12) in additive manufacturing, *Applied Sciences* 9(6) (2019) 1060.
[29] S. Nasiri, M.R. Khosravani, Machine learning in predicting mechanical behavior of additively manufactured parts, *Journal of Materials Research and Technology* 14 (2021) 1137–1153.
[30] P.J. McGinn, Thin-film processing routes for combinatorial materials investigations-a review, *ACS Combinatorial Science* 21(7) (2019) 501–515.
[31] J.D. Hewes, Economic Impact of Combinatorial Materials Science on Industry and Society [1], In Potyrailo, R.A., Amis, E.J. (eds) *High-Throughput Analysis*, Springer: Boston, 2003, pp. 15–30.
[32] R.A. Potyrailo, E.J. Amis, Elements of High-Throughput Analysis in Combinatorial Materials Science, In Potyrailo, R.A., Amis, E.J. (eds) *High-Throughput Analysis*, Springer: Boston, 2003, pp. 1–13.
[33] T. Greathouse, Creation and Characterization of Pulsed Laser Deposition Ag-Cu-Ni, Ag-Cu-Ti Combinatorial Gradient Alloy Films, South Dakota School of Mines and Technology, 2022.
[34] Courtesy of V. Kandadai, Combinatorial Synthesis of Cu-Ni-Ti Thin films using Pulsed laser Deposition Process., South Dakota School of Mines and Technology, 2022.
[35] D. Ma, B. Grabowski, F. Körmann, J. Neugebauer, D. Raabe, Ab initio thermodynamics of the CoCrFeMnNi high entropy alloy: Importance of entropy contributions beyond the configurational one, *Acta Materialia* 100 (2015) 90–97.
[36] C. Zhang, F. Zhang, S. Chen, W. Cao, Computational thermodynamics aided high-entropy alloy design, *Jom* 64(7) (2012) 839–845.
[37] A.J.S.F. Tapia, D. Yim, H.S. Kim, B.-J. Lee, An approach for screening single phase high-entropy alloys using an in-house thermodynamic database, *Intermetallics* 101 (2018) 56–63.
[38] H. Khakurel, M. Taufique, A. Roy, G. Balasubramanian, G. Ouyang, J. Cui, D.D. Johnson, R. Devanathan, Machine learning assisted prediction of the Young's modulus of compositionally complex alloys, *Scientific Reports* 11(1) (2021) 1–10.
[39] K.S. Novoselov, A. Mishchenko, A. Carvalho, A.H. Castro Neto, 2D materials and van der Waals heterostructures, *Science* 353(6298) (2016) aac9439.
[40] X.-L. Qi, S.-C. Zhang, Topological insulators and superconductors, *Reviews of Modern Physics* 83(4) (2011) 1057–1110.
[41] N.P. Armitage, E.J. Mele, A. Vishwanath, Weyl and Dirac semimetals in three-dimensional solids, *Reviews of Modern Physics* 90(1) (2018) 015001.
[42] L. Savary, L. Balents, Quantum spin liquids: A review, *Reports on Progress in Physics* 80(1) (2017) 016502.
[43] S. Mühlbauer, B. Binz, F. Jonietz, C. Pfleiderer, A. Rosch, A. Neubauer, R. Georgii, P. Böni, Skyrmion lattice in a chiral magnet, *Science* 323(5916) (2009) 915–919.

[44] C. Nayak, S.H. Simon, A. Stern, M. Freedman, S. Das Sarma, Non-Abelian anyons and topological quantum computation, *Reviews of Modern Physics* 80(3) (2008) 1083–1159.
[45] V. Stanev, K. Choudhary, A.G. Kusne, J. Paglione, I. Takeuchi, Artificial intelligence for search and discovery of quantum materials, *Communications Materials* 2(1) (2021) 105.
[46] A. Seko, T. Maekawa, K. Tsuda, I. Tanaka, Machine learning with systematic density-functional theory calculations: Application to melting temperatures of single- and binary-component solids, *Physical Review B* 89(5) (2014) 054303.
[47] O. Isayev, D. Fourches, E.N. Muratov, C. Oses, K. Rasch, A. Tropsha, S. Curtarolo, Materials cartography: Representing and mining materials space using structural and electronic fingerprints, *Chemistry of Materials* 27(3) (2015) 735–743.
[48] J.C. Snyder, M. Rupp, K. Hansen, K.-R. Müller, K. Burke, Finding density functionals with machine learning, *Physical Review Letters* 108(25) (2012) 253002.
[49] M. Yu, S. Yang, C. Wu, N. Marom, Machine learning the Hubbard U parameter in DFT+U using Bayesian optimization, *npj Computational Materials* 6(1) (2020) 180.
[50] S.R. Xie, G.R. Stewart, J.J. Hamlin, P.J. Hirschfeld, R.G. Hennig, Functional form of the superconducting critical temperature from machine learning, *Physical Review B* 100(17) (2019) 174513.
[51] H.C. Po, A. Vishwanath, H. Watanabe, Symmetry-based indicators of band topology in the 230 space groups, *Nature Communications* 8(1) (2017) 50.
[52] K. Choudhary, K.F. Garrity, F. Tavazza, High-throughput discovery of topologically non-trivial materials using spin-orbit spillage, *Scientific Reports* 9(1) (2019) 1–8.
[53] G.R. Schleder, C.M. Acosta, A. Fazzio, Exploring two-dimensional materials thermodynamic stability via machine learning, *ACS Applied Materials Interfaces* 12(18) (2020) 20149–20157.
[54] P. Villars, J.C. Phillips, Quantum structural diagrams and high-Tc superconductivity, *Physical Review B* 37(4) (1988) 2345–2348.
[55] H. Liang, V. Stanev, A.G. Kusne, I. Takeuchi, CRYSPNet: Crystal structure predictions via neural networks, *Physical Review Materials* 4(12) (2020) 123802.
[56] C.J. Court, J.M. Cole, Auto-generated materials database of Curie and Néel temperatures via semi-supervised relationship extraction, *Scientific Data* 5(1) (2018) 180111.
[57] Y. Yamaji, T. Yoshida, A. Fujimori, M. Imada, Hidden self-energies as origin of cuprate superconductivity revealed by machine learning, *Physical Review Research* 3(4) (2021).
[58] N. Andrejevic, J. Andrejevic, B.A. Bernevig, N. Regnault, F. Han, G. Fabbris, T. Nguyen, N.C. Drucker, C.H. Rycroft, M. Li, Machine-learning spectral indicators of topology, *Advanced Materials* 34(49) (2022) 2204113.

12 The Future of Data Science in Materials Science

Parvathi Chundi, Vidya Bommanapally, and Venkataramana Gadhamshetty

12.1 INTRODUCTION

Machine leaning (ML) methods have made great contributions to 2D materials science and engineering as evidenced by the collection of works presented in earlier chapters. ML technologies are transforming the way scientists design materials by managing the complexity of vast space of options that need to explored. Yet there are several challenges that must be addressed while working with ML methods. Most ML methods such as neural networks need large quantities of training data that is of high quality so that millions of parameters can be tuned to obtain an accurate model. With small size datasets, these techniques may result in overfitting. For example, building a semantic segmentation model requires thousands of images in the training dataset. Such large datasets may not be easy to obtain from the 2D materials domain, where large training datasets may need expensive manual processes and specialized equipment to collect the data. Even in cases where large training datasets are available, the dataset needs to be properly labeled and should be largely free of noise. For ML tasks such as semantic segmentation and object recognition, labeling a dataset to obtain training and testing datasets can be tedious. For an image semantic segmentation task, each pixel in an image in the dataset must be assigned a class. The most popular dataset for the semantic segmentation task, the CoCo dataset (https://opencv.org/introduction-to-the-coco-dataset/), contains 1.5M labeled images with 80 categories including 'car', 'motorcycle', 'stop sign', etc., and the images in this dataset do not need special expertise to label the dataset. However, labeling datasets in the 2D Materials domain needs domain expertise (e.g., distinguishing oxidized 2D material surface from its pristine counterpart in an image, and labeling the pixels accordingly) and may not be readily available. Moreover, multiple experts may need to label an image to account for human biases, and these multiple labels for each pixel need to be reconciled to decide which label is the correct one for each pixel in an image. Finally, once an ML model is obtained, the modeled results must be carefully analyzed by the domain experts to plan next steps, such as learn the model for a different task (image classification may be changed to object detection), prepare the training data again for obtaining a better model (this is in case if the model accuracy is unacceptable to the domain scientists), or make a plan to validate the model observations in the laboratory.

DOI: 10.1201/9781003132981-12

For ML methods to be effective for designing materials and discovering their properties, scientists must embrace two central notions—data management and validation of ML results in the lab and ultimately in a real-world setting. In this chapter, we present some research directions that will enhance the effectiveness of ML-driven materials discovery. Machine learning, as a subfield of computer science, is growing leaps and bounds, and many of these advances will not be covered in this chapter. Instead, we select a few relevant directions that are based on the relevant case studies conducted by the authors in their recent research projects (e.g., National Science Foundation OIA # 1920954, 1849206).

12.2 LEARNING WITH SMALL TRAINING DATASETS

For ML-assisted materials design to be successful, we need to contend with the issue of small training datasets. In this section, we describe some of the popular ways in which this issue can be dealt with.

12.2.1 Data Augmentation

When we have a small dataset, we may be able to augment it with artificially generated data with similar properties as the original dataset. This process is called *data augmentation* [*1*]. As an example, for image-based ML tasks, standard data augmentation methods include augmenting the dataset with random crops, zooms, and mirror-image flips of the images from the original dataset. For text datasets, augmented data can be generated by inserting random characters at random locations in the documents in the original dataset. The training samples obtained from augmentation will carry the labels over from the original sample they were generated from. Data augmentation has been shown to significantly improve the performance of ML models [2,3]. This is because the data augmentation methods generate a new training instance with properties similar to an instance from the original dataset. Data augmentation is a popular technique that is commonly employed in ML pipelines currently.

For engineering and health domains, there is a tremendous amount of domain knowledge in terms of physical and chemical laws which can be useful to generate data for augmentation purposes. Domain scientists have traditionally relied on domain-knowledge-driven simulation models for gaining a better understanding of the physical and chemical phenomena and for discovering new hypotheses. Even though simulation models make simplifying assumptions of the physical phenomena, they have been shown to capture the laws of a domain well enough to simulate different types of scenarios satisfactorily. In engineering domains, simulation models can be used to generate synthetic data wherever possible, combine it with data collected from the laboratory experiments, and use the data as training data for model building. The predictions from an ML model can be used to enhance the simulation model as well.

As discussed in the earlier chapters, recent works have demonstrated the seamless use of ML algorithms for accelerating the discovery of 2D materials. Such algorithms have already been implemented as online web tools for use by broader communities [4]. As readers may have recognized, these tools have been primarily built using theoretical data sets (e.g., geometrical structures) that can be obtained from the existing databases (e.g., Inorganic Crystal Structure Database and Crystallography Open Database). The screening criteria in these tools are also often based on theoretical values (e.g., binding energies <130 meV Å^{-2}, which can be derived using Einstein's Theory of Relativity calculations). However, generating experimental datasets that describe performances of 2D materials can be cumbersome. For instance, 2D materials are being explored as next-generation protective coatings for controlling microbiologically influenced corrosion (MIC) [5]. Considering the extensive effort for designing, developing, synthesizing, and characterizing the performance of a brand new 2D material, the overall process can take several years of time. If one were to test 1000 promising 2D materials with 1000 different microbial species (each representing one biotechnological application), the time required to complete the performance assessment is significant. Generating experimental datasets on MIC prevention performance of the 2D coatings is a complex, expensive, and laborious process. These constraints also restrict the duration of these MIC tests to few weeks, which cannot adequately help determine their performances (i.e., service lives) that are expected to last for several years. This situation also has forced many life cycle assessment (LCA) modeling to rely upon the assumed service lives of these coatings while quantifying their potential sustainability benefits [6,7]. Some of these issues can be alleviated using data augmentation methods. For instance, a recent study by authors' group leveraged deep learning methods (e.g., variation autoencoder, generative adversarial network (GAN) models) for addressing issues with lack of adequate experimental datasets required to predict the electrochemical performances of MIC-resistant graphene coatings [8].

12.2.2 Semisupervised Learning

Semisupervised learning approaches have been shown to be effective in alleviating the need for large, labeled training datasets. These approaches learn a high-level structure from the unlabeled data and combine the learned structure with a small amount of data for a given ML task to learn a model for that task. Note that semisupervised approaches need a large amount of high-quality data for building a model. However, only a small portion of it needs to be labeled.

An obvious semisupervised learning approach is *self-training* where the small, labeled dataset is used to train a model M and use that model to infer labels for the unlabeled samples in the larger dataset. The labels inferred from M are typically referred to as *pseudo-labels* to differentiate them from the ground-truth labels obtained from domain experts. Then a new model is obtained using some combination of the small, labeled dataset and the data with pseudo-labels. This process is repeated some number of times depending on the user or some convergence metric. Another variation of self-training is to generate pseudo-labels for random training samples and include them in the training set for next model. Although self-training can build models using small-size training datasets, it suffers from what is known as confirmation-bias [9] where incorrect predictions in pseudo-labels inferred for the unlabeled samples can make the accuracy of models built using the repeated model training worse over time.

There are many ways of making the unlabeled data useful when a model with small amount of labeled data. These include semisupervised learning approaches that include entropy minimization, consistency regularization, etc. Please see [1] for a detailed description of these techniques. Active learning is a form of semisupervised learning where the goal is to use as few training samples as possible for learning a model with reasonable accuracy. Informally, an active learning-based learning algorithm queries an authoritative source—an human expert or a function over the labeled dataset—to learn the correct prediction for a given sample. See [10] for more details.

12.2.3 Transfer Learning

Transfer learning is employed when an existing deep neural network model can be used for a new yet a similar problem. As an example, in a recent study, transfer learning was used to learn representations of microstructures and then used the resulting model to discover the underlying annealing conditions [11]. Learning microstructures is typically referred to as the *source* task, and the task of identifying the annealing conditions is referred to as the *target* task. Transfer learning transfers information from the source dataset to the target dataset via a shared set of parameters [1].

In transfer learning, a large source dataset is used to learn a model for the source task. This model is then *fine-tuned* on a small dataset belonging to the target task. Transfer learning assumes that the training instances for the source and target task are the same (e.g., both are RGB images or numeric vectors, etc.) or that the training instances of the source task can be easily converted to the target data format. However, the instance labels for the source task and target task can be different. In order to use the model for the source task for the target task, weights between the output layer and the last hidden layer of the model for the source task are *fine-tuned* using the training instances of the target task (for details, see Decost). Therefore, transfer learning leverages the large source training dataset for building a deep neural network and uses this model for a target task by modifying only a part of is using the small target dataset.

12.2.4 Few-Shot Learning

Few-shot learning [12] refers to ML algorithms that can learn to predict from very few labeled training samples, much like humans do. If a model can learn from a single labeled sample, then it is called *one-shot learning*, and if no labeled samples are needed, then it is called *zero-shot* learning. Here, we illustrate how the few-shot classification method works. Few-shot classification method is given an abundant training sample for base classes and is asked to learn predict previously unseen classes using a limited amount of labeled samples. Few-shot classification approaches are usually evaluated using *C-Way N-shot* classification in which a model is expected to classify *C* classes using *N* training samples for each class where N and C are small.

Few-shot classification can employ transfer learning to build a model from the abundantly available base class data and then fine-tune the model using the labeled data available for the C (previously unseen, unique) classes. Few-shot classification can also use *meta-learning* which means *learning to learn*. For more details on this, please see [1]

12.3 PHYSICS-INSPIRED NEURAL NETWORKS

Physics-inspired neural networks (PiNNs) were inspired by the challenges of collecting large amounts of data needed for employing deep neural networks in complex biological and engineering domains. Typical deep neural networks use the training data to identify a nonlinear function that maps a training instance, which is usually a high-dimensional vector, to a label accurately. There is an abundance of prior domain knowledge in biological and engineering fields that is not considered by a typical neural network while learning the nonlinear map. Not incorporating the domain knowledge has been shown to lead to predictions that are inconsistent with the existing domain knowledge [13]. Having large amounts of training data that reflects the domain completely can mitigate these inconsistent predictions which is expensive or even impossible in biological/engineering domains.

Originally, automatic differentiation was incorporated into a deep neural network to obtain a PiNN. The process of learning the nonlinear map between input vectors and the set of labels in a PiNN is constrained to obey any symmetry, invariance, or conversation principles that underlie the training data, where the principles are captured as nonlinear partial differential equations. We can characterize the incorporation of prior domain knowledge into a deep neural network in three ways [14]: (1) Physics-guided neural networks (PgNNs) use off-the-shelf deep learning networks to construct an appropriate mapping from input vectors and labels which are collected from computations and experiments and curated to ensure compliance with the domain's rules and knowledge. (2) PiNNs use loss functions consisting of residuals of physics equations and boundary constraints to build a model that satisfies the domain constraints. PgNNs suffer from lack of robustness and generalizability whereas PiNNs are not suitable for emerging domains where the differential equations that govern the complex dynamics underlying the domain are not fully understood. (3) So, physics-encoded neural networks have been proposed where the prior knowledge is encoded into the core architecture of a deep neural network.

Please see [14] for a great comparison of the capabilities of PgNNs, PiNNs, and PeNNs. PgNNs can be used to learn mappings from sparse data to discover latent dependencies among the input data points and for interpolation and still need large datasets as their learning really does not incorporate any rules of the domain. PiNNs can be used to discover latent dependencies, potential boundary conditions, etc., from a training dataset; however, the loss functions in these networks can destabilize the learning process. PeNNs can make complex extrapolations based on the input data, how they suffer from low convergence rates. Nonetheless, PgNNs, PiNNs, and PeNNs have expanded the deep learning network applications to complex scientific and engineering applications greatly.

12.4 DIGITAL TWINS

The concept of a *digital twin* is not new and has been around at least since 2003 [15] and formalized in a paper by the National Aeronautical Space Administration (NASA) in 2012 [16]. For our purposes, the following definition would serve well—'A Digital Twin is a virtual instance of a physical system (twin) that is continually updated with the latter's performance, maintenance, and health status data throughout the physical system's life cycle' [17]. With recent advances in AI, simulation, and data management, building a digital twin of a physical process/asset is becoming a reality. Because of tight coupling, a digital twin of a physical asset/process can be used to make predictions about how the physical process will evolve under different conditions.

12.5 DATA-CENTRIC ARTIFICIAL INTELLIGENCE

Traditionally, for an ML task (such as regression, segmentation, etc.), it starts with a dataset containing labeled, training instances, and produces the best model for the given dataset, i.e., the one that generalizes best on the test dataset. For learning the best model for the given dataset, one may try different ML algorithms, different deep learning architectures, and tune the hyperparameters (parameters of the ML algorithms) to obtain the best model. This approach is said to be model-centric artificial intelligence as improving performance on AI task focuses on improvement of the model. In contrast, the data-centric artificial intelligence (DCAI) is about AI algorithms that understand and, if needed, modify the data, so that AI models can be improved. DCAI focuses on systematically changing the dataset so that the model performance on an AI task can be improved [18,19]. The difference between model-centric and data-centric AI approaches as explained is visually shown in Figure 12.1.

As ML tasks are essential for any organization, production machine learning platforms have become necessary for supporting ML tasks [20]. Production machine learning platforms support continuous data collection and model building. As new data and observations become available in emerging domains such as 2D materials engineering, these need to be incorporated into ML models to enhance their predictive power. Although ML models perform well on the training and testing data used for building the model, they tend to perform poorly on new data. Therefore, it is important to build new models continuously.

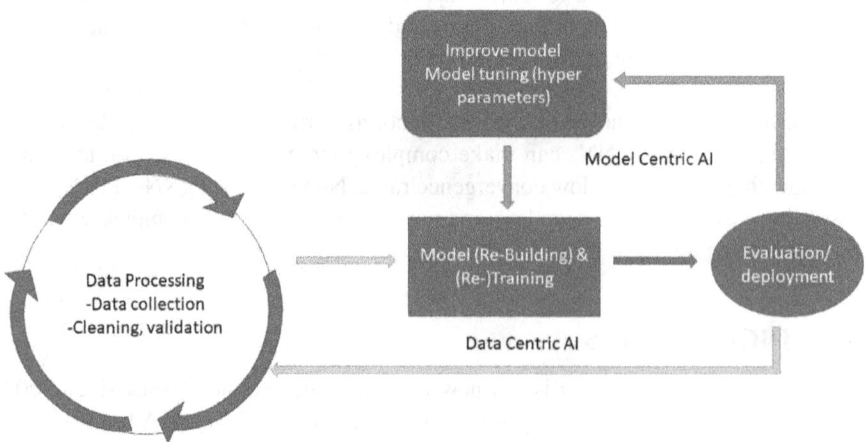

FIGURE 12.1 Data Centric AI vs Model centric AI.

Building an ML model once requires several tasks to prepare the data: explore the dataset to understand what it contains, fix fundamental errors such as missing values, and transform the data into a format needed for the ML technique. As new data become available or generated by experiments, the model must be updated to reflect the new data. For some ML tasks, the training data needed for an ML task may be available in different data sources and in different formats and will need additional data operations to collect the data. The quality of a training dataset also plays a major role in the quality of the ML model learned from that training dataset. The quality of training data must be sufficient to ensure that the ML model learned from that training dataset is robust (i.e., will not produce erroneous results) and unbiased (i.e., will not always produce the same outcome). Therefore, DCAI provides a fuller view of all technologies needed to build a production machine learning platform to build accurate and effective ML models with low cost and high efficiency.

The following issues are central to DCAI and are explained more in the following sections.

1. Data collection and cleaning
2. Robust and fair model training
3. Continuous learning

12.5.1 Data Collection

Data collection issues for building an ML model include finding the right type of data that is suitable for the task at hand. If one wants to build a model for identifying stop signs in an image, the training set must consist of an image with and without stop signs as well images with stop signs of different sizes and seen from different angles. This kind of data may not be readily available in large quantities. The relevant data may need to be *discovered* and labeled properly, and perhaps added to some existing data. *Data discovery* is the process of searching for data relevant to the ML task at hand. Data discovery methods search for data in data storages of an organization or over the web. For a 2D materials expert, data may be available in experimental and observational data storages, publicly available data banks, and research publications. All these data storages may have diverse data which must be processed to discover the attributes and values that are relevant to the ML task. Data discovery tools such as the Google Data Search can be used to search for datasets that are relevant to the ML task. For tabular datasets (such as CSV files containing experimental measurements/settings, descriptions of material attributes, etc.), the data discovery tools work especially well because these tools can suggest other features and records that can enhance the original dataset. Most public data sources such as PubMed offer functionality (APIs) to download relevant datasets whereas data repositories offer downloadable datasets.

Once the data are collected, it must be cleaned to make sure that each instance in the dataset is complete (no missing values) and contains only valid values. Depending on the type of data in the training dataset, the operations to clean the data collected and make sure it is valid will be different. For example, for research publication dataset where each training instance is one paper abstract, data cleaning operations make sure that all instances have titles, publication dates, etc. Data validation operations make sure that the values in the publication date for all instances in the dataset are of valid format, and the number of pages is a positive integer, etc.

The data collection and validation tasks are more complex for supervised learning tasks. In this case, it is important that the training dataset does not have *selection bias*, i.e., the dataset used for training is drawn from the distribution of the domain where the model will be used for prediction. Training datasets suffer from selection bias if the dataset does not reflect instances as the domain evolves or if it does not reflect all classes/categories present in the domain. This may happen if the data are collected in a biased manner, for example, sales revenue data collected only during the holidays, or images of a brain from brain cancer patients, etc., and these situations must be avoided so that the training dataset is free of selection bias. In terms of how much training data is to be collected for a supervised learning task, one can use statistical methods.

For supervised learning tasks, each item in the dataset needs to be labeled as well. If there are existing labels, semisupervised learning methods can be used to predict labels as explained in Section 12.2.2. Otherwise, each training instance must be assigned a label by a human annotator. The goal of the data labeling process is to assign a label to each data instance with high confidence. Since human annotators may not always agree on a label (whether an X-ray image shows disease) for a given data instance, each instance in the dataset may be assigned a label by multiple annotators and then is assigned one of those labels as the final label after curation. One simple method to assign a label for an instance is to simply choose the label assigned by most human annotators (if it exists) or choose a label computed as a function of the individual annotations (such as average or maximum value of the individual annotations). We can also define functions to assign a confidence score for each label assigned to a data instance or a confidence score for each annotator as well.

12.5.2 Robust and Fair Model Training

Machine learning tasks build models from training data that describe the behavior of some domain phenomenon. Once the data are collected, cleaned, and labeled, the next step is to build an ML model that learns a function from the training data to predict some value or property about the phenomenon. Even after careful cleaning and preparing the data to train an ML model, there are no guarantees that the cleaned data are, in fact, free from noisy and missing instances and features. Often times, some behaviors of the phenomenon (e.g., properties of a metal at an extremely low temperature) are missing entirely from the training data or some features that are a part of capturing the phenomenon are missing (e.g., the following feature—*number of days an item was on sale* is important to predict the revenue for that item). For these reasons, an important question that arises is 'can we build a *robust* model from the training data, one that can learn to predict from the training data despite the noisy and/or missing records or features in the training data?'.

Noisy training data (data containing noisy instances where an instance contains wrong values for its features) are typically thought to be a *data poisoning* attack on the training dataset by an *adversary*. *Adversarial training* can be used to improve the robustness of a model by using modified objective functions to learn a model that predicts different classes for clean data and poisoned data. Other method to learn a robust model in the presence of noisy features is to learn an additional model with reduced set of features. A poisoned instance may be assigned a different label by the original model and the reduced model resulting in robust training.

Adversarial machine learning [21] is a popular area of research that develops methods for robust model learning in the presence of the attack paradigms as described below:

1. **Training-Time Adversarial Attach (Backdoor Attack)**: It aims to generate a model, an adversarial model, such that it performs well on data that is not poisonous (i.e., clean) while predicting an adversarial sample as belonging to an adversarial class.
2. **Deployment-Time Adversarial Attack (Weight Attack)**: Given a benign model deployed in a hardware device, the attacker aims at slightly modifying the model parameters in memory so that obtain an adversarial model so that adversarial inputs or some benign samples are labeled as adversarial sample, whereas other benign samples are labeled as ground truth labels.
3. **Inference Time Adversarial Attack**: Given a benign model, the attacker aims at modifying a benign sample to obtain a corresponding adversarial sample such that the prediction is different with a ground-truth label or the same with an adversarial label.

Now, let us talk about noisy labels which are very common because typically manual methods are used to label datasets, and therefore, it is common to have missing and/or incorrect labels. Here, we assume that the training instance contains correct feature values but a wrong label. Sometimes, the human annotators simply disagree on which label to assign for a training instance. Robust training in the presence of noisy labels received a lot of attention because it is a commonly occurring problem in several domains. There are techniques for robust training for every step of the training procedure. Various sample-selection techniques are proposed for choosing a subset of the training dataset that will lead to robust training. Different neural architectures, and loss computing and loss adjustment functions, as well as robust regularization functions have been proposed for robust model building [see references for more information].

Semisupervised and unsupervised approaches are typically used to deal with missing labels. Semisupervised approaches assume that the dataset contains clean labeled data together with unlabeled (or incorrectly labeled) data. Methods such as Mean-Teacher [9] and MixMatch [22] are used to build models in these cases. For unlabeled data, techniques such as self-supervised learning and generative models are used [9,22].

We now focus on model fairness where biased data may cause a model to be discriminating. Here, the goal is to address bias in training data. Data bias can be addressed by preprocessing the training data. Here, data can be *repaired* to reduce bias or use the available data to generate more unbiased data using generative models such as generative adversarial networks or GANs. Model fairness can be added as a constraint to the model's learning function, which may not always be feasible. In cases where data or model cannot be repaired to ensure fairness, the model predictions may be postprocessed to mitigate biased predictions which is not generally advisable.

The training methods for robust and fair model training address different data flaws. There are recent efforts to combine the two training methodologies to build models that are both robust and fair. Fair training can be made more robust by addressing the scenarios where the attributes that contribute to unfairness are noisy or missing. Similarly, robust model training can improve fairness by using adversarial training, removing anomalies, and spurious features. Emerging techniques such as FR-Train provides a framework that combines a classifier, a discriminator for fairness, and a discriminator for robustness to build a model that is both fair and robust.

12.5.3 Continuous Learning

Given a data collection, one can learn to build many models from it to predict the phenomena in several different scenarios. However, as new data become available from experiments and other data sources, it is important to monitor the data continuously to check if the old training datasets and models can be improved and/or new models are needed for predicting new phenomena. *Lifelong learning* (LL) has been proposed to build models that learn as humans do [23] which *retain learned knowledge from previous tasks and use it to help future learning.* A simple method for life-long learning is the leader clustering algorithm which works as follows. Suppose we have clusters over all previous instances. When a new instance arrives, we add it to one of the previously computed clusters that the new instance is most similar to. If no such previous cluster exists, we create a new cluster with the new instance. Note that in this case, there is no constraint on the number of clusters. Semisupervised learning method is the most appropriate method for continuous supervised learning. We use the model and the already available labeled data to label the newly generated instances.

It is important to identify, as the new data become available, if it is significantly different from the old data, i.e., the statistical distribution of the new data is different than the old data. If that is the case, it is important to rebuild the models from scratch instead of tweaking them to include the new data. Based on the knowledge shared by the tasks, the LL approaches' knowledge can be of two types [23]. Global knowledge is where the tasks share a global latent structure and the same is used for the new task. Local knowledge is where each task has a specific local knowledge and that required for the new task can be chosen from them. The categorization tasks can also be done from the type used in the LL approaches. Independent tasks are those where the tasks are learned independent of each other yet can share some latent information. Dependent tasks usually add a new class each time with a new task in continuous supervised learning and hence depend on previous tasks. The architecture hence has to accommodate the previous knowledge of the tasks rather than a simple data, task, learner, and output model of isolated (regular) machine learning paradigms.

The main components of a generic LL approach consist of a knowledge store, knowledge learner, or the new task model learner [23,24]. The components and flow

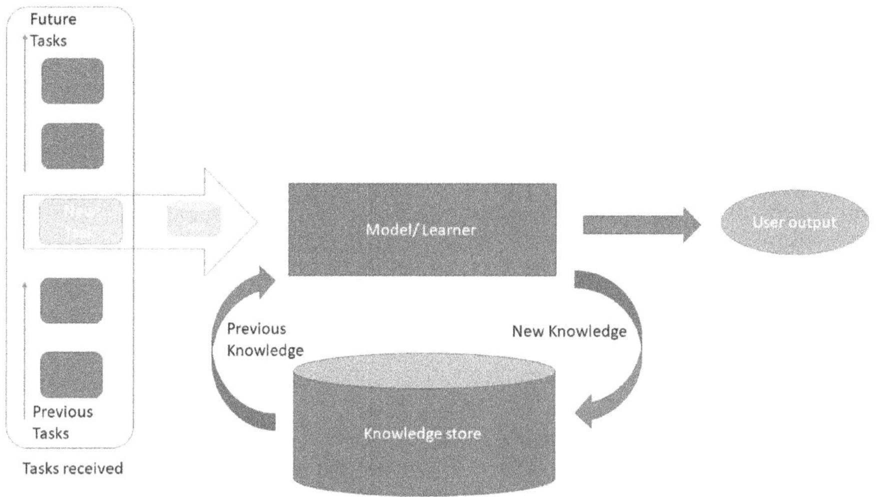

FIGURE 12.2 Architecture of Lifelong machine learning system.

of the process are shown in the architecture in Figure 12.2. The knowledge learnt from the past tasks is stored into the knowledge store, and the knowledge required for the new tasks is acquired from the same. The new knowledge learnt for the new task from the current data is then stored in the knowledge base for future tasks. User output is the required task output like a model or results of the current task. The architectures may vary with additional components as the setting demands in different learning scenarios. The knowledge base holds the past knowledge such as models, any relevant meta knowledge from the data, output of the tasks such as new classes generated and learnt from the previous tasks.

Life-long learning research can be grouped into four main areas—supervised, unsupervised, semisupervised, and reinforcement learning [23,25,26]. LL supervised learning has been designed for classification, object detection, sentiment classification NLP problems, and cumulative learning, where old classifiers are updated with new classes using neural networks. The LL unsupervised learning research has mostly focused on topic modeling and information extraction problems. The topics extracted in previous tasks from multiple domains can be used for future tasks and stored into the knowledge base. Semisupervised learning setting for LL has been studied as a never-ending language learner and has been used for information extraction tasks actively performing the extraction by crawling the web to build large knowledge bases from various domains. LL reinforcement learning has enabled learning from multiple and dynamic environments incorporating previous tasks and cross-domain experience into agents improving their decision-making.

With increased progress in LL, to achieve human-like learning capabilities by learning from continuous data shifts and previous knowledge, there are challenges LL entails that need to be addressed [23,27]. LL requires to store large amounts of knowledge from previous tasks continually to assist future tasks and is one of the main challenges in LL approaches. It would be infeasible to save such huge knowledge and hence might require further processing to determine what kind of knowledge should be stored. Part of such processing might be to decide if the learned knowledge is important that might be useful for future tasks and discard unimportant information. Also minimizing any errors in the knowledge is another challenge in LL approach. New tasks may create errors in the knowledge and would propagate to all the future tasks corrupting the knowledge and require techniques to mitigate such errors. Apart from storing the knowledge, representing such huge data is also inevitable and hence requires knowledge representation techniques to address the challenge.

12.6 GPT MODELS

With the introduction of transformer models, natural language-processing models have seen astonishing improvement, revolutionizing the AI community by altering the dynamics of NLP models usage on the internet. Open AI's GPT (generative pretrained transformers) models belong under such a category. BERT (bidirectional encoder representations from transformers)-pretrained foundation models are trained on large datasets and applied to various downstream tasks, often trained for a specific kind of downstream task separately [28]. GPT models have approached this task in two stages: (1) unsupervised pretraining, leveraging large unlabeled text corpus for training the models rather than using fully labeled datasets for discriminative training of specific tasks. (2) Supervised fine-tuning, GPT models learned from stage 1 are further fine-tuned requiring only minimal labeled data for specific downstream tasks. This architecture has not only observed a boost in performance traits but also leveraged large unlabeled data resources from various domains available in abundance, making it a semisupervised learning paradigm. This has improved the generalization capabilities of GPT models for a wide range of tasks such as summarization, classification, translation, sentiment analysis, question answering, and more. In order for the model to adapt to these varieties of tasks, task-specific input transformations are performed by converting input data into token sequences. GPT models have also shown improved performance in modeling long-range structures in the data. They have also demonstrated zero-shot performance on various tasks.

The first GPT models were pretrained on BooksCorpus containing 7000 unpublished books [29], on a 12-layer transformer architecture as shown in Figure 12.3, comprising decoder-only transformer and masked self-attention heads, 768 encoded tokens dimension for word embedding, trained with 117M parameters. Inspired by the performance of the model trained on large dataset and the architecture, GPT-2 was modeled to be trained on larger dataset with even more parameters attributing to a larger model. With its high-capacity language model, GPT-2 achieved significant improvement in zero-shot task transfer performance setting a baseline. GPT-2 was constructed with 48 layers model [30], with a word embedding vector dimension of 1600, trained with 1.5B parameters on WebText dataset of 40GB with over 8 million documents. Unlike GPT, GPT-2 is not fine-tuned for any specific task rather used in zero-shot manner.

These language models provided a foundation inspiring to build models with larger capacity, trained on larger datasets from different domains to generalize on multiple tasks in a zero-shot setting, scaling the language models toward large

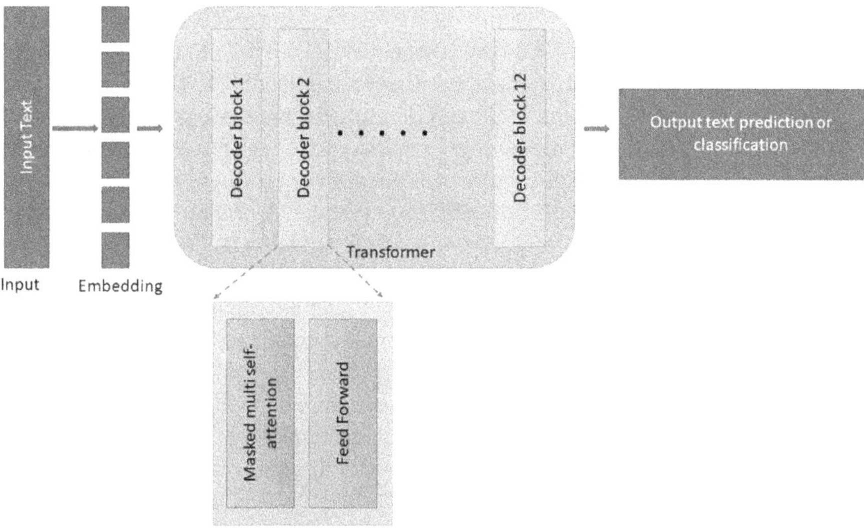

FIGURE 12.3 Architecture of GPT model with 12-layer decoder only transformer.

language models like GPT-3. This scaling has been observed to achieve state-of-the-art performances in NLP tasks including question-answering, translation, on-the-fly reasoning in a few-shot, and zero-shot setting. GPT-3 is an autoregressive model built with 96 layers [31], increased embedding vector dimension of 12888, trained with 175B parameters on five datasets including Common crawl, WebText2, Books1, Books2, and Wikipedia. Though GPT-3 has achieved impressive performance on different language model datasets and tasks in quality text generation for news articles, closed-book question answering, LAMBDA dataset testing long-range dependencies, translation tasks, reading comprehension, common sense reasoning tasks, still comes with weaknesses [31,32]. GPT-3 faces limitations in text synthesis while synthesizing long sentences by generating repeated sequences. Also, the algorithmic limitation where all the tokens are weighed equally leading to inability in differentiating important from unimportant is some of its challenges.

While the first two GPT models have seen their applications in text completion, classification, and text-generation tasks, GPT-3 model also found its application in conversational AI achieving human-like responses in generating social media content and chatbots. GPT models have also been used in material science for material composition generation [33]. GPT-3 models have also been applied in biomedical and healthcare domain as an automatic agent providing instant customer service through conversational AI answering trivial questions [32]. GPT-3 and its later version GPT-3.5 have also been effectively used in extracting and comprehending research papers from material science and engineering related databases, and also in interpretative and predictive material science applications including protein folding, molecular property prediction, and material discovery [34,35].

ChatGPT is one of the large language models that have attracted enormous attention of the world for its reasoning, text-generation capabilities in generating human-like text [36]. *ChatGPT* is built on GPT-3.5 version and is a sibling model of InstructGPT. InstructGPT [37] is a language model trained to follow instructions and provide response with a reinforcement learning human feedback. The input prompts used for the supervised fine-tuning of the GPT-3 model are those previously submitted to OpenAI API. A reward model is used to incorporate the reinforcement element where the responses generated by the fine-tuned model to these user prompts are rated by humans and are used to train the reward model. A proximity policy optimization (PPO) is used as the reinforcement learning model policy of maximizing the reward. ChatGPT differs from InstructGPT in terms of data collection setup. Within a short span of introduction of ChatGPT, it has found enormous applications in the field of education, healthcare, literature, and many more [38]. ChatGPT has been used for summarizing research papers, in providing experiential learning for students [38], common sense reasoning, software development, and translation being some of many applications observed. GPT-4 [39] model has found to exceed ChatGPT in its reasoning capabilities providing better responses with increased accuracy.

12.7 FUTURE DIRECTIONS IN USING ML FOR 2D MATERIALS

Here, we focus on digital twin technology (DTT) as an example for discussing recommended future directions to enable the use of ML for 2D materials research. DDTs have been effectively used for capturing geometrical features and component-level performances of many engineered systems [40]. Despite these successes, there exist knowledge gaps regarding whether DTTs can be utilized for accelerating 2D materials discovery. Like any other advanced materials, 2D materials and their composites also represent complex a physical system, which requires analysis at multiple length (e.g., nano- and microscale) and timescales (e.g., service life, seconds to years). Considering that the 2D materials are characterized by nanoscale dimensions, the analysis become even more complicated because the nanoscale physics for several 2D materials is yet unknown. For biological applications, unveiling the phenotypical and genotypical responses as a function of such nanoscale phenomena is often difficult. To develop digital twins of 2D materials, especially for biotechnology applications, it is important to capture, model, and predict their performances (i.e., changes in structure, process, and performance) under different processing histories (e.g., degradation over time in marine environments). Simultaneously, adequate infrastructure will be required to integrate the 2D materials' processing–structure–property relationships with the biological response at a genetic level (i.e., omics response).

Focused research efforts will be required to capture such integrative responses at different spatial and temporal resolution scales. At a temporal scales, the 2D materials performances will be needed to capture at a shorter timescales (e.g., seconds to minutes, interactions with proteins involved in conditioning phase), relative longer timescales (e.g., days to weeks, biofilm maturation), and longer timescales (e.g., years, macrofouling). It should be noted that the properties of 2D materials would change over time, and hence, the 2D-material–microbe interactions at different timescales may not be necessarily related to each another. At a spatial scale, heterogeneities across the length scale of 2D materials (e.g., defects, dopants, strain, edges, and layer stackings) may result in unpredictable variations. Thus, any DTT of 2D materials, particularly in biological applications, will require a systemic approach that has clear vision for the possible trade-offs in terms of multifunctional requirements at different length and timescales, which will further vary as a function of environmental parameters.

To develop unifying and representative data sets for creating and training DTTs at a systemic level, it is beneficial to have a single expert or group of diverse professionals guided by a single expert. Such a concerted effort should have clear understanding on different 2D materials data modalities, which include physical and morphological (based on microscopy and spectroscopy), mechanical (nanoindentation, micro-/nanomechanical devices, pressurized bulge tests), electrical (current–voltage, Hall effect, and 4-point probe tests), electrochemical properties (cyclic voltammetry, electrochemical impedance spectroscopy), and biological responses (e.g., omics interactions).

Complementary modeling efforts (e.g., DFT and MD simulations) should align faithfully with these experimental measurements at selected time and length scales. Comprehensive mathematical frameworks will also be needed to fuse the disparate materials and biological datasets for assisting development of DTTs. The authors firmly believe that such systemic approaches are possible by embracing convergence research approaches, similar to the one adopted by the authors' group in ongoing research projects (e.g., data-driven materials discovery for bioengineering innovations, NSF OIA # 1920954). This project intentionally blended diverse disciplinary expertise (e.g., biologists, bioinformaticians, computer scientists, corrosion scientists, environmental engineers, and materials scientists) in a concerted and reciprocal manner, where the overall efforts were focused on a single inquiry regarding a grand challenge of exploring infinitesimally thin 2D coatings for addressing vexing challenges caused by sulfate reducing bacterial biofilms. Readers are suggested to review literature to gain a broader understanding on such convergence research approaches [41].

To extend the digital twins to include 2D materials data over a hierarchy of length scales, novel frameworks that tightly integrate the conceptual chassis with the cyberinfrastructure are needed. Readers are encouraged to review such framework suggested by Dingreville and his coworkers. This framework consists of the following key elements: (1) principal component analysis and N-point correlation functions (NPCFs), defined as generating statistical averages over N copies of the phenomena, (2) a two-step Bayesian framework for performance prediction, and (3) a cyberinfrastructure that leverages new material ontologies for managing multimodal materials data [42]. Such foundational elements have been reported to show a promise for extending existing digital twins to incorporate key details of the material over different length scales (i.e., atomistic to macroscale) [42]. In conclusion, although the idea of implementing DTT for materials is at an infancy stage, their potential promises can motivate the readers to explore them through a rigorous R&D effort. Practicing professionals have also started exploring DTTs of materials in realistic applications (e.g., monitoring marine fouling performance of protective coatings in shipping industries). Such DTTs can provide stakeholders with indefinite access to a real time and virtual monitoring of evolution of biofouling, including the early warning monitoring system for onset of biofouling. This will allow the stakeholders to facilitate early interventions and well-informed decisions. To turn the DTTs into a reality, it is also important to embrace advanced data collection technologies (e.g., internet of things and embedded sensors), data security guidelines, and finally, the effective communication strategies (e.g., dashboard) to inform the promising benefits and potential risks. Finally, advanced microscopy and spectroscopy methods are needed to enable remote and viable monitoring of 2D materials properties at a nanoscale.

REFERENCES

[1] Murphy, K. P. (2022). *Probabilistic Machine Learning: An Introduction.* Cambridge: MIT Press.

[2] Kumar, T., Turab, M., Raj, K., Mileo, A., Brennan, R., & Bendechache, M. (2023). Advanced Data Augmentation Approaches: A Comprehensive Survey and Future directions. *arXiv preprint arXiv:2301.02830.*

[3] Shorten, C., & Khoshgoftaar, T. M. (2019). A survey on image data augmentation for deep learning. *Journal of Big Data*, *6*(1), pp. 1–48.

[4] Vahdat, M.T., Agrawal, K.V. and Pizzi, G., 2022. Machine-learning accelerated identification of exfoliable two-dimensional materials. *Machine Learning: Science and Technology, 3*(4), p. 045014.

[5] Chilkoor, G., Shrestha, N., Kutana, A., Tripathi, M., Robles Hernández, F.C., Yakobson, B.I., Meyyappan, M., Dalton, A.B., Ajayan, P.M., Rahman, M.M. and Gadhamshetty, V., 2020. Atomic layers of graphene for microbial corrosion prevention. *ACS Nano, 15*(1), pp. 447-454.

[6] Chilkoor, G., Upadhyayula, V.K., Gadhamshetty, V., Koratkar, N. and Tysklind, M., 2017. Sustainability of renewable fuel infrastructure: a screening LCA case study of anticorrosive graphene oxide epoxy liners in steel tanks for the storage of biodiesel and its blends. *Environmental Science: Processes & Impacts*, 19(2), pp. 141–153.

[7] Upadhyayula, V.K., Meyer, D.E., Gadhamshetty, V. and Koratkar, N., 2017. Screening-level life cycle assessment of graphene-poly (ether imide) coatings protecting unalloyed steel from severe atmospheric corrosion. *ACS Sustainable Chemistry & Engineering, 5*(3), pp. 2656-2667.

[8] Allen, C., Aryal, S., Do, T., Gautum, R., Hasan, M.M., Jasthi, B.K., Gnimpieba, E. and Gadhamshetty, V., 2022. Deep learning strategies for addressing issues with small datasets in 2D materials research: Microbial Corrosion. *Frontiers in Microbiology, 13*, p. 1059123.

[9] Tarvainen, A., & Valpola, H. (2017). Mean teachers are better role models: Weight-averaged consistency targets improve semi-supervised deep learning results. In *Advances in Neural Information Processing Systems, 30*, Curran Associates, Inc.: Long Beach, CA.

[10] Settles, B. (2012). *Active Learning: Synthesis Lectures on Artificial Intelligence and Machine Learning.* Long Island, NY: Morgan & Clay Pool, p. 10, S00429ED1V01Y201207AIM018.

[11] DeCost, B. L., Francis, T., & Holm, E. A. (2017). Exploring the microstructure manifold: image texture representations applied to ultrahigh carbon steel microstructures. *Acta Materialia*, *133*, pp. 30–40.

[12] Ye, H. J., Ming, L., Zhan, D. C., & Chao, W. L. (2022). Few-shot learning with a strong teacher. In *IEEE Transactions on Pattern Analysis and Machine Intelligence*, IEEE.

[13] Raissi, M., Perdikaris, P., & Karniadakis, G. E. (2017). Physics Informed Deep Learning (Part i): Data-Driven Solutions of Nonlinear Partial Differential Equations. *arXiv preprint arXiv:1711.10561.*

[14] Faroughi, S. A., Pawar, N., Fernandes, C., Das, S., Kalantari, N. K., & Mahjour, S. K. (2022). Physics-Guided, Physics-Informed, and Physics-Encoded Neural Networks in Scientific Computing. *arXiv preprint arXiv:2211.07377.*
[15] Grieves, M. (2014). Digital twin: manufacturing excellence through virtual factory replication. *White Paper, 1*(2014), pp. 1–7.
[16] Fuller, A., Fan, Z., Day, C., & Barlow, C. (2020). Digital twin: enabling technologies, challenges and open research. *IEEE Access, 8*, pp. 108952–108971.
[17] Madni, A. M., Madni, C. C., & Lucero, S. D. (2019). Leveraging digital twin technology in model-based systems engineering. *Systems, 7*(1), p. 7.
[18] Whang, S. E., Roh, Y., Song, H., & Lee, J. G. (2023). Data collection and quality challenges in deep learning: a data-centric ai perspective. *The VLDB Journal, 32*, pp. 1–23.
[19] Polyzotis, N., & Zaharia, M. (2021). What Can Data-Centric AI Learn from Data and ML Engineering?. *arXiv preprint arXiv:2112.06439.*
[20] Polyzotis, N., Roy, S., Whang, S. E., & Zinkevich, M. (2018). Data lifecycle challenges in production machine learning: a survey. *ACM SIGMOD Record, 47*(2), pp. 17–28.
[21] Wu, B., Liu, L., Zhu, Z., Liu, Q., He, Z., & Lyu, S. (2023). Adversarial Machine Learning: A Systematic Survey of Backdoor Attack, Weight Attack and Adversarial Example. *arXiv preprint arXiv:2302.09457.*
[22] Berthelot, D., Carlini, N., Goodfellow, I., Papernot, N., Oliver, A., & Raffel, C. A. (2019). Mixmatch: A holistic approach to semi-supervised learning. In *Advances in Neural Information Processing Systems, 32*. Vancouver: Curran Associates, Inc.
[23] Chen, Z., & Liu, B. (2018). *Lifelong Machine Learning. Synthesis Lectures on Artificial Intelligence and Machine Learning.* Rafael, CA: Morgan and Claypool Publishers.
[24] Wang, L., Zhang, X., Su, H., & Zhu, J. (2023). A Comprehensive Survey of Continual Learning: Theory, Method and Application. *arXiv preprint arXiv:2302.00487.*
[25] Liu, B. (2017). Lifelong machine learning: a paradigm for continuous learning. *Frontiers of Computer Science,* 11, pp. 359–361.
[26] Silver, D. L., Yang, Q., & Li, L. (2013, March). Lifelong machine learning systems: Beyond learning algorithms. In *2013 AAAI Spring Symposium Series,* Palo Alto, CA.
[27] Hong, X., Guan, S. U., Man, K. L., & Wong, P. W. (2020). Lifelong machine learning architecture for classification. *Symmetry, 12*(5), p. 852.
[28] Zhou, C., Li, Q., Li, C., Yu, J., Liu, Y., Wang, G., ... & Sun, L. (2023). A Comprehensive Survey on Pretrained Foundation Models: A History from BERT to ChatGPT. *arXiv preprint arXiv:2302.09419.*
[29] Radford, A., Narasimhan, K., Salimans, T., & Sutskever, I. (2018). Improving language understanding by generative pre-training, OpenAI.
[30] Radford, A., Wu, J., Child, R., Luan, D., Amodei, D., & Sutskever, I. (2019). Language models are unsupervised multitask learners. *OpenAI blog, 1*(8), p. 9.
[31] Brown, T., Mann, B., Ryder, N., Subbiah, M., Kaplan, J. D., Dhariwal, P., et al. (2020). Language models are few-shot learners. *Advances in Neural Information Processing Systems, 33*, pp. 1877–1901.
[32] Zong, M., & Krishnamachari, B. (2022). A Survey on GPT-3. *arXiv preprint arXiv:2212.00857.*
[33] Polak, M. P., Modi, S., Latosinska, A., Zhang, J., Wang, C. W., Wang, S., et al. Morgan, D. (2023). Flexible, Model-Agnostic Method for Materials Data Extraction from Text Using General Purpose Language Models. *arXiv preprint arXiv:2302.04914.*
[34] Hu, Y., & Buehler, M. J. (2023). Deep language models for interpretative and predictive materials science. *APL Machine Learning, 1*(1), p. 010901.
[35] Fu, N., Wei, L., Song, Y., Li, Q., Xin, R., Omee, S. S., Dong, R., Siriwardane, E. M., & Hu, J. (2023). Material transformers: deep learning language models for generative materials design. *Machine Learning: Science and Technology*, 4(1), p. 015001. https://doi.org/10.1088/2632-2153/acadcd

[36] Introducing ChatGPT. (n.d.). Retrieved from https://openai.com/blog/chatgpt
[37] Ouyang, L., Wu, J., Jiang, X., Almeida, D., Wainwright, C., Mishkin, P., et al. (2022). Training language models to follow instructions with human feedback. *Advances in Neural Information Processing Systems, 35*, pp. 27730–27744.
[38] Shahriar, S., & Hayawi, K. (2023). Let's Have a Chat! A Conversation with ChatGPT: Technology, Applications, and Limitations. *arXiv preprint arXiv:2302.13817.*
[39] GPT-4. (n.d.). Retrieved from https://openai.com/product/gpt-4
[40] Lim, J., Perullo, C. A., Milton, J., Whitacre, R., Jackson, C., Griffin, C., … & Lieuwen, T. C. (2021, June). The EPRI gas turbine digital twin–a platform for operator focused integrated diagnostics and performance forecasting. In *Turbo Expo: Power for Land, Sea, and Air.* American Society of Mechanical Engineers. Vol. 84966, p. V004T09A009..
[41] Peek, L., Tobin, J., Adams, R.M., Wu, H. and Mathews, M.C., 2020. A framework for convergence research in the hazards and disaster field: the natural hazards engineering research infrastructure CONVERGE facility. *Frontiers in Built Environment*, 6, p. 110.
[42] Kalidindi, S.R., Buzzy, M., Boyce, B.L. and Dingreville, R., 2022. Digital twins for Materials. *Frontiers in Materials, 9*, p. 48.

Index

Note: *Italic* page numbers refer to figures.

For Product Safety Concerns and Information please contact our EU representative GPSR@taylorandfrancis.com
Taylor & Francis Verlag GmbH, Kaufingerstraße 24, 80331 München, Germany

www.ingramcontent.com/pod-product-compliance
Lightning Source LLC
Chambersburg PA
CBHW060400220326
41598CB00023B/2975